Graduate Texts in Mathematics 251

T0255045

For other titles published in this series, go to
http://www.springer.com/series/136

Robert A. Wilson

The Finite Simple Groups

 Springer

Professor Robert A. Wilson
Queen Mary, University of London
School of Mathematical Sciences
Mile End Road
London E1 4NS
UK
R.A.Wilson@qmul.ac.uk

ISSN 0072-5285
ISBN 978-1-4471-2527-3   e-ISBN 978-1-84800-988-2
DOI 10.1007/978-1-84800-988-2
Springer London Dordrecht Heidelberg New York

British Library Cataloguing in Publication Data
A catalogue record for this book is available from the British Library

Mathematics Subject Classification (2000): 20D

Cover design: SPI Publisher Services

Printed on acid-free paper

Springer is part of Springer Science+Business Media (www.springer.com)

Preface

This book is intended as an introduction to all the finite simple groups. During the monumental struggle to classify the finite simple groups (and indeed since), a huge amount of information about these groups has been accumulated. Conveying this information to the next generation of students and researchers, not to mention those who might wish to apply this knowledge, has become a major challenge.

With the publication of the two volumes by Aschbacher and Smith [12, 13] in 2004 we can reasonably regard the proof of the Classification Theorem for Finite Simple Groups (usually abbreviated CFSG) as complete. Thus it is timely to attempt an overview of all the (non-abelian) finite simple groups in one volume. For expository purposes it is convenient to divide them into four basic types, namely the alternating, classical, exceptional and sporadic groups.

The study of alternating groups soon develops into the theory of permutation groups, which is well served by the classic text of Wielandt [170] and more modern treatments such as the comprehensive introduction by Dixon and Mortimer [53] and more specialised texts such as that of Cameron [19]. The study of classical groups via vector spaces, matrices and forms encompasses such highlights as Dickson's classic book [48] of 1901, Dieudonné's [52] of 1955, and more modern treatments such as those of Taylor [162] and Grove [72]. The complete collection of groups of Lie type (comprising the classical and exceptional groups) is beautifully exposed in Carter's book [21] using the simple complex Lie algebras as a starting point. And sporadic attempts have been made to bring the structure of the sporadic groups to a wider audience—perhaps the most successful book-length introduction being that of Griess [69]. But no attempt has been made before to bring within a single cover an introductory overview of all the finite simple groups (unless one counts the 'Atlas of finite groups' [28], which might reasonably be considered to be an overview, but is certainly not introductory).

The remit I have given myself, to cover all of the finite simple groups, gives both advantages and disadvantages over books with more restricted subject matter. On the one hand it allows me to point out connections, for example between exceptional behaviour of generic groups and the existence of sporadic groups. On the other hand it prevents me from proving everything in as much detail as the reader may desire. Thus the reader who wishes to understand everything in this book will have to do a lot of homework in filling in gaps, and following up references to more complete treatments. Some of the exercises are specifically designed to fill in gaps in the proofs, and to develop certain topics beyond the scope of the text.

One unconventional feature of this book is that Lie algebras are scarcely mentioned. The reasons for this are twofold. Firstly, it hardly seems possible to improve on Carter's exposition [21] (although this book is now out of print and secondhand copies change hands at astronomical prices). And secondly, the alternative approach to the exceptional groups of Lie type via octonions deserves to be better known: although real and complex octonions have been extensively studied by physicists, their finite analogues have been sadly neglected by mathematicians (with a few notable exceptions). Moreover, this approach yields easier access to certain key features, such as the orders of the groups, and the generic covering groups.

On the other hand, not all of the exceptional groups of Lie type have had effective constructions outside Lie theory. In the case of the family of large Ree groups I provide such a construction for the first time, and give an analogous description of the small Ree groups. The importance of the octonions in these descriptions led me also to a new octonionic description of the Leech lattice and Conway's group, and to an ambition, not yet realised, to see the octonions at the centre of the construction of all the exceptional groups of Lie type, and many of the sporadic groups, including of course the Monster.

Complete uniformity of treatment of all the finite simple groups is not possible, but my ideal (not always achieved) has been to begin by describing the appropriate geometric/algebraic/combinatorial structure, in enough detail to calculate the order of its automorphism group, and to prove simplicity of a clearly defined subquotient of this group. Then the underlying geometry/algebra/combinatorics is further developed in order to describe the subgroup structure in as much detail as space allows. Other salient features of the groups are then described in no particular order.

This book may be read in sequence as a story of all the finite simple groups, or it may be read piecemeal by a reader who wants an introduction to a particular group or family of groups. The latter reader must however be prepared to chase up references to earlier parts of the book if necessary, and/or make use of the index. Chapters 4 and 5 are largely (but not entirely) independent of each other, but both rely heavily on Chapters 2 and 3. The sections of Chapter 4 are arranged in what I believe to be the most appropriate order pedagogically, rather than logically or historically, but could be read in a different order. For example, one could begin with Section 4.3 on $G_2(q)$ and proceed via triality (Section 4.7) to $F_4(q)$ (Section 4.8) and $E_6(q)$ (Section 4.10), postponing the twisted groups until later. The ordering of sections in Chapter 5 is traditional, but a more avant-garde approach might begin with J_1 (Section 5.9.1), and follow this with the exotic incarnation of (the double cover of) J_2 as a quaternionic reflection group (in the first few parts of Section 5.6), and/or the octonionic Leech lattice (Section 5.6.12). But one cannot go far in the study of the sporadic groups without a thorough understanding of M_{24}.

I was introduced to the weird and wonderful world of finite simple groups by a course of lectures on the sporadic simple groups given by John Conway in Cambridge in the academic year 1978–9. During that course, and the following three years when he was my Ph.D. supervisor, he taught me most of what subsequently appeared in the 'Atlas of finite groups' [28], and a large part of what now appears in this book. I am of course extremely indebted to him for this thorough initiation.

Especial thanks go also to my former colleague at the University of Birmingham, Chris Parker, who, early in 2003, fuelled by a couple of pints of beer, persuaded me there was a need for a book of this kind, and volunteered to write half of it; who persuaded the Head of School to let us teach a two-semester course on finite simple groups in 2003–4; who developed the original idea into a detailed project plan; and who then quietly left me to get on and write the book. It is not entirely his fault that the book which you now have in your hands bears only a superficial resemblance to that original plan: I excised the chapters which he was going to write, on Lie algebras and algebraic groups, and shovelled far more into the other chapters than we ever anticipated. At the same time the planned 150 pages grew to nearly 300. Indeed, the more I wrote, the more I became aware of how much I had left out. I would need at least another 300 pages to do justice to the material, but one has to stop somewhere. I apologise to those readers who find that I stopped just at the point where they started to get interested.

Several colleagues have read substantial parts of various drafts of this book, and made many valuable comments. I particularly thank John Bray, whose keen nose for errors and assiduousness in sorting out some of the finer points has improved the accuracy and reliability of the text enormously; John Bradley, whose refusal to accept woolly arguments helped me tighten up the exposition in many places; and Peter Cameron whose comments on some early draft chapters have led to significant improvements, and whose encouragement has helped to keep me working on this book.

I owe a great deal also to my students for their careful reading of various versions of parts of the text, and their uncovering of countless errors, some minor, some serious. I used to tell them that if they had not found any errors, it was because they had not read it properly. I hope that this is now less true than it used to be. I thank in particular Jonathan Ward, Johanna Rämö, Simon Nickerson, Nicholas Krempel, and Richard Barraclough, and apologise to any whose names I have inadvertently omitted.

It is a truism that errors remain, and the fault (if fault there be), human nature. By convention, the responsibility is mine, but in fact that is unrealistic.

As Gauss himself said, "In science and mathematics we do not appeal to authority, but rather *you are responsible for what you believe.*" Nevertheless, I shall endeavour to maintain a web-site of corrections that have been brought to my attention, and will be grateful for notification of any further errors that you may find.

Thanks go also to Karen Borthwick at Springer-Verlag London for her gentle but persistent pressure, and to the anonymous referees for their enthusiasm for this project and their many helpful suggestions. I am grateful to Queen Mary, University of London, for their initially relatively light demands on me when I moved there in September 2004, which left me time to indulge in the pleasures of writing. It is entirely my own fault that I did not finish the book before those demands increased to the point where only a sabbatical would suffice to bring this project to a conclusion. I am therefore grateful to Jianbei An and the University of Auckland, and John Cannon and the University of Sydney, for providing me with time, space, and financial support during the last six months which enabled me, among other things, to sign off this book.

London   *Robert Wilson*
June 2009

Contents

1

Introduction

1.1 A brief history of simple groups

The study of (non-abelian) finite simple groups can be traced back at least
as far as Galois, who around 1830 understood their fundamental significance
as obstacles to the solution of polynomial equations by radicals (square roots,
cube roots, etc.). From the very beginning, Galois realised the importance of
classifying the finite simple groups, and knew that the alternating groups A_n
are simple for $n \geqslant 5$, and he constructed (at least) the simple groups $\mathrm{PSL}_2(p)$
for primes $p \geqslant 5$.

Every finite group G has a *composition series*

$$1 = G_0 \lhd G_1 \lhd G_2 \lhd \cdots \lhd G_{n-1} \lhd G_n = G \qquad (1.1)$$

where each group is normal in the next, and the series cannot be refined any
further: in other words, each G_i/G_{i-1} is *simple*. The Jordan–Hölder theorem
states that the set of *composition factors* G_i/G_{i-1} (counting multiplicities) is
independent of the choice of composition series. Thus if any composition series
contains a non-abelian composition factor then they all do. Galois's theorem
states that a polynomial equation in one variable has a solution by radicals
if and only if the corresponding 'Galois group' has a composition series with
cyclic factors.

The 19th century saw slow progress in finite group theory until 1870, in
which year appeared Camille Jordan's 'Traité des substitutions' [104], followed
in 1872 by the publication of Sylow's theorems. The former contains constructions of the simple groups we now call $\mathrm{PSL}_n(p)$, while the latter provides the
first tools for classifying simple groups. And we must not forget the extraordinary paper of Mathieu [131] from 1861 in which he constructs the groups
we now know as the sporadic groups M_{11} and M_{12}, and which was followed
by another paper [132] in 1873 constructing M_{22}, M_{23} and M_{24}.

It was not until the dawn of the 20th century that a well-developed theory
of the finite classical groups began to emerge, most notably in the work of

R.A. Wilson, *The Finite Simple Groups*,
Graduate Texts in Mathematics 251,
© Springer-Verlag London Limited 2009

L. E. Dickson. In large part this work was inspired by Killing's classification of complex simple Lie algebras (if that is not an oxymoron) into the types A_n, B_n, C_n, D_n, G_2, F_4, E_6, E_7 and E_8. Dickson constructed finite simple groups analogous to all of these except F_4, E_7 and E_8, for every finite field (with a small number of exceptions which are not simple).

One wonders why Dickson did not go on to construct finite simple groups of the remaining three types. It seems extraordinary that it was another fifty years before Chevalley provided a uniform construction of all these groups, in his famous 1955 paper [23]. The types A_n, B_n, C_n and D_n give rise to the classical groups $\mathrm{PSL}_{n+1}(q)$ (linear), $\mathrm{P\Omega}_{2n+1}(q)$ (orthogonal, odd dimension), $\mathrm{PSp}_{2n}(q)$ (symplectic) and $\mathrm{P\Omega}_{2n}^+(q)$ (orthogonal, even dimension, plus type). But where were the unitary groups, and $\mathrm{P\Omega}_{2n}^-(q)$? Very soon it was realised that these could be obtained by 'twisting' the Chevalley construction. Using the unitary groups as a model, Steinberg, Tits and Hertzig independently constructed two new families ${}^3D_4(q)$ and ${}^2E_6(q)$.

Soon afterwards, Suzuki and Ree saw how to 'twist' the groups of types B_2, G_2 and F_4, provided the characteristic of the field was 2, 3, or 2 respectively. And that seemed to be that. There was a feeling in the early 1960s (or so I am told) that probably all the finite simple groups had been discovered, and all that remained was to prove this.

Meanwhile, other parts of group theory had been developing by leaps and bounds. The Feit–Thompson paper [59] of 1963 proved the monumental result that every finite group of odd order is soluble, or to put it another way, every non-abelian finite simple group has even order. Thus every nonabelian finite simple group contains an element of order 2 (an *involution*) and soon the seemingly outrageous notion began to take root, that one could prove by induction that all finite simple groups were known. Thompson again provided the base case for the induction by classifying the minimal simple groups.

And so, in the early 1960s, the attempt to get a complete classification of the finite simple groups began in earnest. But it turned out to be a lot harder than some people had predicted. More or less the first case to try to eliminate after Thompson's work (at least logically, if not historically) was the case of a simple group with involution centraliser $C_2 \times A_5$. Janko's construction of such a group in 1964 sent shock-waves throughout the group theory community. Suddenly it seemed that the classification project might not be so easy after all. Maybe there were still hundreds, thousands, infinitely many simple groups left to find?

In the decade that followed, a further twenty 'sporadic' (the term originally used by Burnside to describe the Mathieu groups) simple groups were discovered, and then the supply suddenly dried up. By 1980 there was a general feeling that the classification of finite simple groups was almost complete, and there were probably no more finite simple groups to find. Not that that has stopped some people from continuing to look. Announcements were made that the proof was almost complete, and (premature) predictions of the imminent death of group theory filled the air.

1.2 The Classification Theorem

The *Classification Theorem for Finite Simple Groups* (traditionally abbreviated CFSG, conveniently forgetting that what is important is not so much the *Classification*, as the *Theorem*) states that every finite simple group is isomorphic to one of the following:

 (i) a cyclic group C_p of prime order p;
 (ii) an alternating group A_n, for $n \geqslant 5$;
(iii) a classical group:

linear:	$\mathrm{PSL}_n(q)$,	$n \geqslant 2$, except $\mathrm{PSL}_2(2)$ and $\mathrm{PSL}_2(3)$;
unitary:	$\mathrm{PSU}_n(q)$,	$n \geqslant 3$, except $\mathrm{PSU}_3(2)$;
symplectic:	$\mathrm{PSp}_{2n}(q)$,	$n \geqslant 2$, except $\mathrm{PSp}_4(2)$;
orthogonal:	$\mathrm{P\Omega}_{2n+1}(q)$,	$n \geqslant 3$, q odd;
	$\mathrm{P\Omega}_{2n}^+(q)$,	$n \geqslant 4$;
	$\mathrm{P\Omega}_{2n}^-(q)$,	$n \geqslant 4$

 where q is a power p^a of a prime p;
(iv) an exceptional group of Lie type:

$$G_2(q), q \geqslant 3; F_4(q); E_6(q); {}^2E_6(q); {}^3D_4(q); E_7(q); E_8(q)$$

 where q is a prime power, or

$$ {}^2B_2(2^{2n+1}), n \geqslant 1; {}^2G_2(3^{2n+1}), n \geqslant 1; {}^2F_4(2^{2n+1}), n \geqslant 1 $$

 or the Tits group ${}^2F_4(2)'$;
 (v) one of 26 sporadic simple groups:
 - the five Mathieu groups M_{11}, M_{12}, M_{22}, M_{23}, M_{24};
 - the seven Leech lattice groups Co_1, Co_2, Co_3, McL, HS, Suz, J_2;
 - the three Fischer groups Fi_{22}, Fi_{23}, Fi_{24}';
 - the five Monstrous groups $\mathbb{M}$, $\mathbb{B}$, Th, HN, He;
 - the six pariahs J_1, J_3, J_4, O'N, Ly, Ru.

Conversely, every group in this list is simple, and the only repetitions in this list are:

$$
\begin{aligned}
\mathrm{PSL}_2(4) &\cong \mathrm{PSL}_2(5) \cong A_5; \\
\mathrm{PSL}_2(7) &\cong \mathrm{PSL}_3(2); \\
\mathrm{PSL}_2(9) &\cong A_6; \\
\mathrm{PSL}_4(2) &\cong A_8; \\
\mathrm{PSU}_4(2) &\cong \mathrm{PSp}_4(3).
\end{aligned}
\tag{1.2}
$$

It is the chief aim of this book to explain, as far as space allows, the *statement* of CFSG. Thus we seek to introduce all the finite simple groups, to provide concrete constructions whenever possible, to calculate the orders of the groups, prove simplicity, and study their actions on various natural geometrical or combinatorial objects to the point where much of the subgroup structure is revealed. In so doing, we prove a substantial part (though by no means all) of the converse part of CFSG, that is, we prove the existence of many of the finite simple groups. (In the literature on CFSG the word 'construction' is generally used in this technical sense of 'existence proof', but in this book I shall often use it in the weaker sense of building, possibly without proof, an object which is in fact the group in question.) On the other hand, there is nothing at all here about the proof of the main part of CFSG, that is, the non-existence of any other finite simple groups.

1.3 Applications of the Classification Theorem

If one has (or if many people have) spent decades classifying certain objects, one is apt to forget just why one started the project in the first place. Predictions of the death of group theory in 1980 were the pronouncements of just such amnesiacs. But group theorists did not spend such an enormous amount of effort classifying simple groups just in order to put them in a glass case and admire them.

The first serious applications of CFSG were in permutation group theory. For example, the classical problem of classifying multiply-transitive groups, which had led Mathieu to the discovery of the first five sporadic groups, was easily solved: essentially, there are no others. With more work, one can classify 2-transitive groups. The O'Nan–Scott Theorem was the first result aimed towards a classification of primitive permutation groups. It reduced this problem (for a fixed degree n) to the problem of classifying maximal subgroups of index n in the almost simple groups, rather than in arbitrary groups.

This emphasised the need which was already felt, to have a good classification of the maximal subgroups of the simple groups. For individual groups it is possible to obtain complete explicit lists of maximal subgroups. For example, the maximal subgroups of the sporadic groups can be obtained by (often formidable) calculations: all except the Monster have been completed in this way. For families of groups it is not always possible to obtain such an explicit answer. In the case of the alternating groups, the O'Nan–Scott theorem lists some maximal subgroups explicitly, and says that any other maximal subgroup is almost simple, acting primitively. A theorem of Liebeck, Praeger and Saxl [120] tells us exactly when a subgroup of the latter type is *not* maximal, but it is impossible to give an explicit list of the rest.

A similar programme for classifying the maximal subgroups of the classical groups began with the publication of Aschbacher's paper [5] on the subject in 1984. For small dimensions explicit lists of all the maximal subgroups are known, going back to the classification of the maximal subgroups of $PSL_2(q)$ more than a century ago by L. E. Dickson and others [48]. With the benefit of exhaustive lists of representations of quasisimple groups in dimensions up to 250, due to Hiss and Malle [79] and Lübeck [124], there is some prospect that eventually there will be explicit lists of maximal subgroups for classical groups in these dimensions. Similarly, one would hope to be able to do the same for the exceptional groups of Lie type: so far, complete lists are available for five of the ten families.

1.4 Remarks on the proof of the Classification Theorem

There has been much debate about whether CFSG deserves to be called a theorem, and this debate has contributed to the philosophical arguments about what a theorem is, what a proof is, what mathematics is (or are) and how we recognise them when we see them. I believe most mathematicians are pragmatic in their daily professional lives, and do not expect to reach the Platonic ideal of a perfect proof which confers absolute certainty on a result. Certainly some mathematicians who argue most vociferously for the absolute nature of proof are amongst those whose own proofs often fall short of this ideal. Thus my own point of view is that it is ultimately meaningless to argue about whether a written (or spoken) argument 'is' or 'is not' a 'proof'. One can only really argue about the degree of certainty we derive from the argument.

The twentieth century saw announcements of solutions of many long-standing difficult problems in mathematics, including besides the CFSG, also the four-colour problem, Fermat's Last Theorem, the Poincaré conjecture, and others. It is natural, and necessary, to greet these announcements with a healthy degree of scepticism, as not all of them have stood up to the test of time. But in most cases a gradual process of expert scrutiny, tidying up and correcting minor (or major) errors leads eventually to a general acceptance that the problem in question has indeed been solved. On the other hand, it is impossible in practice to satisfy the mathematician's desire for absolute certainty. After all, we are only human and therefore fallible. We make mistakes, which sometimes lie hidden for years.

So what of the CFSG? Has it indeed been proved? Certainly the process of collating the various parts of the proof, filling in the gaps and correcting errors, has taken longer than anyone expected when the imminent completion of the proof was announced around 1980. The project by Gorenstein, Lyons and Solomon [66] to write down the whole proof in one place is still in progress: six of a projected eleven volumes have been published so far. The so-called 'quasithin' case is not included in this series, but has been dealt with in two volumes, totalling some 1200 pages, by Aschbacher and Smith [12, 13]. Nor do they consider the problem of existence and uniqueness of the 26 sporadic simple groups: fortunately this is not in the slightest doubt. So by now most parts of the proof have been gone over by many people, and re-proved in different ways. Thus the likelihood of catastrophic errors is much reduced, though not completely eliminated.

1.5 Prerequisites

I have tried in this book to keep the prerequisites to a minimum, but I do assume a familiarity with abstract group theory up to the level of Sylow's theorems and the Jordan–Hölder theorem, as well as the basics of linear algebra, such as can be found in Kaye and Wilson [105], and a reasonable mathematical maturity. In a few of the proofs I also need to assume a basic knowledge of representation theory, such as can be found in James and Liebeck [98], although this is not necessary for most of the text. In the chapter on sporadic groups I shall from time to time use basic properties of graphs, codes, lattices and other mathematical objects, but I hope that these can be picked up from the context. For the record, here is a summary of roughly what I assume as background in group theory. (Don't read this unless you need to!)

Groups, subgroups and cosets

A *group* is a (finite) set G with an *identity* element 1, a (binary) *multiplication* $x.y$ (or xy) and a (unary) *inverse* x^{-1} satisfying the *associative law* $(xy)z = x(yz)$, the *identity laws* $x1 = 1x = x$ and the *inverse laws* $xx^{-1} = x^{-1}x = 1$ for all $x, y, z \in G$ (and the *closure laws* $xy \in G$ and $x^{-1} \in G$ which we take for granted). It is *abelian* if $xy = yx$ for all $x, y \in G$, *non-abelian* otherwise. A *subgroup* is a subset H closed under multiplication and inverses. (It is sufficient to check $xy^{-1} \in H$ for all $x, y \in H$.) *Left cosets* of H in G are subsets $gH = \{gh \mid h \in H\}$ and *right cosets* are $Hg = \{hg \mid h \in H\}$. The left (or right) cosets all have the same size, and partition G, so that $|G| = |H||G : H|$ (*Lagrange's Theorem*), where $|G|$ is the *order* of G, i.e. the number of elements in G, and $|G : H|$ is the *index* of H in G, i.e. the number of left (or right) cosets. The *order* of an element $g \in G$ is the order n of the *cyclic group* $\langle g \rangle = \{1, g, g^2, \ldots, g^{n-1}\}$ (denoted C_n) that it *generates*, and the *exponent* of G is the lowest common multiple of the orders of the elements, that is the smallest positive integer e such that $g^e = 1$ for all $g \in G$.

Homomorphisms and quotient groups

A *homomorphism* is a map $\phi : G \to H$ which preserves the multiplication, $\phi(xy) = \phi(x)\phi(y)$ (from which it follows that $\phi(1) = 1$ and $\phi(x^{-1}) = \phi(x)^{-1}$). The *kernel* of ϕ is $\ker \phi = \{g \in G \mid \phi(g) = 1\}$, and is a subgroup which satisfies $g(\ker \phi) = (\ker \phi)g$, i.e. its left and right cosets are equal (such a subgroup N is called *normal*, written $N \trianglelefteq G$, or $N \triangleleft G$ if also $N \neq G$). An *isomorphism* is a bijective homomorphism, i.e. one satisfying $\ker \phi = \{1\}$ and $\phi(G) = H$: in this case we write $G \cong H$.

If N is a normal subgroup of G, the *quotient* group G/N has elements xN (for all $x \in G$) and group operations $(xN)(yN) = (xy)N$, and $(xN)^{-1} = x^{-1}N$. The *first isomorphism theorem* states that if $\phi : G \to H$ is a homomorphism then the image of ϕ, $\phi(G) \cong G/\ker \phi$ (and the isomorphism is given by $\phi(x) \mapsto x(\ker \phi)$).

The normal subgroups of G/N are in one-to-one correspondence with the normal subgroups K of G which contain N, and the *second isomorphism theorem* is $(G/N)/(K/N) \cong G/K$. If H is any subgroup of G, and N is any normal subgroup of G, then $HN = \{xy \mid x \in H, y \in N\}$ is a subgroup of G and $N \cap H$ is a normal subgroup of H, and the *third isomorphism theorem* is $HN/N \cong H/(N \cap H)$.

Simple groups and composition series

A group S is *simple* if it has exactly two normal subgroups (1 and S). In particular, an abelian group is simple if and only if it has prime order. A series

$$1 = G_0 \lhd G_1 \lhd G_2 \lhd \cdots \lhd G_{n-1} \lhd G_n = G \tag{1.3}$$

for a group G is called a *composition series* if all the factors G_i/G_{i-1} are simple (and they are then called *composition factors*).

The *fourth isomorphism theorem* (or *Zassenhaus's butterfly lemma*) states that if $X \lhd Y \leqslant G$ and $A \lhd B \leqslant G$ then

$$\frac{(Y \cap B)X}{(Y \cap A)X} \cong \frac{(Y \cap B)}{(Y \cap A)(X \cap B)} \cong \frac{(Y \cap B)A}{(X \cap B)A}.$$

Hence any two series for G have isomorphic *refinements*, and by induction on the length of a composition series, any two composition series for a finite group have the same composition factors, counted with multiplicities (the *Jordan–Hölder Theorem*). A *normal series* is one in which all terms G_i are normal in G, and if it has no proper refinements it is called a *chief series*, and its factors G_i/G_{i-1} *chief factors*.

Soluble groups

A group is *soluble* if it has a composition series with abelian (hence cyclic of prime order) composition factors. A *commutator* is an element $x^{-1}y^{-1}xy$, denoted $[x, y]$, and the subgroup generated by all commutators $[x, y]$ of elements $x, y \in G$ is the *commutator subgroup* (or *derived subgroup*), written $[G, G]$ or G'. Writing $G^{(0)} = G$ and $G^{(n)} = (G^{(n-1)})'$, it follows that G is soluble if and only if $G^{(n)} = 1$ for some n. Also G/N is abelian if and only if N contains G', so G/G' is the *largest abelian quotient* of G.

Group actions and conjugacy classes

The *right regular representation* of a group G is the identification of each element $g \in G$ with the permutation $x \mapsto xg$ of the elements of G. This shows that every finite group is isomorphic to a group of permutations (*Cayley's theorem*). If G is a group of permutations on a set Ω, and $a \in \Omega$, the *stabiliser* of a is the subgroup H consisting of all permutations in G which map a to itself. Then Lagrange's theorem can be re-interpreted as the *orbit–stabiliser theorem*, that $|G|/|H|$ equals the number of images of a under G (i.e. the *length* of the *orbit* of a).

Now let G act on itself by conjugation, $g : x \mapsto g^{-1}xg$, so that the orbits are the *conjugacy classes* $[x] = \{g^{-1}xg \mid g \in G\}$, and the stabiliser of x is the *centraliser* of x, $C_G(x) = \{g \in G \mid g^{-1}xg = x\}$. In particular, the conjugacy classes partition G, and their sizes divide the order of G. An element x is in a conjugacy class of size 1 if and only if x *commutes* with every element of G, i.e. $x \in Z(G) = \{y \in G \mid g^{-1}yg = y$ for all $g \in G\}$, the *centre* of G, which is a normal subgroup of G.

p-groups and nilpotent groups

A finite group is called a *p-group* if its order is a power of the prime p (and so by Lagrange's Theorem all its elements have order some power of p). Every conjugacy class in G has p^a elements for some a, and $\{1\}$ is a conjugacy class, so there are at least p conjugacy classes of size 1, and $Z(G)$ has order at least p. Define $Z_1(G) = Z(G)$ and $Z_n(G)/Z_{n-1}(G) = Z(G/Z_{n-1}(G))$, so that if G is a p-group then $Z_n(G) = G$ for some n. A group with this property is called *nilpotent* (of *class* at most n), and the series

$$1 = Z_0(G) \lhd Z_1(G) \lhd Z_2(G) \lhd \cdots$$

is called the *upper central series*.

The *direct product* $G_1 \times \cdots \times G_k$ of groups $G_1, \ldots, G_k$ is defined on the set $\{(g_1, \ldots, g_k) \mid g_i \in G_i\}$ by the group operations $(g_1, \ldots, g_k)(h_1, \ldots, h_k) = (g_1 h_1, \ldots, g_k h_k)$ and $(g_1, \ldots, g_k)^{-1} = (g_1^{-1}, \ldots, g_k^{-1})$. A finite group is nilpotent if and only if it is a direct product of p-groups.

Abelian groups

If m and n are coprime, then $C_m \times C_n \cong C_{mn}$. Hence in any finite abelian group there is an element whose order is equal to the exponent of the group. Indeed, every finite abelian group is isomorphic to a group $C_{n_1} \times C_{n_2} \times \cdots \times C_{n_r}$ with n_i dividing n_{i-1} for all $2 \leqslant i \leqslant r$. Conversely, the integers n_i are uniquely determined by the group.

Sylow's theorems

If G is a finite group of order $p^k n$, where p is prime and n is not divisible by p, then the *Sylow theorems* state that

(i) G has subgroups of order p^k;
(ii) these *Sylow p-subgroups* are all conjugate; and
(iii) the number s_p of Sylow p-subgroups satisfies $s_p \equiv 1 \bmod p$. (Note also that, by the orbit–stabiliser theorem, s_p is a divisor of n).

To prove the first statement, let G act by right multiplication on all subsets of G of size p^k: since the number of these subsets is not divisible by p, there is a stabiliser of order divisible by p^k, and therefore equal to p^k. To prove the second statement, and also to prove that any p-subgroup is contained in a Sylow p-subgroup, let any p-subgroup Q act on the right cosets Pg of any Sylow p-subgroup P by right multiplication: since the number of cosets is not divisible by p, there is an orbit $\{Pg\}$ of length 1, so $PgQ = Pg$ and gQg^{-1} lies inside P. To prove the third statement, let a Sylow p-subgroup P act by conjugation on the set of all the other Sylow p-subgroups: the orbits have length divisible by p, for otherwise P and Q are distinct Sylow p-subgroups of $N_G(Q)$, which is a contradiction.

An important corollary of Sylow's theorems is the *Frattini argument*: if $N \lhd G$ and P is a Sylow p-subgroup of N, then $G = N_G(P)N$.

Automorphism groups

An *automorphism* of a group G is an isomorphism of G with itself. The set of all automorphisms of G forms a group under composition, and is denoted $\mathrm{Aut}(G)$. The *inner* automorphisms are the automorphisms $\phi_g : x \mapsto g^{-1}xg$, for $g \in G$. These form a subgroup $\mathrm{Inn}(G)$ of $\mathrm{Aut}(G)$. Indeed, if $\alpha \in \mathrm{Aut}(G)$, then $\alpha^{-1}\phi_g\alpha = \phi_{g\alpha}$, so that $\mathrm{Inn}(G)$ is a normal subgroup of $\mathrm{Aut}(G)$. Now $\phi_{gh} = \phi_g\phi_h$, and $\phi_g = \phi_h$ if and only if $gh^{-1} \in Z(G)$, so the map ϕ defined by $\phi : g \mapsto \phi_g$ is a homomorphism from G onto $\mathrm{Inn}(G)$ with kernel $Z(G)$. Therefore $\mathrm{Inn}(G) \cong G/Z(G)$ and, in particular, if $Z(G) = 1$ then $G \cong \mathrm{Inn}(G)$. The *outer automorphism group* of G is $\mathrm{Out}(G) = \mathrm{Aut}(G)/\mathrm{Inn}(G)$.

1.6 Notation

Unfortunately there is no general consensus on notation for simple groups, extensions of groups, and so on. In this book I shall usually, but not always, follow the notation of the 'Atlas of finite groups' [28]. The main exception is for orthogonal groups, where Atlas notation is most likely to be misunderstood, and I prefer to follow Dieudonné [52]. Notation for simple groups is given in Section 1.2. Extensions of groups are written in one of the following ways: $A \times B$ denotes a direct product, with normal subgroups A and B; also $A{:}B$ denotes a semidirect product (or split extension), with a normal subgroup A and a subgroup B; and $A{\cdot}B$ denotes a non-split extension, with a normal subgroup A and quotient B, but no subgroup B; finally $A.B$ or just AB denotes an unspecified extension.

The expression $[n]$ denotes an (unspecified) group of order n, while n or C_n denotes (usually) a cyclic group of order n. If p is prime, p^n denotes an elementary abelian group of order p^n, i.e. a direct product of n copies of C_p. Often I shall use q^n (where q is a power of p) also to denote an elementary abelian p-group, although this is not standard Atlas notation. This includes the case $n = 1$.

1.7 How to read this book

Above all, this book should be read *slowly*. In order to cover a lot of ground, the text proceeds at a fast pace, and therefore demands the undivided attention of the reader. I believe that the reader will gain a lot more by mastering a small section than by skimming the whole book. (I have tried to make it as easy as possible to dip into the book and read one section which is of interest.) There are plenty of exercises collected at the end of each chapter. These range from routine calculations to consolidate and test the basic material, to substantial projects which go beyond the material presented in the book. The serious reader will of course attempt at least a selection of these exercises.

Within each chapter, the material gets significantly harder from beginning to end. Therefore at first reading one might profitably give up on each chapter when the going gets tough, and start again with the more elementary parts of the next chapter.

Gaps in the proofs are of varying sizes. A small gap, which we hope the reader will be able to plug with a minimum (or a modicum) of effort, is signalled by a phrase such as 'it is easy to see that'. A large gap, or complete absence of proof, which might require considerable effort, or reference to the literature, is signalled by a phrase such as 'it turns out that', or 'in fact'. The word 'shape' is used frequently to indicate that the notation is imprecise, but that it would take up too much space to make it precise.

Chapters 2 and 3 are reasonably 'complete' and for the most part 'elementary', and could be covered at advanced undergraduate level. The material in Chapters 4 and 5 is harder, and I have found it necessary to cover this in less detail, which makes it more suitable for project work than for a traditional lecture course.

2

The alternating groups

2.1 Introduction

The most familiar of the (finite non-abelian) simple groups are the alternating
groups A_n, which are subgroups of index 2 in the symmetric groups S_n. In
this chapter our main aims are to define these groups, prove they are simple,
determine their outer automorphism groups, describe in general terms their
subgroups, and construct their covering groups. At the end of the chapter we
briefly introduce reflection groups as a generalisation of the symmetric groups,
as they play an important role not only in the theory of groups of Lie type, but
also in the construction of many sporadic groups, as well as in the elucidation
of much exceptional behaviour of low-dimensional classical groups.

By way of introduction we bring in the basic concepts of permutation
group theory, such as k-transitivity and primitivity, before presenting one
of the standard proofs of simplicity of A_n for $n \geqslant 5$. Then we prove that
$\mathrm{Aut}(A_n) \cong S_n$ for $n \geqslant 7$, while for $n = 6$ there is an exceptional outer
automorphism of S_6. The subgroup structure of A_n and S_n is described by
the O'Nan–Scott Theorem, which we state and prove after giving a detailed
description of the subgroups which arise in that theorem.

Next we move on to the covering groups, and construct the Schur double
covers $2 \cdot A_n$ for all $n \geqslant 4$. We also construct the exceptional triple covers $3 \cdot A_6$
and $3 \cdot A_7$ (and hence $6 \cdot A_6$ and $6 \cdot A_7$), but make no attempt to prove the fact
that there are no other covers. Finally, we define and prove the Coxeter presentation for S_n on the fundamental transpositions $(i, i+1)$, as an introduction
to reflection groups in general. We state Coxeter's classification theorem for
real reflection groups, and the crystallographic restriction, for future use.

2.2 Permutations

We first define the *symmetric group* $\mathrm{Sym}(\Omega)$ on a set Ω as the group of all
permutations of that set. Here a *permutation* is simply a bijection from the

R.A. Wilson, *The Finite Simple Groups*,
Graduate Texts in Mathematics 251,
© Springer-Verlag London Limited 2009

set to itself. If Ω has cardinality n, then we might as well take $\Omega = \{1, \ldots, n\}$. The resulting symmetric group is denoted S_n, and called *the* symmetric group of degree n.

Since a permutation π of Ω is determined by the images $\pi(1)$ (n choices), $\pi(2)$ ($n-1$ choices, as it must be distinct from $\pi(1)$), $\pi(3)$ ($n-2$ choices), and so on, we see that the number of permutations is $n(n-1)(n-2)\ldots 2.1 = n!$ and therefore $|S_n| = n!$.

A permutation π may conveniently be written simply as a list of the images $\pi(1), \ldots, \pi(n)$ of the points in order, or more explicitly, as a list of the points $1, \ldots, n$ with their images $\pi(1), \ldots, \pi(n)$ written underneath them. For example, $\begin{pmatrix} 1 & 2 & 3 & 4 & 5 \\ 1 & 5 & 2 & 3 & 4 \end{pmatrix}$ denotes the permutation fixing 1, and mapping 2 to 5, 3 to 2, 4 to 3, and 5 to 4. If we draw lines between equal numbers in the two rows, the lines cross over each other, and the crossings indicate which pairs of numbers have to be interchanged in order to produce this permutation. In this example, the line joining the 5s crosses the 4s, 3s and 2s in that order, indicating that we may obtain this permutation by first swapping 5 and 4, then 5 and 3, and finally 5 and 2. A single interchange of two elements is called a *transposition*, so we have seen how to write any permutation as a product of transpositions. Of course, for any given permutation there are many ways of doing this.

2.2.1 The alternating groups

An alternative interpretation of this picture is to read it from bottom to top, and record the *positions* of the strings that are swapped. In this example, we first swap the second and third strings, then the third and fourth, and finally the fourth and fifth. Thus the second string moves to the fifth position, the third string moves to the second position, and so on. In this way we have written our permutation as a product of swaps of *adjacent* strings. Moreover, the product of two permutations can be expressed by concatenating any two corresponding lists of swapping strings.

But if we write the identity permutation as a product of such transpositions, and the line connecting the is crosses over the line connecting the js, then they must cross back again: thus the number of crossings for the identity element is even. It follows that if π is written in two different ways as a product of such transpositions, then either the number of transpositions is even in both cases, or it is odd in both cases. Therefore the map ϕ from S_n onto the group $\{\pm 1\}$ of order 2 defined by $\phi(\pi) = 1$ whenever π is the product of an even number of transpositions, is a (well-defined) group homomorphism. As ϕ is onto, its kernel is a normal subgroup of index 2, which we call the *alternating* group of degree n. It has order $\frac{1}{2}n!$, and its elements are called the *even* permutations. The other elements of S_n are the *odd* permutations. (An alternative proof that A_n has index 2 in S_n can be found in Exercise 2.1.)

The notation for permutations as functions on the left (where $\pi\rho$ means ρ followed by π) is unfortunately inconsistent with the normal convention for permutations that $\pi\rho$ means π followed by ρ. Therefore we adopt a different notation, writing a^π instead of $\pi(a)$, to avoid this confusion. We then have $a^{\pi\rho} = \rho(\pi(a))$, and permutations are read from left to right, rather than right to left as for functions.

2.2.2 Transitivity

Given a group H of permutations, i.e. a subgroup of a symmetric group S_n, we are interested in which points can be mapped to which other points by elements of the group H. If every point can be mapped to every other point, we say H is *transitive* on the set Ω. In symbols, this is expressed by saying that for all a and b in Ω, there exists $\pi \in H$ with $a^\pi = b$. In any case, the set $\{a^\pi \mid \pi \in H\}$ of points reachable from a is called the *orbit* of H containing a. It is easy to see that the orbits of H form a partition of the set Ω.

More generally, if we can simultaneously map k points wherever we like, the group is called *k-transitive*. This means that (for $k \leqslant n$) for every list of k distinct points $a_1, \ldots, a_k$ and every list of k distinct points $b_1, \ldots, b_k$ there exists an element $\pi \in H$ with $a_i^\pi = b_i$ for all i. In particular, 1-transitive is the same as transitive.

For example, it is easy to see that the symmetric group S_n is k-transitive for all $k \leqslant n$, and that the alternating group A_n is k-transitive for all $k \leqslant n-2$.

It is obvious that if $H \neq 1$ is k-transitive then H is $(k-1)$-transitive, and is therefore m-transitive for all $m \leqslant k$. There is however a concept intermediate between 1-transitivity and 2-transitivity which is of interest in its own right. This is the concept of primitivity, which is best explained by defining what it is not.

2.2.3 Primitivity

A *block system* for a subgroup H of S_n is a partition of Ω preserved by H; that is, a set of mutually disjoint non-empty subsets of Ω whose union is Ω. We call the elements of the partition *blocks*. In other words, if two points a and b are in the same block of the partition, then for all elements $\pi \in H$, the points a^π and b^π are also in the same block as each other. There are two block systems which are always preserved by every group: one is the partition consisting of the single block Ω; at the other extreme is the partition in which every block consists of a single point. These are called the *trivial* block systems. A non-trivial block system is often called a *system of imprimitivity* for the group H. If $n \geqslant 3$ then any group which has a system of imprimitivity is called *imprimitive*, and any non-trivial group which is not imprimitive is called *primitive*. (It is usual also to say that S_2 is primitive, but that S_1 is neither primitive nor imprimitive.)

It is obvious that

$$\text{if } H \text{ is primitive, then } H \text{ is transitive.} \tag{2.1}$$

For, if $H \neq 1$ is not transitive, then the orbits of H form a system of imprimitivity for H, so H is not primitive. On the other hand, there exist plenty of transitive groups which are not primitive. For example, in S_4, the subgroup H of order 4 generated by $\begin{pmatrix} 1 & 2 & 3 & 4 \\ 2 & 1 & 4 & 3 \end{pmatrix}$ and $\begin{pmatrix} 1 & 2 & 3 & 4 \\ 3 & 4 & 1 & 2 \end{pmatrix}$ is transitive, but preserves the block system $\{\{1,2\},\{3,4\}\}$. It also preserves the block systems $\{\{1,3\},\{2,4\}\}$ and $\{\{1,4\},\{2,3\}\}$. Of course, in any block system for a transitive imprimitive group, all the blocks have the same size.

Another important basic result about primitive groups is that

$$\text{every 2-transitive group is primitive.} \tag{2.2}$$

For, if H is imprimitive, we can choose three distinct points a, b and c such that a and b are in the same block, while c is in a different block. (This is possible since the blocks have at least two points, and there are at least two blocks.) Then there can be no element of H taking the pair (a, b) to the pair (a, c), so it is not 2-transitive.

2.2.4 Group actions

Suppose that G is a subgroup of S_n acting transitively on Ω. Let H be the stabiliser of the point $a \in \Omega$, that is, $H = \{g \in G \mid a^g = a\}$. Then the points of Ω are in natural bijection with the (right) cosets Hg of H in G. This bijection is given by $Hx \leftrightarrow a^x$. It is left as an exercise for the reader (see Exercise 2.2) to prove that this is a bijection. In particular, $|G : H| = n$.

We can turn this construction around, so that given any subgroup H in G, we can let G act on the right cosets of H according to the rule $(Hx)^g = Hxg$. Numbering the cosets of H from 1 to n, where $n = |G : H|$, we obtain a permutation action of G on these n points, or in other words a group homomorphism from G to S_n. If this homomorphism is injective, we say G acts *faithfully*.

2.2.5 Maximal subgroups

This correspondence between transitive group actions on the one hand, and subgroups on the other, permits many useful translations between combinatorial properties of Ω and group-theoretical properties of G. For example, a primitive group action corresponds to a maximal subgroup, where a subgroup H of G is called *maximal* if there is no subgroup K with $H < K < G$. More precisely:

Proposition 2.1. *Suppose that the group G acts transitively on the set Ω, and let H be the stabiliser of $a \in \Omega$. Then G acts primitively on Ω if and only if H is a maximal subgroup of G.*

Proof. We prove both directions of this in the contrapositive form. First assume that H is not maximal, and choose a subgroup K with $H < K < G$. Then the points of Ω are in bijection with the (right) cosets of H in G. Now the cosets of K in G are unions of H-cosets, so correspond to sets of points, each set containing $|K : H|$ points. But the action of G preserves the set of K-cosets, so the corresponding sets of points form a system of imprimitivity for G on Ω.

Conversely, suppose that G acts imprimitively, and let Ω_1 be the block containing a in a system of imprimitivity. Since G is transitive, it follows that the stabiliser of Ω_1 acts transitively on Ω_1, but not on Ω. Therefore this stabiliser strictly contains H and is a proper subgroup of G, so H is not maximal.

2.2.6 Wreath products

The concept of imprimitivity leads naturally to the idea of a *wreath product* of two permutation groups. Recall the *direct product*

$$G \times H = \{(g, h) \mid g \in G, h \in H\} \tag{2.3}$$

with identity element $1_{G \times H} = (1_G, 1_H)$ and group operations

$$\begin{aligned}
(g_1, h_1)(g_2, h_2) &= (g_1 g_2, h_1 h_2), \\
(g, h)^{-1} &= (g^{-1}, h^{-1}).
\end{aligned} \tag{2.4}$$

Recall also the *semidirect product* $G{:}H$ or $G{:}_\phi H$, where $\phi : H \to \operatorname{Aut}(G)$ describes an action of H on G. We define $G{:}H = \{(g, h) \mid g \in G, h \in H\}$ with identity element $1_{G{:}H} = (1_G, 1_H)$ and group operations

$$\begin{aligned}
(g_1, h_1)(g_2, h_2) &= (g_1 g_2^{\phi(h_1^{-1})}, h_1 h_2), \\
(g, h)^{-1} &= ((g^{-1})^{\phi(h)}, h^{-1}).
\end{aligned} \tag{2.5}$$

Now suppose that H is a permutation group acting on $\Omega = \{1, \ldots, n\}$. Define $G^n = G \times G \times \cdots \times G = \{(g_1, \ldots, g_n) \mid g_i \in G\}$, the direct product of n copies of G, and let H act on G^n by permuting the n subscripts. That is $\phi : H \to \operatorname{Aut}(G^n)$ is defined by

$$\phi(\pi^{-1}) : (g_1, \ldots, g_n) \mapsto (g_{1^\pi}, \ldots, g_{n^\pi}). \tag{2.6}$$

Then the *wreath product* $G \wr H$ is defined to be $G^n{:}_\phi H$. For example, if $H \cong S_n$ and $G \cong S_m$ then the wreath product $S_m \wr S_n$ can be formed by taking n copies of S_m, each acting on one of the sets $\Omega_1, \ldots, \Omega_n$ of size m, and then permuting the subscripts $1, \ldots, n$ by elements of H. This gives an imprimitive action of $S_m \wr S_n$ on $\Omega = \bigcup_{i=1}^n \Omega_i$, preserving the partition of Ω into the Ω_i. More generally, any (transitive) imprimitive group can be embedded in a wreath product: if the blocks of a system of imprimitivity for G are $\Omega_1, \ldots, \Omega_k$, then clearly all the Ω_i have the same size, and G is a subgroup of $\operatorname{Sym}(\Omega_1) \wr S_k$.

2.3 Simplicity

2.3.1 Cycle types

An alternative notation for a permutation π is obtained by considering the *cycles* of π. These are defined by taking an element $a \in \Omega$, which maps under π to a^π: this in turn maps to a^{π^2}, which maps to a^{π^3} and so on. Because Ω is finite, eventually we get a repetition $a^{\pi^j} = a^{\pi^k}$ and therefore $a^{\pi^{j-k}} = a$. Thus the first time we get a repetition is when we get back to the start of the cycle, which can now be written $(a, a^\pi, a^{\pi^2}, \ldots, a^{\pi^{k-1}})$, where k is the *length* of the cycle. Repeating this with a new element b not in this cycle, we get another cycle of π, disjoint from the first. Eventually, we run out of elements of Ω, at which point π is written as a product of disjoint cycles.

The *cycle type* of a permutation is simply a list of the lengths of the cycles, usually abbreviated in some way. Thus the identity has cycle type (1^n) and a transposition has cycle type $(2, 1^{n-2})$. Note, incidentally, that a cycle of *even* length is an *odd* permutation, and vice versa. Thus a permutation is even if and only if it has an even number of cycles of even length.

If $\rho \in S_n$ is another permutation, then $\pi^\rho = \rho^{-1}\pi\rho$ maps a^ρ via a and a^π to $a^{\pi\rho}$. Therefore each cycle $(a, a^\pi, a^{\pi^2}, \ldots, a^{\pi^{k-1}})$ of π gives rise to a corresponding cycle $(a^\rho, a^{\pi\rho}, a^{\pi^2\rho}, \ldots, a^{\pi^{k-1}\rho})$ of π^ρ. So the cycle type of π^ρ is the same as the cycle type of π. Conversely, if π and π' are two permutations with the same cycle type, we can match up the cycles of the same length, say $(a, a^\pi, a^{\pi^2}, \ldots, a^{\pi^{k-1}})$ with $(b, b^{\pi'}, b^{\pi'^2}, \ldots, b^{\pi'^{k-1}})$. Now define a permutation ρ by mapping a^{π^j} to $b^{\pi'^j}$ for each integer j, and similarly for all the other cycles, so that $\pi' = \pi^\rho$. Thus two permutations are conjugate in S_n if and only if they have the same cycle type.

By performing the same operation to conjugate a permutation π to itself, we find the centraliser of π. Specifically, if π is an element of S_n of cycle type $(c_1^{k_1}, c_2^{k_2}, \ldots, c_r^{k_r})$, then the centraliser of π in S_n is a direct product of r groups $C_{c_i} \wr S_{k_i}$ (see Exercise 2.25).

2.3.2 Conjugacy classes in the alternating groups

Next we determine the conjugacy classes in A_n. The crucial point is to determine which elements of A_n are centralised by odd permutations. Given an element g of A_n, and an odd permutation ρ, either g^ρ is conjugate to g by an element π of A_n or it is not. In the former case, g is centralised by the odd permutation $\rho\pi^{-1}$, while in the latter case, every odd permutation maps g into the same A_n-conjugacy class as g^ρ, and so no odd permutation centralises g.

If g has a cycle of even length, it is centralised by that cycle, which is an odd permutation. Similarly, if g has two cycles of the same odd length (possibly of length 1!), it is centralised by an element ρ which interchanges the two cycles: but then ρ is the product of an odd number of transpositions, so is an odd permutation.

On the other hand, if g does not contain an even cycle or two odd cycles of the same length, then it is the product of disjoint cycles of distinct odd lengths, and every element ρ centralising g must map each of these cycles to itself. The first point in each cycle can be mapped to an arbitrary point in that cycle, but then the images of the remaining points are determined. Thus we obtain all such elements ρ as products of powers of the cycles of g. In particular ρ is an even permutation.

This proves that g is centralised by no odd permutation if and only if g is a product of disjoint cycles of distinct odd lengths. It follows immediately that the conjugacy classes of A_n correspond to cycle types if there is a cycle of even length or there are two cycles of equal length, whereas a cycle type consisting of distinct odd lengths corresponds to two conjugacy classes in A_n.

For example, in A_5, the cycle types of even permutations are (1^5), $(3, 1^2)$, $(2^2, 1)$, and (5). Of these, only (5) consists of disjoint cycles of distinct odd lengths. Therefore there are just five conjugacy classes in A_5.

2.3.3 The alternating groups are simple

It is easy to see that a subgroup H of G is normal if it is a union of whole conjugacy classes in G. The group G is *simple* if it has precisely two normal subgroups, namely 1 and G. Every non-abelian simple group G is *perfect*, i.e. $G' = G$.

The numbers of elements in the five conjugacy classes in A_5 are 1, 20, 15, 12 and 12 respectively. Since no proper sub-sum of these numbers including 1 divides 60, there can be no subgroup which is a union of conjugacy classes, and therefore A_5 is a simple group.

We now prove by induction that A_n is simple for all $n \geqslant 5$. The induction starts when $n = 5$, so we may assume $n > 5$. Suppose that N is a non-trivial normal subgroup of A_n, and consider $N \cap A_{n-1}$, where A_{n-1} is the stabiliser in A_n of the point n. This is normal in A_{n-1}, so by induction is either 1 or A_{n-1}. In the second case, $N \geqslant A_{n-1}$, so contains all the elements of cycle type $(3, 1^{n-3})$ and $(2^2, 1^{n-4})$ (since it is normal). But it is easily seen that every even permutation is a product of such elements, so $N = A_n$. Therefore we can assume that $N \cap A_{n-1} = 1$, which means that every non-identity element of N is fixed-point-free (i.e. fixes no points). Thus $|N| \leqslant n$, for if $x, y \in N$ map the point 1 to the same point then xy^{-1} fixes 1 so is trivial.

But N must contain a non-trivial conjugacy class of elements of A_n, and it is not hard to show that if $n \geqslant 5$ then there is no such class with fewer than n elements. We leave this verification as an exercise (Exercise 2.10). This contradiction proves that N does not exist, and so A_n is simple. (An alternative proof of simplicity of A_n is given in Exercise 3.4.)

2.4 Outer automorphisms

2.4.1 Automorphisms of alternating groups

If $n \geqslant 4$ then A_n has trivial centre, so that $A_n \cong \mathrm{Inn}(A_n) \trianglelefteq \mathrm{Aut}(A_n)$. Moreover, each element of S_n induces an automorphism of A_n, by conjugation in S_n, so S_n is (isomorphic to) a subgroup of $\mathrm{Aut}(A_n)$. It turns out that for $n \geqslant 7$ it is actually the whole of $\mathrm{Aut}(A_n)$. We prove this next. (I am grateful to Chris Parker for supplying this argument.)

First we observe that, since $(a, b, c)(a, b, d) = (a, d)(b, c)$, the group A_n is generated by its 3-cycles. Indeed, it is generated by the 3-cycles $(1, 2, 3)$, $(1, 2, 4)$, $\ldots$, $(1, 2, n)$. Also note that for $n \geqslant 5$, A_n has no subgroup of index k less than n—for if it did there would be a homomorphism from A_n onto a transitive subgroup of A_k, contradicting the fact that A_n is simple. We next prove:

Lemma 2.2. *If $n \geqslant 7$ and $A_{n-1} \cong H \leqslant A_n$, then H is the stabiliser of one of the n points on which A_n acts.*

Proof. By the above remark, H cannot act on a non-trivial orbit of length less than $n-1$, so if it is not a point stabiliser then it must act transitively on the n points. For $n = 7$ this is impossible, as 7 does not divide the order of A_6. For $n > 8$, each element of H which corresponds to a 3-cycle of A_{n-1} centralises a subgroup isomorphic to A_{n-4}, with $n - 4 \geqslant 5$, so again by the above remark this subgroup must have an orbit of at least $n - 4$ points. Therefore the '3-cycles' of H can move at most four points, so must act as 3-cycles on the n points. The same is true for $n = 8$, as the 3-cycles centralise A_5, which contains $C_2 \times C_2$, whereas the elements of cycle type $(3^2, 1^2)$ do not centralise $C_2 \times C_2$ in A_8.

Now the elements of H corresponding to $(1, 2, 3)$ and $(1, 2, 4)$ in A_{n-1} generate a subgroup isomorphic to A_4, and therefore map to cycles (a, b, c) and (a, b, d) in A_n. Similarly, the elements corresponding to $(1, 2, j)$ must all map to (a, b, x). It follows that the images (a, b, x) of the $n - 3$ generating elements of H together move exactly $n - 1$ points. Therefore H is one of the point stabilisers isomorphic to A_{n-1}, as required.

Now we are ready to prove the theorem:

Theorem 2.3. *If $n \geqslant 7$ then $\mathrm{Aut}(A_n) \cong S_n$.*

Proof. Any automorphism of A_n permutes the subgroups, and in particular permutes the n subgroups isomorphic to A_{n-1}. But these subgroups are in natural one-to-one correspondence with the n points of Ω, and therefore any automorphism acts as a permutation of Ω, so is an element of S_n.

The theorem is also true for $n = 5$ and for $n = 4$ (see Exercise 2.16).

2.4.2 The outer automorphism of S_6

Of all the symmetric groups, S_6 is perhaps the most remarkable. One manifestation of this is its exceptional outer automorphism. This is an isomorphism from S_6 to itself which does not correspond to a permutation of the underlying set of six points. What this means is that there is a completely different way for S_6 to act on six points.

To construct a non-inner automorphism ϕ of S_6 we first note that ϕ must map the point stabiliser S_5 to another subgroup $H \cong S_5$. However, H cannot fix any of the six points on which S_6 acts, so H must be transitive on these six points.

Thus our first job is to construct a transitive action of S_5 on six points. This may be obtained in a natural way as the action of S_5 by conjugation on its six Sylow 5-subgroups. (If we wish to avoid using Sylow's theorems at this point we can simply observe that the 24 elements of order 5 belong to six cyclic subgroups $\langle (1, 2, x, y, z) \rangle$, and that these are permuted transitively by conjugation by elements of S_5.)

Going back to S_6, we have now constructed our transitive subgroup H of index 6. Thus S_6 acts naturally (and transitively) on the six cosets Hg by right multiplication. More explicitly, we can define a group homomorphism $\phi : S_6 \to \mathrm{Sym}(\{Hg \mid g \in S_6\}) \cong S_6$. The kernel of ϕ is trivial, since S_6 has no non-trivial normal subgroups of index 6 or more. Hence ϕ is a group isomorphism, i.e. an automorphism of S_6.

But ϕ is not an inner automorphism, because it maps the transitive subgroup H to the stabiliser of the trivial coset H, whereas inner automorphisms preserve transitivity. [A more sophisticated version of this construction is given in Section 3.3.5 in the discussion of $\mathrm{PSL}_2(5)$. An alternative construction of the outer automorphism of S_6 is given in Section 4.2.]

This is the only outer automorphism of S_6, in the sense that S_6 has index 2 in its full automorphism group. Indeed, we can prove the stronger result that the outer automorphism group of A_6 has order 4. For any automorphism maps the 3-cycles to elements of order 3, which are either 3-cycles or products of two disjoint 3-cycles. Therefore it suffices to show that any automorphism which maps 3-cycles to 3-cycles is in S_6. But this follows by the same argument as in Lemma 2.2 and Theorem 2.3 (see Exercise 2.18).

2.5 Subgroups of S_n

There are a number of more or less obvious subgroups of the symmetric groups. In order to simplify the discussion it is usual to (partly) classify the maximal subgroups first, and to study arbitrary subgroups by looking at them as subgroups of the maximal subgroups. In this section we describe some important classes of (often maximal) subgroups, and prove maximality in a few cases.

The converse problem, of showing that any maximal subgroup is in one of these classes, is addressed in Section 2.6.

The first two classes are the intransitive subgroups and the transitive imprimitive subgroups. The other four are types of maximal primitive subgroups of S_n which are 'obvious' to the experts, and are generally labelled the primitive wreath product, affine, diagonal, and almost simple types. We shall not prove that any of these are maximal, and indeed sometimes they are not.

2.5.1 Intransitive subgroups

If H is an intransitive subgroup of S_n, then it has two or more orbits on the underlying set of n points. If these orbits have lengths $n_1, \ldots, n_r$, then H is a subgroup of the subgroup $S_{n_1} \times \cdots \times S_{n_r}$ consisting of all permutations which permute the points in each orbit, but do not mix up the orbits. If $r > 2$, then we can mix up all the orbits except the first one, to get a group $S_{n_1} \times S_{n_2 + \cdots + n_r}$ which lies between H and S_n. Therefore, in this case H cannot be maximal.

On the other hand, if $r = 2$, we have the subgroup $H = S_k \times S_{n-k}$ of S_n, and it is quite easy to show this is a maximal subgroup, as long as $k \neq n - k$. For, we may as well assume $k < n - k$, and that the factor S_k acts on $\Omega_1 = \{1, 2, \ldots, k\}$, while the factor S_{n-k} acts on $\Omega_2 = \{k + 1, \ldots, n\}$. If g is any permutation not in H, let K be the subgroup generated by H and g. Our aim is to show that K contains all the transpositions of S_n, and therefore is S_n.

Now g must move some point in Ω_2 to a point in Ω_1, but cannot do this to all points in Ω_2, since $|\Omega_2| > |\Omega_1|$. Therefore we can choose $i, j \in \Omega_2$ with $i^g \in \Omega_1$ and $j^g \in \Omega_2$. Then $(i, j) \in H$ so $(i^g, j^g) \in H^g \leqslant K$. Conjugating this transposition by elements of H we obtain all the transpositions of S_n (except those which are already in H), and therefore $K = S_n$. This implies that H is a maximal subgroup of S_n. Note that we have now completely classified the intransitive maximal subgroups of S_n, so any other maximal subgroup must be transitive. For example, the intransitive maximal subgroups of S_6 are S_5 and $S_4 \times S_2$.

2.5.2 Transitive imprimitive subgroups

In the case when $k = n - k$, this proof breaks down, and in fact the subgroup $S_k \times S_k$ is not maximal in S_{2k}. This is because there is an element h in S_{2k} which interchanges the two orbits of size k, and normalises the subgroup $S_k \times S_k$. For example we may take $h = (1, k + 1)(2, k + 2) \cdots (k, 2k)$. Indeed, what we have here is the wreath product of S_k with S_2. This can be shown to be a maximal subgroup of S_{2k} by a similar method to that used above (see Exercise 2.27).

More generally, if we partition the set of n points into m subsets of the same size k (so that $n = km$), then the wreath product $S_k \wr S_m$ can act on this partition: the *base group* $S_k \times \cdots \times S_k$ consists of permutations of each of the

m subsets separately, while the wreathing action of S_m acts by permuting the m orbits of the base group. It turns out that this subgroup is maximal in S_n also (see Exercise 2.28). Thus we obtain a list of all the transitive imprimitive maximal subgroups of S_n. These are the groups $S_k \wr S_m$ where $k > 1$, $m > 1$ and $n = km$. For example, the transitive imprimitive maximal subgroups of S_6 are $S_2 \wr S_3$ (preserving a set of three blocks of size 2, for example generated by the three permutations $(1,2)$, $(1,3,5)(2,4,6)$ and $(3,5)(4,6)$) and $S_3 \wr S_2$ (preserving a set of two blocks of sise 3, for example generated by the three permutations $(1,2,3)$, $(1,2)$ and $(1,4)(2,5)(3,6)$).

2.5.3 Primitive wreath products

We have completely classified the imprimitive maximal subgroups of S_n, so all the remaining maximal subgroups of S_n must be primitive. To see an example of a primitive subgroup of S_n, consider the case when $n = k^2$, and arrange the n points in a $k \times k$ array. Let one copy of S_k act on this array by permuting the columns around, leaving each row fixed as a set. Then let another copy of S_k act by permuting the rows around, leaving each column fixed as a set. These two copies of S_k commute with each other, so generate a group $H \cong S_k \times S_k$. Now H is imprimitive, as the rows form one system of imprimitivity, and the columns form another. But if we adjoin the permutation which reflects in the main diagonal, so mapping rows to columns and vice versa, then we get a group $S_k \wr S_2$ which turns out to be primitive. For example, there is a primitive subgroup $S_3 \wr S_2$ in S_9, which however turns out not to be maximal. In fact the smallest case which is maximal is the subgroup $S_5 \wr S_2$ in S_{25}.

Generalising this construction to an m-dimensional array in the case when $n = k^m$, with $k > 2$ and $m > 1$, we obtain a primitive action of the group $S_k \wr S_m$ on k^m points. To make this more explicit, we identify Ω with the Cartesian product $\Omega_1{}^m$ of m copies of a set Ω_1 of size k, and let an element $(\pi_1, \ldots, \pi_m)$ of the base group $S_k{}^m$ act by

$$(a_1, \ldots, a_m) \mapsto (a_1{}^{\pi_1}, \ldots, a_m{}^{\pi_m}) \qquad (2.7)$$

for all $a_i \in \Omega_1$. The wreathing action of $\rho^{-1} \in S_m$ is then given by the natural action permuting the coordinates, thus:

$$\rho^{-1} : (a_1, \ldots, a_m) \mapsto (a_{1\rho}, \ldots, a_{m\rho}). \qquad (2.8)$$

This action of the wreath product is sometimes called the *product action*, to distinguish it from the imprimitive action on km points described in Section 2.5.2 above. We shall not prove maximality of these subgroups in S_n or A_n, although they are in fact maximal in A_n if $k \geqslant 5$ and k^{m-1} is divisible by 4, and maximal in S_n if $k \geqslant 5$ and k^{m-1} is not divisible by 4.

2.5.4 Affine subgroups

The affine groups are essentially the symmetry groups of vector spaces. Let p be a prime, and let $\mathbb{F}_p = \mathbb{Z}/p\mathbb{Z}$ denote the field of order p (for more on finite fields see Section 3.2). Let V be the vector space of k-tuples of elements of $\mathbb{F}_p$. Then V has p^k elements, and has a symmetry group which is the semidirect product of the group of translations $t_a : v \mapsto v + a$, by the general linear group $\mathrm{GL}_k(p)$ consisting of all invertible $k \times k$ matrices over $\mathbb{F}_p$. This group, sometimes denoted $\mathrm{AGL}_k(p)$, and called the *affine general linear group*, acts as permutations of the vectors, so is a subgroup of S_n where $n = p^k$. The translations form a normal subgroup isomorphic to the additive group of the vector space, which is isomorphic to a direct product of k copies of the cyclic group C_p. In other words it is an *elementary abelian* group of order p^k, which we denote E_{p^k}, or simply p^k. With this notation, $\mathrm{AGL}_k(p) \cong p^k{:}\mathrm{GL}_k(p)$.

An example of an affine group is the group $\mathrm{AGL}_3(2) \cong 2^3{:}\mathrm{GL}_3(2)$, which acts as a permutation group on the 8 vectors of $\mathbb{F}_2{}^3$, and so embeds in S_8. Indeed, it is easy to check that all its elements are even permutations, so it embeds in A_8. Another example is $\mathrm{AGL}_1(7) \cong 7{:}6$ which is a maximal subgroup of S_7. Note however that its intersection with A_7 is a group $7{:}3$ which is *not* maximal in A_7. These groups $7{:}6$ and $7{:}3$ are examples of *Frobenius groups*, which are by definition transitive non-regular permutation groups in which the stabiliser of any two points is trivial. Other examples of Frobenius groups are the *dihedral* groups $D_{2n} \cong n{:}2$ of symmetries of the regular n-gon.

2.5.5 Subgroups of diagonal type

The diagonal type groups are less easy to describe. They are built from a non-abelian simple group T, and have the shape

$$T^k.(\mathrm{Out}(T) \times S_k) \cong (T \wr S_k).\mathrm{Out}(T). \tag{2.9}$$

Here there is a normal subgroup $T \wr S_k$, extended by a group of outer automorphisms which acts in the same way on all the k copies of T. This group contains a subgroup $\mathrm{Aut}(T) \times S_k$ consisting of a diagonal copy of T (i.e. the subgroup of all elements $(t, \dots, t)$ with $t \in T$), extended by its outer automorphism group and the permutation group. This subgroup has index $|T|^{k-1}$, so the permutation action of the group on the cosets of this subgroup gives an embedding of the whole group in S_n, where $n = |T|^{k-1}$.

The smallest example of such a group is $(A_5 \times A_5){:}(C_2 \times C_2)$ acting on the cosets of a subgroup $S_5 \times C_2$. This group is the semidirect product of $A_5 \times A_5 = \{(g, h) \mid g, h \in A_5\}$ by the group $C_2 \times C_2$ of automorphisms generated by $\alpha : (g, h) \mapsto (g^\pi, h^\pi)$, where π is the transposition $(1, 2)$, and $\beta : (g, h) \mapsto (h, g)$. The point stabiliser is the centraliser of β, generated by α, β and $\{(g, g) \mid g \in A_5\}$. Therefore an alternative way to describe the action of the group on 60 points is as the action by conjugation on the 60 conjugates of β.

2.5.6 Almost simple groups

Finally, there are the almost simple primitive groups. A group G is called *almost simple* if it satisfies $T \leqslant G \leqslant \mathrm{Aut}(T)$ for some simple group T. Thus it consists of a simple group, possibly extended by adjoining some or all of the outer automorphism group. If M is any maximal subgroup of G, then the permutation action of G on the cosets of M is primitive, so G embeds as a primitive subgroup of S_n, where $n = |G : M|$. The class of almost simple maximal subgroups of S_n is chaotic in general, and to describe them completely would require complete knowledge of the maximal subgroups of all almost simple groups—a classic case of reducing an impossible problem to an even harder one!

However, a result of Liebeck, Praeger and Saxl [120] states that (subject to certain technical conditions) every such embedding of G in S_n is maximal unless it appears in their explicit list of exceptions. It is also known that as n tends to infinity, for almost all values of n there are no almost simple maximal subgroups of S_n or A_n.

2.6 The O'Nan–Scott Theorem

The O'Nan–Scott theorem gives us a classification of the maximal subgroups of the alternating and symmetric groups. Roughly speaking, it tells us that every maximal subgroup of S_n or A_n is of one of the types described in the previous section. It does not tell us exactly what the maximal subgroups are, but it does provide a first step towards writing down the list of maximal subgroups of A_n or S_n for any particular reasonable value of n.

Theorem 2.4. *If H is any proper subgroup of S_n other than A_n, then H is a subgroup of one or more of the following subgroups:*

 (i) an intransitive group $S_k \times S_m$, where $n = k + m$;
 (ii) an imprimitive group $S_k \wr S_m$, where $n = km$;
 (iii) a primitive wreath product, $S_k \wr S_m$, where $n = k^m$;
 (iv) an affine group $\mathrm{AGL}_d(p) \cong p^d{:}\mathrm{GL}_d(p)$, where $n = p^d$;
 (v) a group of shape $T^m.(\mathrm{Out}(T) \times S_m)$, where T is a non-abelian simple group, acting on the cosets of a subgroup $\mathrm{Aut}(T) \times S_m$, where $n = |T|^{m-1}$;
 (vi) an almost simple group acting on the cosets of a maximal subgroup of index n.

Note that the theorem does not assert that all these subgroups are maximal in S_n, or in A_n. This is a rather subtle question. As we noted in Section 2.5.6, the last category of subgroups also requires us to know all the maximal subgroups of all the finite simple groups, or at least those of index n. In practice, this means that we can only ever hope to get a *recursive* description of the maximal subgroups of A_n and S_n.

In view of the fundamental importance of the O'Nan–Scott Theorem, we shall give a proof. However, this proof is not easy, and could reasonably be omitted at a first reading.

2.6.1 General results

In this section we collect a number of general facts about (finite) groups which will be useful in the proof of the O'Nan–Scott theorem, as well as being of more general importance. Throughout this section we assume that H acts faithfully on a set Ω.

Lemma 2.5. *Every non-trivial normal subgroup N of a primitive group H is transitive.*

Proof. Otherwise the orbits of N form a system of imprimitivity for H.

A normal subgroup N of a group H is called *minimal* if $N \neq 1$ and N contains no normal subgroup of H except 1 and N.

Lemma 2.6. *Any two distinct minimal normal subgroups N_1 and N_2 of any group H commute.*

Proof. By normality, $[N_1, N_2] \leqslant N_1 \cap N_2 \trianglelefteq H$, so by minimality

$$[N_1, N_2] = N_1 \cap N_2 = 1.$$

A subgroup K of a group N is called *characteristic* if it is fixed by all automorphisms of N. The following is obvious:

Lemma 2.7. *If K is characteristic in N and N is normal in H then K is normal in H.*

A group $K \neq 1$ is called *characteristically simple* if K has no proper non-trivial characteristic subgroups. Thus Lemma 2.7 is saying that any minimal normal subgroup of H is characteristically simple.

Lemma 2.8. *If K is characteristically simple then it is a direct product of isomorphic simple groups.*

Proof. If T is any minimal normal subgroup of K, then so is T^α for any $\alpha \in \mathrm{Aut}K$. So by the proof of Lemma 2.6 either $T^\alpha = T$ or $T \cap T^\alpha = 1$. In the latter case $TT^\alpha = T \times T^\alpha$ is a direct product. Since K is characteristically simple, it is generated by all the T^α. By induction we obtain that K is a direct product of a certain number of such T^α. But then any normal subgroup of T is normal in K, so by minimality of T, T is simple.

Corollary 2.9. *Every minimal normal subgroup N of a finite group H is a direct product of isomorphic simple groups (not necessarily non-abelian).*

Proof. By minimality, N is characteristically simple.

A group N is called *regular* on Ω if for each pair of points a and b in Ω, there is exactly one element of N mapping a to b. In particular N is transitive and $|N| = |\Omega|$, and every non-identity element of N is fixed-point-free.

Lemma 2.10. *If H is primitive, and N is a non-trivial normal subgroup of H, then either $C_H(N)$ is trivial, or $C_H(N)$ is regular and $|C_H(N)| = |\Omega|$.*

Proof. Clearly $C_H(N)$ is normal in H, so by Lemma 2.5, if $C_H(N) \neq 1$ then both N and $C_H(N)$ are transitive. Moreover, if $1 \neq x \in C_H(N)$ has any fixed points, then the set of fixed points of x is preserved by N. This contradiction implies that every element of $C_H(N)$ is fixed-point-free. This means that $C_H(N)$ is regular.

This has a number of important consequences.

Corollary 2.11. *If H is primitive, and N_1 and N_2 are non-trivial normal subgroups of H, and $[N_1, N_2] = 1$, then $N_2 = C_H(N_1)$ and vice versa. In particular, H contains at most two minimal normal subgroups, and if it has an abelian normal subgroup then it has only one minimal normal subgroup.*

Proof. By Lemma 2.5, N_1 is transitive, and by Lemma 2.10, $C_H(N_2)$ is regular. But $N_1 \subseteq C_H(N_2)$, and therefore N_1 and $C_H(N_2)$ have the same order and are equal.

Corollary 2.12. *With the same notation, $N_1 \cong N_2$.*

Proof. The result is trivial if $N_1 = N_2$, so assume $N_1 \neq N_2$, and therefore $N_1 \cap N_2 = 1$. Fix a point $x \in \Omega$, and let K be the stabiliser of x in the group $N_1 N_2$. Then $K \cap N_1 = K \cap N_2 = 1$ as N_1 and N_2 are regular. Therefore $KN_1 = KN_2 = N_1 N_2$, and by the third isomorphism theorem

$$K \cong K/(K \cap N_1) \cong KN_1/N_1 = N_2 N_1/N_1 \cong N_2/(N_1 \cap N_2) \cong N_2$$

and similarly $K \cong N_1$.

Lemma 2.13. *Suppose that H is primitive and N is a non-trivial normal subgroup of H. Let K be the stabiliser in H of a point. Then $KN = H$.*

Proof. By Lemma 2.5, N is transitive, so the result follows by the orbit–stabiliser theorem.

The following result is called the *Dedekind modular law* and although it is very easy to prove it is surprisingly useful.

Lemma 2.14. *If K, X and N are subgroups of a group G and $X \leqslant N$, then $N \cap (KX) = (N \cap K)X$.*

Proof. It is obvious that $(N \cap K)X \leqslant N \cap (KX)$. Conversely, if $k \in K$ and $x \in X$ satisfy $kx \in N$, then also $k \in N$, so $kx \in (N \cap K)X$.

A subgroup X of H is called *K-invariant* if $K \leqslant N_H(X)$.

Lemma 2.15. *Suppose that H is primitive and N is a minimal normal subgroup of H. Let K be the stabiliser in H of a point. Then $K \cap N$ is maximal among K-invariant proper subgroups of N.*

Proof. If $K \cap N < X < N$ and $K \leqslant N_H(X)$ then KX is a subgroup of H. Moreover, X contains elements (of N) not in K, so $K < KX$; and H contains elements (of N) not in X, so $N \cap (KX) = (N \cap K)X = X < N$ and therefore $KX < KXN = KN = H$. This contradicts the maximality of K in H.

2.6.2 The proof of the O'Nan–Scott Theorem

With this preparation we are ready to embark on the proof of the theorem. Let H be a subgroup of S_n not containing A_n, and let N be a minimal normal subgroup of H. Let K be the stabiliser in H of a point.

Reduction to the case N unique and non-abelian.

Certainly H is either intransitive (giving case (i) of the theorem), or transitive imprimitive (giving case (ii) of the theorem), or primitive. So we may assume from now on that H is primitive.

If N is abelian, then by Corollary 2.9 it is an elementary abelian p-group, and by Lemma 2.10 it acts regularly, and by Corollary 2.11, $N = C_H(N)$. Therefore H is affine (case (iv) of the theorem).

Otherwise, all minimal normal subgroups of H are non-abelian. If there is more than one minimal normal subgroup, say N_1 and N_2, then by Corollaries 2.11 and 2.12 $N_1 \cong N_2$ and both N_1 and N_2 act regularly on Ω.

Thus N_1 and N_2 act in the same way on the n points, so there is an element x of S_n conjugating N_1 to N_2. Moreover, by Corollary 2.11, $N_2 = C_H(N_1)$. Therefore x conjugates N_2 to N_1, and $\langle H, x \rangle$ has a unique minimal normal subgroup $N = N_1 \times N_2$. So this case reduces to the case when there is a unique minimal normal subgroup.

The case N unique and non-abelian.

From now on, we can assume that H has a unique minimal normal subgroup, N, which is non-abelian. If N is simple, then $C_H(N) = 1$ and so we are in case (vi) of the theorem. Otherwise, N is non-abelian, non-simple, say $N = T_1 \times \cdots \times T_m$ with $T_i \cong T$ simple for all $1 \leqslant i \leqslant m$, and $m > 1$, and H permutes the T_i transitively by conjugation. Also, by Lemma 2.13 we have $KN = H$, where K is the stabiliser in H of a point.

For each i, let K_i be the image of $K \cap N$ under the natural projection from N to T_i. In particular, $K \cap N \leqslant K_1 \times \cdots \times K_m$. We divide into two cases: either $K_i \neq T_i$ for some i (and therefore for all i) or $K_i = T_i$ for all i.

Case 1: $K_i \neq T_i$.

Now K normalises $K_1 \times \cdots \times K_m$, so that, in this case, by maximality (Lemma 2.15), we have $K \cap N = K_1 \times \cdots \times K_m$, and K permutes the K_i transitively since $H = KN$. Let k be the index of K_i in T_i. Then H is evidently contained in the group $S_k \wr S_m$ acting in the product action (case (iii) of the theorem).

Case 2: $K_i = T_i$.

This is the hardest case. For the purposes of this proof define the *support* of an element $(t_1, \ldots, t_m) \in T_1 \times \cdots \times T_m = N$ to be the set $\{i \mid t_i \neq 1\}$. Let Ω_1 be a minimal (non-empty) subset of $\{1, \ldots, m\}$ such that $K \cap N$ contains an element whose support is Ω_1. Then the subgroup of all elements of $K \cap N$ with support Ω_1 still maps onto T_i, for each $i \in \Omega_1$, since it maps onto a normal subgroup and T_i is simple. Now if Ω_2 is another such set, intersecting Ω_1 non-trivially, then there are elements x and y in $K \cap N$ such that $[x, y] \neq 1$ has support contained in $\Omega_1 \cap \Omega_2$. Minimality of Ω_1 implies that $\Omega_1 \cap \Omega_2 = \Omega_1$. In other words, Ω_1 is a block in a block system invariant under K, and therefore under $H = KN$.

The blocks cannot have size 1, for then K contains N, a contradiction. Now

$$n = |\Omega| = |H : K| = |N : N \cap K| \tag{2.10}$$

since $H = KN$ and $KN/N \cong K/N \cap K$. If the block system is non-trivial, with l blocks of size k, say, and $l > 1$, then $N \cong T^{kl}$ and $N \cap K \cong T^l$ so $n = |T|^{(k-1)l}$. Thus we see that H lies inside $S_r \wr S_l$, in its product action, where $r = |T|^{k-1}$. This is case (iii) of the theorem again.

Otherwise, the block system is trivial, so $K \cap N$ is a diagonal copy of T inside $T_1 \times \cdots \times T_m$, and we can choose our notation such that it consists of the elements $(g, g, \ldots, g)$ for all $g \in T$. Also $n = |T|^{m-1}$, and the n points can be identified with the n conjugates of $K \cap N$ by elements of N. The largest subgroup of S_n preserving this setup is as in case (v) of the theorem. This concludes the proof of the O'Nan–Scott Theorem.

2.7 Covering groups

2.7.1 The Schur multiplier

We have seen that the alternating groups arise as (normal) subgroups of the symmetric groups, such that $S_n/A_n \cong C_2$. They also arise as *quotients* of bigger groups $2 \cdot A_n$, by a subgroup C_2, so that $2 \cdot A_n/C_2 \cong A_n$. Under the natural quotient map, each element π of A_n comes from two elements of $2 \cdot A_n$. We label these two elements $+\pi$ and $-\pi$, but it must be understood that there is no canonical choice of which element gets which sign: your choice of signs may be completely different from mine. To avoid confusion, we write the cycles in $2 \cdot A_n$ with square brackets instead of round ones.

This group $2 \cdot A_n$ is called the *double cover* of A_n. More generally, if G is any finite group, we say that $\widetilde{G}$ is a *covering group* of G if $Z(\widetilde{G}) \leqslant \widetilde{G}'$ and $\widetilde{G}/Z(\widetilde{G}) \cong G$. If the centre has order 2, 3, etc., the covering group is often referred to as a *double*, *triple*, etc., cover as appropriate. It turns out that every finite perfect group G has a unique maximal covering group $\widehat{G}$, with the property that every other covering group is a quotient of $\widehat{G}$. This is called the *universal cover*, and its centre is called the *Schur multiplier* of G. On the other hand, if G is not perfect there may be more than one maximal covering group. For example, the group $C_2 \times C_2$ has four: one is isomorphic to the quaternion group Q_8 and the other three are isomorphic to the dihedral group D_8. [The *quaternion group* Q_8 consists of the elements $\pm 1, \pm i, \pm j, \pm k$ and the multiplication is given by $i^2 = j^2 = k^2 = -1$, $ij = -ji = k$, $jk = -kj = i$, and $ki = -ik = j$.]

2.7.2 The double covers of A_n and S_n

To define $2 \cdot A_n$ it suffices to define the multiplication. One way to do this is to define a double cover of the whole symmetric group, as follows. First choose arbitrarily the element $[1, 2]$, mapping to the transposition $(1, 2)$ of S_n. Then define all the elements $[i, j]$ inductively by the rule $[i, j]^{\pm \pi} = -[i^\pi, j^\pi]$ if π is an odd permutation. Then define the elements mapping to cycles $(a_i, a_{i+1}, \ldots, a_j)$ by $[a_i, a_{i+1}, \ldots, a_j] = [a_i, a_{i+1}][a_i, a_{i+2}] \cdots [a_i, a_j]$.

Finally, all elements are obtained by multiplying together disjoint cycles in the usual way. However, we must be careful not to permute the cycles, or start a cycle at a different point, as this may change $+\pi$ into $-\pi$. For example, our rules tell us that
$$[1, 2] = [1, 2]^{[1,2]} = -[2, 1]$$
while

$$[1, 2]^{[3,4]} = -[1, 2]$$
$$\Rightarrow [1, 2][3, 4] = -[3, 4][1, 2]. \tag{2.11}$$

Any product can now be computed using these rules, by first writing each cycle as a product of transpositions, and then simplifying. However, it is not obvious that if we compute the same product in two different ways, then we get the same answer. You will have to take this on trust for now. [A proof can be obtained by embedding the symmetric group in an orthogonal group, and using the construction of the double cover of the orthogonal group in Section 3.9.]

For example, consider the case $n = 4$. There are 6 transpositions in S_4, lifting to 12 elements $\pm [i, j]$ in the double cover. These elements square to ± 1, but they are all conjugate, so either they all square to 1 or they all square to -1. Let us suppose for simplicity that $[i, j]^2 = 1$ for all i and j. Next there are some elements like $[1, 2][3, 4]$. We have already seen that $[3, 4][1, 2] = -[1, 2][3, 4]$ and therefore

$$[1, 2][3, 4][1, 2][3, 4] = -[3, 4][1, 2][1, 2][3, 4]$$
$$= [3, 4][3, 4]$$
$$= -1. \tag{2.12}$$

Similarly the elements $\pm[1,3][2,4]$ and $\pm[1,4][2,3]$ square to -1, so together these elements form a copy of the quaternion group of order 8. The method for multiplying elements together can best be demonstrated by an example, such as the following.

$$
\begin{aligned}
[2,1,3][3,1,4] &= [2,1][2,3][3,1][3,4] \\
&= [2,1][3,1]^{[2,3]}[2,3][3,4] \\
&= -[2,1][2,1][2,3][3,4] \\
&= [2,1][2,1][3,2][3,4] \\
&= [3,2,4]
\end{aligned}
$$

In this way we obtain two double covers of S_n: the first one, denoted $2{\cdot}S_n^+$, is the one in which the elements $[i,j]$ have order 2. The second one, in which the elements $[i,j]$ square to the central involution -1, is denoted $2{\cdot}S_n^-$. Both contain the same subgroup of index 2, the double cover $2{\cdot}A_n$ of A_n.

See Section 3.3.1 for an explicit representation of $2{\cdot}S_4$ as $GL_2(3)$. See also Sections 5.6.8 and 5.6.1, and Exercises 2.35 and 2.37, for descriptions of $2{\cdot}A_4$ and $2{\cdot}A_5$ as groups of unit quaternions.

2.7.3 The triple cover of A_6

The double covers of the alternating groups, described in Section 2.7.2, are in fact the only covering groups of A_n for $n \geqslant 8$, and for $n = 4$ or 5, but A_6 and A_7 have exceptional triple covers as well. These can both be seen as the groups of symmetries of certain sets of vectors in complex 6-space. We let $\omega = e^{2\pi i/3} = (-1 + \sqrt{-3})/2$ be a primitive (complex) cube root of unity, and consider the vectors $(0,0,1,1,1,1)$, $(0,1,0,1,\omega,\overline{\omega})$ and their multiples by ω and $\overline{\omega} = \omega^2$. Then take the images of these vectors under the group S_4 of coordinate permutations generated by $(1,2)(3,4)$, $(3,4)(5,6)$, $(1,3,5)(2,4,6)$ and $(1,3)(2,4)$ (that is, the stabiliser in A_6 of the partition $\{\{1,2\},\{3,4\},\{5,6\}\}$, which we shall write as $(12 \mid 34 \mid 56)$ for short). These vectors come in triples of scalar multiples, and there are 15 such triples. Indeed, each triple consists of the three vectors whose zeroes are in a given pair of coordinates.

In addition to the above coordinate permutations, this set of 45 vectors is invariant under other monomial elements (that is, products of permutations and diagonal matrices). For example, it is a routine exercise to verify that the set is invariant under $(1,2,3)\mathrm{diag}(1,1,1,1,\overline{\omega},\omega)$, i.e. the map

$$
(x_1,\ldots,x_6) \mapsto (x_3,x_1,x_2,x_4,\overline{\omega}x_5,\omega x_6). \tag{2.13}
$$

This group G of symmetries now acts on the 6 coordinate positions, inducing all even permutations. Thus we obtain a homomorphism from G onto A_6, whose kernel consists of diagonal matrices. But any element of the kernel must take each of the 45 vectors to a scalar multiple of itself (since it does not move the zeroes), so is a scalar. Hence the kernel is the group of scalars

$\{1, \omega, \overline{\omega}\}$ of order 3. Thus we have constructed a group G of order 1080 with $Z(G) \cong C_3$ and $G/Z(G) \cong A_6$.

Finally notice that $(4, 5, 6)\mathrm{diag}(\omega, \overline{\omega}, 1, 1, 1, 1)$ is also a symmetry, and its commutator with $(1, 2, 3)\mathrm{diag}(1, 1, 1, 1, \overline{\omega}, \omega)$ is $\mathrm{diag}(\omega, \dots, \omega)$, so the scalars are inside G'. Therefore $G = G'$, since A_6 is simple, so G is perfect.

This group, written $3 \cdot A_6$, can be extended to a group $3 \cdot S_6$ by adjoining the map which interchanges the last two coordinates and then replaces every coordinate by its complex conjugate.

This construction of $3 \cdot A_6$ and $3 \cdot S_6$ is of fundamental importance for the sporadic groups, as well as for much of the exceptional behaviour of small classical and exceptional groups. Compare for example Section 5.2.1 on the hexacode, used for constructing the Mathieu group M_{24}, Sections 3.12.2 and 3.12.3 on exceptional covers of $\mathrm{PSU}_4(3)$ and $\mathrm{PSL}_3(4)$, and Section 5.6.8 on the exceptional double cover of $G_2(4)$.

2.7.4 The triple cover of A_7

Now the groups $3 \cdot A_7$ and $3 \cdot S_7$ can be described by extending the above set of 45 vectors to a set of 63 by adjoining the 18 images of $(2, 0, 0, 0, 0, 0)$ under $3 \cdot A_6$. There are now some new symmetries, such as

$$
\frac{1}{2}
\begin{pmatrix}
2 & 0 & 0 & 0 & 0 & 0 \\
0 & 0 & 1 & 1 & 1 & 1 \\
0 & 1 & 0 & 1 & \omega & \overline{\omega} \\
0 & 1 & 1 & 0 & \overline{\omega} & \omega \\
0 & 1 & \overline{\omega} & \omega & 1 & 0 \\
0 & 1 & \omega & \overline{\omega} & 0 & 1
\end{pmatrix}.
\tag{2.14}
$$

Indeed there are just 7 'coordinate frames' consisting of 6 mutually orthogonal vectors (up to scalar multiplication) from the set of 63. We label the standard coordinate frame with the number 7, and the frame given by the rows of the above matrix with the number 1. Similary we obtain frames 2 to 6 containing the vectors $(0, 2, 0, 0, 0, 0)$, $\dots$, $(0, 0, 0, 0, 0, 2)$ respectively. With this numbering, the matrix (2.14) corresponds to the permutation $(1, 7)(5, 6)$ of A_7.

This gives a map from our group onto the group A_7 of permutations of the 7 coordinate frames. The kernel K of this map fixes each of the 7 frames, and therefore fixes the intersection of every pair of frames. But each of the 21 triples $\{v, \omega v, \overline{\omega} v\}$ of vectors is the intersection of two frames, so each triple is fixed by K, and the argument given above for $3 \cdot A_6$ shows that K consists of scalars.

Therefore the group G we have constructed satisfies $K = Z(G) \cong C_3$ and $G/Z(G) \cong A_7$, as well as $G = G'$. This group is denoted $3 \cdot A_7$.

2.8 Coxeter groups

2.8.1 A presentation of S_n

We have looked at the symmetric groups as groups of permutations of points, but for many purposes we want to use linear algebra, so it is convenient to consider the points as basis vectors in a vector space. More formally, let $V = \mathbb{R}^n$ be the canonical real vector space of dimension n, and let $\{e_1, \ldots, e_n\}$ be the canonical basis, so that $e_1 = (1, 0, 0, \ldots, 0)$ and so on. Let S_n act on this vector space by permuting the basis vectors in the natural way, that is $e_i{}^\pi = e_{i^\pi}$. (Here we write linear maps as superscripts rather than as functions, for conformity with our notation for permutations. Note that π^{-1}, not π, maps the general vector $\sum_{i=1}^n \lambda_i e_i$ to $\sum_{i=1}^n \lambda_{i^\pi} e_i$.)

Now the transpositions (i, j) have a single eigenvalue -1, and all other eigenvalues 1: specifically, $e_i - e_j$ is an eigenvector with eigenvalue -1, while $e_i + e_j$ and e_k for $i \neq k \neq j$ are linearly independent eigenvectors with eigenvalue 1. Such linear maps are called *reflections*, because they fix a space of dimension $n - 1$, and 'reflect' across this subspace by negating all vectors in the orthogonal 1-space. We call S_n a *reflection group* since it is generated by these reflections.

The symmetric group S_n can be generated by the $n-1$ *fundamental* transpositions $(i, i + 1)$, or by the corresponding *fundamental reflections* acting on V. Obviously, these reflections have order 2, and commute if they are not adjacent, while if they are adjacent, their product $(i, i + 1)(i + 1, i + 2) = (i, i + 2, i + 1)$ has order 3. What is not so obvious is that these relations are all that you need to work in S_n. To make this more precise, we define a *presentation* for a group G, written $G \cong \langle X \mid R \rangle$, to consist of a set X of *generators* and a set R of *relations*, which are equations in the generators, sufficient to define the entire multiplication table of G. For example the dihedral group has a presentation

$$D_{2n} \cong \langle a, b \mid a^2 = b^2 = (ab)^n = 1 \rangle. \tag{2.15}$$

Thus we are asserting that S_n is defined by the so-called *Coxeter presentation*

$$\langle r_1, \ldots, r_{n-1} \mid r_i{}^2 = 1, (r_i r_j)^2 = 1 \text{ if } |i - j| > 1, (r_i r_{i+1})^3 = 1 \rangle. \tag{2.16}$$

To prove this, by induction on n, note first that it works for $n = 2$. Now assume $n > 2$, and we prove that the subgroup H generated by $r_1, \ldots, r_{n-2}$ has index at most n. But $r = r_{n-1}$ commutes with $K = \langle r_1, \ldots, r_{n-3} \rangle$, which by induction has index at most $n - 1$ in H. Therefore r has at most $n - 1$ images under conjugation by H. Moreover, every element in $H \cup HrH$ is in one of the cosets Hx, where x is either the identity or one of these (at most) $n - 1$ conjugates of r.

We need to show that $G = H \cup HrH$. To do this we show that

$$(H \cup HrH)g \subseteq H \cup HrH \tag{2.17}$$

for all $g \in G$. Then, as $1 \in H \cup HrH$, we have

$$G \subseteq H \cup HrH. \tag{2.18}$$

Plainly, if $g \in H$, then the claim (2.17) holds. So, as $G = \langle H, r \rangle$, we may assume that $g = r$. By induction we assume that $H = K \cup KqK$ where $q = r_{n-2}$. We have $Hr \subseteq HrH$ and, as every element of K commutes with r

$$
\begin{aligned}
HrHr &= Hr(K \cup KqK)r \\
&= H(Kr^2 \cup KrqrK) \\
&= H(K \cup KqrqK) \\
&\subseteq H(K \cup HrH) \\
&= H \cup HrH. \tag{2.19}
\end{aligned}
$$

This proves the claim.

2.8.2 Real reflection groups

The idea of a reflection group, introduced in Section 2.8.1 for the symmetric groups, turns out to be extremely important in many areas of mathematics. The finite real reflection groups were investigated and completely classified by Coxeter. We do not have space here to prove this classification, so we shall merely state it.

Every finite reflection group in n-dimensional real orthogonal space (so-called *Euclidean space*) can be generated by n reflections, and is defined by the angles between the reflecting vectors (by which we mean vectors in the -1-eigenspaces of the generating reflections). Notice that two reflections generate a dihedral group: if the order of this group is $2k$, then the reflecting vectors can be chosen to be at an angle $\pi - \pi/k$ to each other—that is, as near to being opposite as possible. Indeed, it turns out that we can always choose the generating reflections so that every pair has this property, in an essentially unique way.

We draw a diagram consisting of nodes representing the n generating (or *fundamental*) reflections, joined by edges labelled k whenever the product of the two reflections has order $k > 2$. Any labels which are 3 are usually omitted, for simplicity.

If this diagram is disconnected, it means that all the reflections in one component commute with all the reflections in all the other components, so the reflection group is a direct product of smaller reflection groups. Thus we only really need to describe the connected components of the diagrams. The ones which occur are shown in Table 2.1. The last column of this table gives some indication of the structure of the corresponding reflection groups, which will be explained in more detail in Section 3.12.4.

We saw in Section 2.8.1 that the diagram for A_n gives a presentation for the group S_{n+1}, by taking abstract generators of order 2 corresponding to the nodes of the diagram, and specifying that their products have order 2

Table 2.1. The Coxeter diagrams

Name	Diagram	Group
A_n $(n \geqslant 1)$	•——• $\cdots$ •——•	S_{n+1}
B_n $(n \geqslant 2)$	•——• $\cdots$ •——$\overset{4}{}$•	$C_2 \wr S_n$
D_n $(n \geqslant 4)$	•——• $\cdots$ •—⊥—•	$C_2^{n-1} S_n$
E_6	•——•——•——•——• (with node above)	$SO_6^-(2)$
E_7	•——•——•——•——•——• (with node above)	$SO_7(2) \times 2$
E_8	•——•——•——•——•——•——• (with node above)	$2 {\cdot} SO_8^+(2)$
F_4	•——•——$\overset{4}{}$•——•	$2^{1+4}{:}(S_3 \times S_3)$
H_3	•——•——$\overset{5}{}$•	$2 \times A_5$
H_4	•——•——•——$\overset{5}{}$•	$2 {\cdot}(A_5 \times A_5){:}2$
$I_2(k)$ $(k \geqslant 5)$	•——$\overset{k}{}$•	D_{2k}

if the nodes are not joined, and order 3 if the nodes are joined. In fact this generalises to all the Coxeter groups, where an edge labelled k denotes that the corresponding product of reflections has order k. Thus for example the diagram of type H_3 in Table 2.1 indicates that $2 \times A_5$ has a presentation

$$\langle a, b, c \mid a^2 = b^2 = c^2 = (ac)^2 = (ab)^3 = (bc)^5 = 1 \rangle. \tag{2.20}$$

2.8.3 Roots, root systems, and root lattices

In some contexts it turns out that only some of these reflection groups arise. For example, the dihedral groups are the symmetry groups of the regular k-gons, but these k-gons only tessellate the plane if $k = 3$, 4, or 6. If we put this restriction onto all the non-abelian dihedral groups in our diagram (this is called the *crystallographic condition*), we obtain a shorter list of groups. In this context, only the labels 3 (usually omitted), 4 and 6 arise, and it is usual (following Dynkin) to replace an edge labelled 4 by a double edge, and an edge labelled 6 by a triple edge.

   The importance of the crystallographic condition is that by choosing the lengths of the reflecting vectors suitably, the $\mathbb{Z}$-span of these vectors is a discrete subgroup of the additive group $\mathbb{R}^n$, as well as spanning $\mathbb{R}^n$ as a vector space. Such a subgroup is called a *lattice*. We call the reflecting vectors *roots*, and the lattice is the corresponding *root lattice*. The term *root system* is used here to denote the set of roots as a subset of the ambient Euclidean space, although in the literature it often has a more abstract definition. The roots corresponding to the vertices of the diagram are called the *fundamental* or *simple* roots.

For vertices joined by a single edge, corresponding to the symmetries of a triangle, we can take the vectors defining the reflections to be all the same length. For vertices joined by a double edge, corresponding to symmetries of a square, the vectors may be taken as the vertices of the square together with the midpoints of the edges: the fundamental reflections must then be one of each type, so that one vector is $\sqrt{2}$ times as long as the other. We put an arrow on the double edge pointing from the long vector to the short one. [Dynkin originally used open circles for the long roots, and filled circles for the short roots.] Now the diagram B_n comes in two varieties: B_n with the arrow pointing outwards and C_n with the arrow pointing inwards.

Similarly in the case of triple edges, corresponding to symmetries of a regular hexagon, one vector is $\sqrt{3}$ times as long as the other, and again we put an arrow pointing from the long vector to the short one.

The root systems are of types A_n $(n \geqslant 1)$, B_n $(n \geqslant 2)$, C_n $(n \geqslant 3)$, D_n $(n \geqslant 4)$, E_6, E_7, E_8, F_4 and G_2, this last being another name for $I_2(6)$. The corresponding diagrams are called *Dynkin diagrams*.

2.8.4 Weyl groups

The crystallographic reflection groups arise in many contexts, where they are usually called Weyl groups. The Weyl group of type A_n is just the symmetric group S_{n+1}, while that of type B_n (and C_n) is $S_2 \wr S_n$ and that of type D_n is a subgroup of index two in the latter. For $G_2 = I_2(6)$ we get a dihedral group of order 12, and for F_4 a group of order 1152. The Weyl groups of types E_6, E_7 and E_8 are especially interesting groups which we shall meet again later. Writing $W(X_n)$ for the Weyl group of type X_n, we have the following descriptions of exceptional Weyl groups in terms of orthogonal groups (see Chapter 3).

$$\begin{aligned}
W(F_4) &\cong GO_4^+(3) \cong 2^{1+4}{:}(S_3 \times S_3), \\
W(E_6) &\cong GO_6^-(2) \cong U_4(2){:}2, \\
W(E_7) &\cong GO_7(2) \times 2 \cong Sp_6(2) \times 2, \\
W(E_8) &\cong 2{\cdot}GO_8^+(2) \cong 2{\cdot}\Omega_8^+(2){:}2.
\end{aligned} \tag{2.21}$$

In a similar vein, the reflection group of type H_4 is a subgroup of index 2 in $GO_4^+(5)$. More details concerning the exceptional Weyl groups are given in Section 3.12.4.

Further reading

For a comprehensive modern treatment of permutation groups at an introductory level I would recommend the book 'Permutation groups' by Dixon and Mortimer [53]. There one can find a detailed proof of the O'Nan–Scott Theorem, and a construction of the Mathieu groups (see Chapter 5) by building up one step at a time from $PSL_3(4)$ via M_{22} and M_{23} to M_{24}, and from $PSL_2(9)$ via M_{11} to M_{12}. An older classic which still has a lot to offer is Wielandt's book 'Finite permutation groups' [170]. Another classic text is Passman's book 'Permutation groups' [145], which develops the subject from the beginning with the study of multiply-transitive groups as one of its principal aims. Highlights are elucidation of the structure of Frobenius groups, that is, transitive permutation groups in which the stabiliser of two points is trivial but the stabiliser of one point is not (or more generally, 3/2-transitive groups, defined as transitive groups in which all non-trivial orbits of the point stabiliser have the same length), and a construction of the Mathieu groups by Witt's method. Another more modern advanced treatment, which covers a variety of diverse topics, including the O'Nan–Scott Theorem, and infinite permutation groups, is 'Permutation groups' by Cameron [19].

For a more specialised treatment of the symmetric groups and their representation theory, see 'The representation theory of the symmetric group' by James and Kerber [97], or 'The symmetric group' by Sagan [151]. A full and approachable account of the classification of finite real reflection groups (i.e. Coxeter groups) is given by Benson and Grove in 'Finite reflection groups' [73]. For a more advanced treatment of Coxeter groups and related topics see Humphreys 'Reflection groups and Coxeter groups' [85]. For presentations in general see Coxeter and Moser 'Generators and relations for discrete groups' [40] or Johnson 'Presentations of groups' [103].

Exercises

2.1. For a permutation $\pi \in S_n$ define

$$\varepsilon(\pi) = \prod_{1 \leqslant i < j \leqslant n} \frac{i - j}{i^\pi - j^\pi} \in \mathbb{Q}.$$

Show that $\varepsilon(\pi) = \pm 1$ and that ε is a group homomorphism from S_n onto $C_2 = \{1, -1\}$. Hence obtain another proof that the sign of a permutation is well-defined.

2.2. Let $G < S_n$ act transitively on $\Omega = \{1, \ldots, n\}$ and let H be the point stabiliser $\{g \in G \mid a^g = a\}$ for some fixed $a \in \Omega$. Prove that $\phi : a^g \mapsto Hg$ is a bijection between Ω and the set $G : H$ of right cosets of H in G.

Prove also that $Hg = \{x \in G \mid a^x = a^g\}$.

2.3. Prove that the orbits of a group H acting on a set Ω form a partition of Ω.

2.4. Show that A_n is not $(n-1)$-transitive on $\{1, 2, \ldots, n\}$.

2.5. Let G act transitively on Ω. Show that the average number of fixed points of the elements of G is 1, i.e.

$$\frac{1}{|G|} \sum_{g \in G} |\{x \in \Omega \mid x^g = x\}| = 1.$$

2.6. Verify that the semidirect product $G{:}_\phi H$ defined in Section 2.2 is a group. Show that the subset $\{(g, 1_H) \mid g \in G\}$ is a normal subgroup isomorphic to G, and that the subset $\{(1_G, h) \mid h \in H\}$ is a subgroup isomorphic to H.

2.7. Suppose that G has a normal subgroup A and a subgroup B satisfying $G = AB$ and $A \cap B = 1$. Prove that $G \cong A{:}_\phi B$, where $\phi : B \to \operatorname{Aut} A$ is defined by $\phi(b) : a \mapsto b^{-1}ab$.

2.8. Prove that if the permutation π on n points is the product of k disjoint cycles (including trivial cycles), then π is an even permutation if and only if $n - k$ is an even integer.

2.9. Determine the number of conjugacy classes in A_8, and write down one element from each class.

2.10. Show that if $n \geqslant 5$ then there is no non-trivial conjugacy class in A_n with fewer than n elements.

2.11. Prove that $\operatorname{Inn}(G) \trianglelefteq \operatorname{Aut}(G)$.

2.12. Write down all the elements of $\operatorname{Aut}(C_2 \times C_2)$. To which well-known group is it isomorphic?

2.13. Calculate $\operatorname{Inn}(G)$ when $G = D_8$. Show that $\operatorname{Aut}(G) \cong D_8$.

2.14. Show that $\operatorname{Aut}(Q_8) \cong S_4$, where $Q_8 = \langle i, j \mid i^2 = j^2 = (ij)^2 \rangle$ is the quaternion group of order 8.

2.15. Show that if p is an odd prime then

$$\operatorname{Aut}(C_{p^n}) \cong C_{p^n - p^{n-1}} \cong C_{p^{n-1}} \times C_{p-1}.$$

2.16. Prove that $\operatorname{Aut}(A_4) \cong S_4$ and $\operatorname{Aut}(A_5) \cong S_5$.

2.17. Use Lemma 2.2 to show that if $n \geqslant 6$ then A_n cannot act transitively on a set of $n + 1$ points.

2.18. Use the argument of Lemma 2.2 and Theorem 2.3 to show that any automorphism of A_6 which maps 3-cycles to 3-cycles is realised by an element of S_6.

2.19. Construct the outer automorphism of S_6 combinatorially, as follows. From the 6 'points', show that there are 15 'duads' (pairs of points), and 15 'synthemes' (partitions of the 6 points into three duads), and 6 'synthematic totals' (partitions of the 15 duads into five synthemes). [Thus S_6 permutes the 6 synthematic totals.]

Show that any two synthematic totals intersect in a unique syntheme, that any partition of the synthematic totals into three pairs determines a unique duad, and that any 'synthematic total' on the synthematic totals corresponds to a point in a natural way.

2.20. Let S_5 act on the 10 unordered pairs $\{a, b\} \subset \{1, 2, 3, 4, 5\}$. Show that this action is primitive. Determine the stabiliser of one of the 10 pairs, and deduce that it is a maximal subgroup of S_5.

2.21. The previous question defines a primitive embedding of S_5 in S_{10}. Show that this S_5 is not maximal in S_{10}.

[Hint: construct a primitive action of S_6 on 10 points, extending this action of S_5.]

2.22. If $k < \frac{n}{2}$, show that the action of S_n on the $\binom{n}{k}$ unordered k-tuples is primitive.

2.23. If G acts k-transitively on $\{1, 2, \ldots, n\}$ for some $k > 1$, and H is the stabiliser of the point n, show that H acts $(k - 1)$-transitively on the subset $\{1, 2, \ldots, n - 1\}$.

2.24. Let G be the group of permutations of 8 points $\{\infty, 0, 1, 2, 3, 4, 5, 6\}$ generated by $(0, 1, 2, 3, 4, 5, 6)$ and $(1, 2, 4)(3, 6, 5)$ and $(\infty, 0)(1, 6)(2, 3)(4, 5)$. Show that G is 2-transitive. Show that the Sylow 7-subgroups of G have order 7, and that their normalisers have order 21. Show that there are just 8 Sylow 7-subgroups, and deduce that G has order 168. Show that G is simple.

2.25. Let x be an element in S_n of cycle type $(c_1{}^{n_1}, \ldots, c_k{}^{n_k})$, where $c_1, \ldots, c_k$ are distinct positive integers. Show that the centraliser of x in S_n is isomorphic to $(C_{c_1} \wr S_{n_1}) \times \cdots \times (C_{c_k} \wr S_{n_k})$.

2.26. Show that if $H \cong \mathrm{AGL}_3(2) \cong 2^3{:}\mathrm{GL}_3(2)$ is a subgroup of S_8, and $K = H^g$ where g is an odd permutation, then H and K are not conjugate in A_8.

2.27. Prove that $S_k \wr S_2$ is maximal in S_{2k} for all $k \geqslant 2$.

2.28. Prove that $S_k \wr S_m$ is maximal in S_{km} for all $k, m \geqslant 2$.

2.29. Prove that the 'diagonal' subgroups of S_n constructed in Section 2.5.5 are primitive.

2.30. Show that if H is abelian and transitive on Ω, then it is regular on Ω.

2.31. Use the O'Nan–Scott theorem to write down as many maximal subgroups of S_5 as you can. Can you prove your subgroups are maximal?

2.32. Do the same for A_5.

2.33. Let G be a simple group, H a maximal subgroup of G, and K a minimal normal subgroup of H. Prove that $H = N_G(K)$ and that K is characteristically simple.

2.34. Use Exercise 2.33 to determine the maximal subgroups of A_5 from first principles.

2.35. The real quaternion algebra $\mathbb{H}$ (see Section 4.3.1) is made by linearising the quaternion group Q_8, identifying the central element i^2 of Q_8 with the real number -1, and extending the multiplication bilinearly. Show that the quaternion $\omega = \frac{1}{2}(-1+i+j+k)$, where $k = ij$, satisfies $\omega^3 = 1$ and $\omega^{-1}i\omega = j$. Deduce that the group generated by i and ω is a double cover of A_4, permuting the four coordinate axes $\langle 1 \rangle$, $\langle i \rangle$, $\langle j \rangle$, $\langle k \rangle$.

2.36. Prove the following presentations:

(i) $\langle x, y \mid x^2 = y^3 = (xy)^3 = 1 \rangle \cong A_4$;
(ii) $\langle x, y \mid x^2 = y^3 = (xy)^4 = 1 \rangle \cong S_4$;
(iii) $\langle x, y \mid x^2 = y^3 = (xy)^5 = 1 \rangle \cong A_5$.

2.37. Show that the subgroup of the unit quaternions generated by i and $1 + \sigma i + \tau j$, where $\sigma = \frac{1}{2}(\sqrt{5} - 1)$ and $\tau = \frac{1}{2}(\sqrt{5} + 1)$, is a double cover of A_5.
  [Hint: show that modulo -1 these elements satisfy the relations given in Exercise 2.36(iii).]

2.38. Use the outer automorphism of S_6 to prove that the two double covers $2{\cdot}S_6^+$ and $2{\cdot}S_6^-$ of S_6 are isomorphic.

2.39. Write down the 15 vectors which are images of $(0,0,1,1,1,1)$ and $(0,1,0,1,\omega,\overline{\omega})$ under the group S_4 generated by the coordinate permutations $(1,2)(3,4)$, $(1,3,5)(2,4,6)$ and $(1,3)(2,4)$.
  Verify that the map $(x_1,\ldots,x_6) \mapsto (x_3,x_1,x_2,x_4,\overline{\omega}x_5,\omega x_6)$ preserves this set of vectors up to scalar multiplication by ω and $\overline{\omega}$.

2.40. Show that the reflection group of type A_3 is the group of symmetries of a regular tetrahedron.

2.41. Show that the reflection group of type B_3 is the group of symmetries of the cube/octahedron, and is isomorphic to $C_2 \times S_4$.

2.42. Show that the reflection group of type H_3 is the group of symmetries of the dodecahedron/icosahedron, and is isomorphic to $C_2 \times A_5$.

2.43. Show that the root system of type B_3 consists of the midpoints of the edges and faces of a cube.

2.44. Show that the root system of type C_3 consists of the vertices and the midpoints of the edges of a regular octahedron.

2.45. Show that the long roots of the B_n root system (or the short roots of the C_n root system) form a root system of type D_n. What type of root system do the short roots of the B_n root system (or the long roots of the C_n root system) form?

3

The classical groups

3.1 Introduction

In this chapter we describe the six families of so-called 'classical' simple groups.
These are the linear, unitary and symplectic groups, and the three families
of orthogonal groups. All may be obtained from suitable matrix groups G
by taking $G'/Z(G')$. Our main aims again are to define these groups, prove
they are simple, and describe their automorphisms, subgroups and covering
groups.

We begin with some basic facts about finite fields, before constructing
the general linear groups and their subquotients $\mathrm{PSL}_n(q)$, which are usually simple. We prove simplicity using Iwasawa's Lemma, and construct some
important subgroups using the geometry of the underlying vector space.
This leads into a description of projective spaces, especially projective lines
and projective planes, setting the scene for proofs of several remarkable isomorphisms between certain groups $\mathrm{PSL}_n(q)$ and other groups, in particular
$\mathrm{PSL}_2(4) \cong \mathrm{PSL}_2(5) \cong A_5$ and $\mathrm{PSL}_2(9) \cong A_6$. Covering groups and automorphism groups are briefly mentioned.

The other families of classical groups are defined in terms of certain 'forms'
on the vector space, so we collect salient facts about these forms in Section 3.4,
before introducing the symplectic groups in Section 3.5. These are in many
ways the easiest of the classical groups to understand. We calculate their orders, prove simplicity, discuss subgroups, covering groups and automorphisms,
and prove the generic isomorphism $\mathrm{Sp}_2(q) \cong \mathrm{SL}_2(q)$ and the exceptional isomorphism $\mathrm{Sp}_4(2) \cong S_6$. Next the unitary groups are treated in much the same
way.

Most of the difficulties in studying the classical groups occur in the orthogonal groups. The first problem is that there are fundamental differences
between the case when the underlying field has characteristic 2, and the other
cases. The second problem is that even when we take the subgroup of matrices with determinant 1, and factor out any scalars, the resulting group is not
usually simple.

R.A. Wilson, *The Finite Simple Groups*,
Graduate Texts in Mathematics 251,
© Springer-Verlag London Limited 2009

To avoid introducing too many complications at once, we first treat the case of orthogonal groups over fields of odd order, defining the three different types and calculating their orders. We use the spinor norm to obtain the (usually) simple groups, and describe some of their subgroup structure. Then we describe the differences that occur when the field has characteristic 2, and use the quasideterminant to obtain the (usually) simple group.

Clifford algebras and spin groups are constructed next, in order to complete the proofs of certain key facts, especially the fact that the spinor norm is well-defined.

We then prove a simple version of the Aschbacher–Dynkin theorem which gives a basic classification of the different types of maximal subgroups in a classical group. It is analogous to the O'Nan–Scott theorem for permutation groups, but is rather harder to prove, and is less explicit in its conclusions. The proof relies heavily on representation theory, however, and can be omitted by those readers without the necessary prerequisites. More explicit versions of the theorem are stated for individual classes of groups, taken largely from the book by Kleidman and Liebeck [108].

The generic isomorphisms of the orthogonal groups all derive from the Klein correspondence, by which $P\Omega_6^+(q) \cong PSL_4(q)$ (and also $P\Omega_6^-(q) \cong PSU_4(q)$ and $P\Omega_5(q) \cong PSp_4(q)$). This correspondence also leads to an easy proof of the exceptional isomorphisms $PSL_4(2) \cong A_8$ and $PSL_3(2) \cong PSL_2(7)$. Finally we discuss some exceptional behaviour of small classical groups, and relate this to the exceptional Weyl groups introduced in Chapter 2.

3.2 Finite fields

All the classical groups are defined in terms of groups of matrices over fields, so before we can study the finite classical groups we need to know what the finite fields are. A *field* is a set F with operations of addition, subtraction, multiplication and division satisfying the usual rules of arithmetic. That is, F has an element 0 such that $(F, +, -, 0)$ is an abelian group, and $F \setminus \{0\}$ contains an element 1 such that $(F \setminus \{0\}, ., /, 1)$ is an abelian group, and $x(y+z) = xy + xz$. It is an easy exercise to show that the subfield F_0 generated by the element 1 in a finite field F is isomorphic to the integers modulo p, for some p, and therefore p is prime (called the *characteristic* of the field). The field F_0 is called the *prime subfield* of F. Moreover, F is a vector space over F_0, since the vector space axioms are special cases of the field axioms. As every finite-dimensional vector space has a basis of n vectors, $v_1, \ldots, v_n$, say, and every vector has a unique expression $\sum_{i=1}^{n} a_i v_i$ with $a_i \in F_0$, it follows that the field F has p^n elements.

Conversely, for every prime p and every positive integer d there is a field of order p^d, which is unique up to isomorphism. [See below.]

The most important fact about finite fields which we need is that the multiplicative group of all non-zero elements is cyclic. For the polynomial ring $F[x]$ over any field F is a Euclidean domain and therefore a unique factorisation domain. In particular a polynomial of degree n has at most n roots. If the multiplicative group $F^\times$ of a field of order q has exponent e strictly less than $q - 1$, then $x^e - 1$ has $q - 1$ roots, which is a contradiction. Therefore the exponent of $F^\times$ is $q - 1$, so, since it is abelian, $F^\times$ contains elements of order $q - 1$, and therefore it is cyclic.

Note also that all elements x of F satisfy $x^q = x$, and so the polynomial $x^q - x$ factorises in $F[x]$ as $\prod_{\alpha \in F}(x - \alpha)$. Moreover, the number of solutions to $x^n = 1$ in F is the greatest common divisor $(n, q - 1)$ of n and $q - 1$. A useful consequence of this for the field of order q^2 is that for every $\mu \in \mathbb{F}_q$ there are exactly $q + 1$ elements $\lambda \in \mathbb{F}_{q^2}$ satisfying $\lambda\overline{\lambda} = \mu$, where $\overline{\lambda} = \lambda^q$.

We now show that fields of order p^d exist and are unique up to isomorphism. Observe that if f is an irreducible polynomial of degree d over the field $\mathbb{F}_p \cong \mathbb{Z}/p\mathbb{Z}$ of order p, then $\mathbb{F}_p[x]/(f)$ is a field of order p^d. Conversely, if F is a field of order p^d, let x be a generator for $F^\times$, so that the minimum polynomial for x over $\mathbb{F}_p$ is an irreducible polynomial f of degree d, and $F \cong \mathbb{F}_p[x]/(f)$. One way to see that such a field exists is to observe that any field of order $q = p^d$ is a splitting field for the polynomial $x^q - x$. (A splitting field for a polynomial f is a field over which f factorises into linear factors.) Splitting fields always exist, by adjoining roots one at a time until the polynomial factorises into linear factors. But then the set of roots of $x^q - x$ is closed under addition and multiplication, since if $x^q = x$ and $y^q = y$ then $(xy)^q = x^q y^q = xy$ and $(x + y)^q = x^q + y^q = x + y$. Hence this set of roots is a subfield of order q, as required. To show that the field of order $q = p^d$ does not depend on the particular irreducible polynomial we choose, suppose that f_1 and f_2 are two such, and $F_i = \mathbb{F}_p[x]/(f_i)$. Since $f_2(t)$ divides $t^q - t$, and $t^q - t$ factorises into linear factors over F_1, it follows that F_1 contains an element y with $f_2(y) = 0$. Hence the map $x \mapsto y$ extends to a field homomorphism from F_2 to a subfield of F_1. Moreover, the kernel is trivial, since fields have no quotient fields, so this map is a field isomorphism, since the fields are finite and have the same order.

If also $f_1 = f_2$ then any automorphism of $F = F_1 = F_2$ has this form, so is defined by the image of x, which must be one of the d roots of f_1. Hence the group of automorphisms of F has order d. On the other hand, the map $y \mapsto y^p$ (for all $y \in F$) is an automorphism of F, called the *Frobenius automorphism*, and has order d. Hence $\mathrm{Aut}(F)$ is cyclic of order d.

3.3 General linear groups

Let V be a vector space of dimension n over the finite field $\mathbb{F}_q$ of order q. The *general linear group* $\mathrm{GL}(V)$ is the set of invertible linear maps from V to itself. Without much loss of generality, we may take V as the vector space $\mathbb{F}_q{}^n$ of n-tuples of elements of $\mathbb{F}_q$, and identify $\mathrm{GL}(V)$ with the group (denoted $\mathrm{GL}_n(q)$) of invertible $n \times n$ matrices over $\mathbb{F}_q$.

There are certain obvious normal subgroups of $G = \mathrm{GL}_n(q)$. For example, the centre, Z say, consists of all the scalar matrices λI_n, where $0 \neq \lambda \in \mathbb{F}_q$ and I_n is the $n \times n$ identity matrix. Thus Z is a cyclic normal subgroup of order $q - 1$. The quotient G/Z is called the *projective general linear group*, and denoted $\mathrm{PGL}_n(q)$.

Also, since $\det(AB) = \det(A)\det(B)$, the determinant map is a group homomorphism from $\mathrm{GL}_n(q)$ onto the multiplicative group of the field, so its kernel is a normal subgroup of index $q-1$. This kernel is called the *special linear group* $\mathrm{SL}_n(q)$, and consists of all the matrices of determinant 1. Similarly, we can quotient $\mathrm{SL}_n(q)$ by the subgroup of scalars it contains, to obtain the *projective special linear group* $\mathrm{PSL}_n(q)$, sometimes abbreviated to $\mathrm{L}_n(q)$. [The alert reader will have noticed that as defined here, $\mathrm{PSL}_n(q)$ is not necessarily a subgroup of $\mathrm{PGL}_n(q)$. However, there is an obvious isomorphism between $\mathrm{PSL}_n(q)$ and a normal subgroup of $\mathrm{PGL}_n(q)$, so we shall ignore the subtle distinction.]

3.3.1 The orders of the linear groups

Now an invertible matrix takes a basis to a basis, and is determined by the image of an ordered basis. The only condition on this image is that the ith vector is linearly independent of the previous ones—but these span a space of dimension $i - 1$, which has q^{i-1} vectors in it, so the order of $\mathrm{GL}_n(q)$ is

$$\begin{aligned}
|\mathrm{GL}_n(q)| &= (q^n - 1)(q^n - q)(q^n - q^2)\cdots(q^n - q^{n-1}) \\
&= q^{n(n-1)/2}(q - 1)(q^2 - 1)\cdots(q^n - 1).
\end{aligned} \tag{3.1}$$

The orders of $\mathrm{SL}_n(q)$ and $\mathrm{PGL}_n(q)$ are equal, being $|\mathrm{GL}_n(q)|$ divided by $q-1$. To obtain the order of $\mathrm{PSL}_n(q)$, we need to know which scalars λI_n have determinant 1. But $\det(\lambda I_n) = \lambda^n$, and the number of solutions to $x^n = 1$ in the field $\mathbb{F}_q$ is the greatest common divisor $(n, q - 1)$ of n and $q - 1$. Thus the order of $\mathrm{PSL}_n(q)$ is

$$|\mathrm{PSL}_n(q)| = \frac{1}{(n, q - 1)} q^{n(n-1)/2} \prod_{i=2}^{n}(q^i - 1). \tag{3.2}$$

The groups $\mathrm{PSL}_n(q)$ are all simple except for the small cases $\mathrm{PSL}_2(2) \cong S_3$ and $\mathrm{PSL}_2(3) \cong A_4$. We shall prove the simplicity of the remaining groups $\mathrm{PSL}_n(q)$ below. First we prove these exceptional isomorphims. For $\mathrm{PSL}_2(2) \cong \mathrm{GL}_2(2)$, and $\mathrm{GL}_2(2)$ permutes the three non-zero vectors of $\mathbb{F}_2{}^2$; moreover, any two of these vectors form a basis for the space, so the action of $\mathrm{GL}_2(2)$ is 2-transitive, and faithful, so $\mathrm{GL}_2(2) \cong S_3$.

Similarly, $\mathrm{GL}_2(3)$ permutes the four 1-dimensional subspaces of $\mathbb{F}_3{}^2$, spanned by the vectors $(1,0)$, $(0,1)$, $(1,1)$ and $(1,-1)$. The action is 2-transitive since the group acts transitively on ordered bases. Moreover, fixing the standard basis, up to scalars, the matrix $\begin{pmatrix} 1 & 0 \\ 0 & -1 \end{pmatrix}$ interchanges the other two 1-spaces, so the action of $\mathrm{GL}_2(3)$ is S_4. The kernel of the action is just the group of scalar matrices, and the matrices of determinant 1 act as even permutations, so $\mathrm{PSL}_2(3) \cong A_4$. Notice that $\mathrm{SL}_2(3)$ is a proper double cover $2{\cdot}A_4$, and similarly $\mathrm{GL}_2(3) \cong 2{\cdot}S_4$ (see Section 2.7.2).

We use the term *linear group* loosely to refer to any of the groups $\mathrm{GL}_n(q)$, $\mathrm{SL}_n(q)$, $\mathrm{PGL}_n(q)$ or $\mathrm{PSL}_n(q)$.

3.3.2 Simplicity of $\mathrm{PSL}_n(q)$

The key to proving simplicity of the finite classical groups is Iwasawa's Lemma, originally proved in [95].

Theorem 3.1. *If G is a finite perfect group, acting faithfully and primitively on a set Ω, such that the point stabiliser H has a normal abelian subgroup A whose conjugates generate G, then G is simple.*

Proof. For otherwise, there is a normal subgroup K with $1 < K < G$, which does not fix all the points of Ω, so we may choose a point stabiliser H with $K \not\leqslant H$, and therefore $G = HK$ since H is a maximal subgroup of G. So any $g \in G$ can be written $g = hk$ with $h \in H$ and $k \in K$, and therefore every conjugate of A is of the form $g^{-1}Ag = k^{-1}h^{-1}Ahk = k^{-1}Ak \leqslant AK$. Therefore $G = AK$ and $G/K = AK/K \cong A/A \cap K$ is abelian, contradicting the assumption that G is perfect.

In order to apply this, for $n \geqslant 2$ we let $\mathrm{SL}_n(q)$ act on the set Ω of 1-dimensional subspaces of $\mathbb{F}_q{}^n$, so that the kernel of the action is just the set of scalar matrices, and we obtain an action of $\mathrm{PSL}_n(q)$ on Ω. Moreover, this action is 2-transitive (since any basis can be mapped to any other basis, up to a scalar multiplication, by an element of $\mathrm{SL}_n(q)$), and therefore primitive.

To study the stabiliser of a point, we might as well take this point to be the 1-space $\langle(1,0,\ldots,0)\rangle$. The stabiliser then consists of all matrices whose first row is $(\lambda,0,\ldots,0)$ for some $\lambda \neq 0$. It is easy to check that the subgroup of matrices of the shape $\begin{pmatrix} 1 & 0_{n-1} \\ v_{n-1} & I_{n-1} \end{pmatrix}$, where v_{n-1} is an arbitrary column vector of length $n - 1$, is a normal abelian subgroup A. Moreover, all non-trivial elements of A are *transvections*, that is, matrices (or linear maps) t such that $t - I_n$ has rank 1 and $(t - I_n)^2 = 0$. By suitable choice of basis (but remembering that the base change matrix must have determinant 1) it is easy to see that every transvection is contained in some conjugate of A.

We have two more things to verify: first, that $\mathrm{SL}_n(q)$ is generated by transvections, and second, that $\mathrm{SL}_n(q)$ is perfect, except for the cases $\mathrm{SL}_2(2)$ and $\mathrm{SL}_2(3)$. The first fact is a restatement of the elementary result that every matrix of determinant 1 can be reduced to the identity matrix by a finite sequence of elementary row operations of the form $r_i \mapsto r_i + \lambda r_j$. To prove the second it suffices to verify that every transvection is a commutator of elements of $\mathrm{SL}_n(q)$. An easy calculation shows that the commutator

$$\left[\begin{pmatrix} 1 & 0 & 0 \\ 1 & 1 & 0 \\ 0 & 0 & 1 \end{pmatrix}, \begin{pmatrix} 1 & 0 & 0 \\ 0 & 1 & 0 \\ 0 & x & 1 \end{pmatrix}\right] = \begin{pmatrix} 1 & 0 & 0 \\ 0 & 1 & 0 \\ -x & 0 & 1 \end{pmatrix}, \tag{3.3}$$

so by suitable choice of basis we see that if $n > 2$ then every transvection is a commutator in $SL_n(q)$. If $n = 2$ and $q > 3$, then $\mathbb{F}_q$ contains a non-zero element x with $x^2 \neq 1$, and then the commutator

$$\left[\begin{pmatrix} 1 & 0 \\ y & 1 \end{pmatrix}, \begin{pmatrix} x & 0 \\ 0 & x^{-1} \end{pmatrix}\right] = \begin{pmatrix} 1 & 0 \\ y(x^2 - 1) & 1 \end{pmatrix}, \tag{3.4}$$

which is an arbitrary element of A.

We can now apply Iwasawa's Lemma, and deduce that $PSL_n(q)$ is simple provided $n > 2$ or $q > 3$.

3.3.3 Subgroups of the linear groups

Here we introduce some of the more important subgroups of the linear groups, including the subgroups B, N and T, and the parabolic subgroups, as well as the Weyl group $W = N/T$. We use this notation and terminology since it is standard in the theory of groups of Lie type (see Carter's book [21]). The standard properties of B and N can be axiomatised into the notion of a 'BN-pair', which led Tits to the even more abstract notion of a building. A fuller discussion of the subgroup structure, incorporating the Aschbacher–Dynkin theorem on maximal subgroups, will be found later, in Section 3.10. We work in $GL_n(q)$ for simplicity, but it is easy to modify the constructions for $SL_n(q)$, $PGL_n(q)$ and $PSL_n(q)$.

The subgroup B of $GL_n(q)$ consisting of all lower-triangular matrices is the *Borel subgroup*, and N is the subgroup of all *monomial* matrices, i.e. matrices with exactly one non-zero entry in each row and column. Then $T = B \cap N$, called the *maximal split torus*, consists of the diagonal matrices, which form a normal subgroup of N. (Indeed, N is the normaliser of T, except when $q = 2$, in which case T is trivial.) The quotient group $W = N/T$ is called the *Weyl group*. It is clear that the torus T is isomorphic to the direct product of n copies of the cyclic group C_{q-1} (or $n-1$ copies in $SL_n(q)$), and that the Weyl group is isomorphic to the symmetric group S_n of all coordinate permutations.

The subgroup U of all lower unitriangular matrices (i.e. matrices having all diagonal entries 1, and all above-diagonal entries 0) is easily seen to be a group of order $q^{n(n-1)/2}$, so it is a Sylow p-subgroup of $GL_n(q)$, where p is the characteristic of $\mathbb{F}_q$. Moreover, B is the semidirect product of U with T, the group of diagonal matrices, so B has order $q^{n(n-1)/2}(q-1)^n$. This group B may also be defined as the stabiliser of the chain of subspaces

$$0 = V_0 < V_1 < V_2 < \cdots < V_n = V \tag{3.5}$$

defined by $V_i = \{(x_1, \ldots, x_i, 0, \ldots, 0)\}$, so that $\dim(V_i) = i$. A chain of subspaces ordered by inclusion is called a *flag*, and if such a chain has a subspace of each possible dimension it is called a *maximal flag*. Thus B is the stabiliser of a maximal flag.

The *parabolic subgroups* are the stabilisers of flags, and the maximal parabolic subgroups are the stabilisers of minimal flags $0 < W < V$, i.e. the stabilisers of subspaces W. If W has dimension k, say, we may choose a basis $\{e_1, \ldots, e_k\}$ for W and extend it to a basis $\{e_1, \ldots, e_n\}$ for V. Then the matrices for elements of the subspace stabiliser have the shape $\begin{pmatrix} A & 0 \\ C & D \end{pmatrix}$, where A and D are non-singular $k \times k$ and $(n-k) \times (n-k)$ matrices and C is an arbitrary $k \times (n-k)$ matrix. The subset Q of matrices of the shape $\begin{pmatrix} I_k & 0 \\ C & I_{n-k} \end{pmatrix}$ is easily checked to be an elementary abelian normal subgroup of order $q^{k(n-k)}$. (Recall from Section 2.5.4 that an abelian group is called *elementary* if it is a direct product of cyclic groups of order p, for some fixed prime p.) The subset L of matrices of the shape $\begin{pmatrix} A & 0 \\ 0 & D \end{pmatrix}$ is a subgroup isomorphic to $\mathrm{GL}_k(q) \times \mathrm{GL}_{n-k}(q)$. Moreover, $L \cap Q = 1$ and LQ is the full stabiliser of W, so this has the structure of a semidirect product $Q{:}L$ (see Exercise 3.5). In the language of Lie theory, Q is the *unipotent radical* and L is a *Levi complement*. It is left as an exercise (see Exercise 3.7) to show that the maximal parabolic subgroups are indeed maximal subgroups.

Interpretations of the Dynkin diagram

The Weyl group S_n in $\mathrm{GL}_n(q)$ may be generated by the coordinate permutations $(1, 2), (2, 3) \ldots, (n-1, n)$, which in turn may be regarded as reflections in the vectors $(1, -1, 0, \ldots, 0), \ldots, (0, \ldots, 0, 1, -1)$ as described in Section 2.8.1. Since these vectors are the fundamental roots of the A_{n-1} root system, we have a (natural) identification with the Weyl group of type A_{n-1}.

The maximal flag $V_1 < V_2 < \cdots < V_{n-1}$ may also be described by the Dynkin diagram, by labelling the kth node V_k. The stabiliser of V_k is a maximal parabolic subgroup, whose Levi complement is $\mathrm{GL}_k(q) \times \mathrm{GL}_{n-k}(q)$, and the Weyl group of this complement is obtained by deleting the node labelled V_k from the A_{n-1} diagram.

Some more subgroups

For each family of classical groups we aim eventually to prove (in Section 3.10) a version of the Aschbacher–Dynkin Theorem, which is analogous to the O'Nan–Scott Theorem (Theorem 2.4) for the alternating and symmetric groups. Just as in that case we had to construct various subgroups as stabilisers of certain structures on the underlying set, so here we have various structures on the underlying vector space V.

We have already seen the stabilisers in $\mathrm{GL}_n(q)$ of proper non-zero subspaces (also known as the maximal parabolic subgroups): if the subspace has dimension k, so codimension $m = n - k$, then the stabiliser is a group of shape

$q^{km}{:}(\mathrm{GL}_k(q) \times \mathrm{GL}_m(q))$, where $k + m = n = \dim V$. These are analogous to the intransitive subgroups of S_n. Notice that the stabiliser of a subspace of codimension k is isomorphic to that of a subspace of dimension k, but they are not conjugate in $\mathrm{GL}_n(q)$ except when $k = n/2$.

Corresponding to the imprimitive subgroups of S_n there are the stabilisers of direct sum decompositions of the space. Thus if $V = V_1 \oplus V_2 \oplus \cdots \oplus V_m$, with $\dim V_i = k$, then there is a group $\mathrm{GL}_k(q) \wr S_m$ stabilising this decomposition. These groups are also called *imprimitive* linear groups.

There are a number of other types of subgroups which we shall construct in Section 3.10, namely tensor product subgroups (somewhat analogous to the primitive wreath products), subgroups of extraspecial type (somewhat analogous to the affine groups), and various types of almost simple subgroups.

3.3.4 Outer automorphisms

It is a fact that the outer automorphism groups of all the classical groups have a uniform description in terms of so-called diagonal, field, and graph automorphisms. The diagonal automorphisms are so called because they are induced by conjugation by diagonal matrices (with respect to a suitable basis). In the case of the linear groups, $\mathrm{PGL}_n(q)$ acts as a group of automorphisms of $\mathrm{PSL}_n(q)$, and the corresponding quotient group $\mathrm{PGL}_n(q)/\mathrm{PSL}_n(q)$ is a cyclic group of order $d = (n, q - 1)$, called the group of diagonal outer automorphisms.

The field automorphisms are, as the name suggests, induced by automorphisms of the underlying field. Recall from Section 3.2 that the automorphism group of the field $\mathbb{F}_q$ of order $q = p^e$ is a cyclic group of order e generated by the Frobenius automorphism $x \mapsto x^p$. This induces an automorphism of $\mathrm{GL}_n(q)$, also of order e, by mapping each matrix entry to its pth power. Taking the semidirect product of $\mathrm{GL}_n(q)$ with this group of field automorphisms gives us a group $\mathrm{\Gamma L}_n(q)$, and correspondingly the extension of $\mathrm{SL}_n(q)$, $\mathrm{PGL}_n(q)$ or $\mathrm{PSL}_n(q)$ by the induced group of field automorphisms is denoted $\mathrm{\Sigma L}_n(q)$, $\mathrm{P\Gamma L}_n(q)$ or $\mathrm{P\Sigma L}_n(q)$. [Notice that this definition of the field automorphisms is basis-dependent. This does not matter much in the present instance, as all the groups just defined are well-defined up to isomorphism. However, the situation in the other classical groups is not so straightforward.]

The graph automorphisms are induced by an automorphism of the Dynkin diagram, but do not appear particularly uniform from a classical point of view. In the case of the linear groups, the graph automorphism is best explained by the classical concept of *duality*. Formally, the *dual space* of a vector space V over a field F is the vector space V^* of linear functions $f^* : V \to F$. Given a basis $\{e_1, \ldots, e_n\}$ of V there is a well-defined *dual basis* $\{e_1^*, \ldots, e_n^*\}$ of V^* defined by $e_i^*(e_i) = 1$ and $e_i^*(e_j) = 0$ if $i \neq j$. If $g \in \mathrm{GL}(V)$ acts on V by

$$e_i \mapsto \sum_{j=1}^{n} g_{ij} e_j \tag{3.6}$$

and on V^* by

$$e_i^* \mapsto \sum_{j=1}^{n} h_{ij} e_j^* \qquad (3.7)$$

then a straightforward calculation shows that the matrix $h = (h_{ij})$ is the transpose of the inverse of the matrix $g = (g_{ij})$. The *duality automorphism* (with respect to these bases) of $\mathrm{GL}(V)$ is the map which replaces each matrix by the transpose of its inverse. This is an automorphism of the general linear group because

$$((AB)^{\top})^{-1} = (B^{\top} A^{\top})^{-1} = (A^{\top})^{-1} (B^{\top})^{-1} \qquad (3.8)$$

and of the special linear group because also

$$\det((A^{\top})^{-1}) = (\det(A^{\top}))^{-1} = (\det(A))^{-1}. \qquad (3.9)$$

For $n = 2$, observe that duality is an inner automorphism of $\mathrm{SL}_2(q)$, induced by conjugation by $\begin{pmatrix} 0 & 1 \\ -1 & 0 \end{pmatrix}$ (see Exercise 3.10). For $n > 2$, however, duality is not inner, even in $\mathrm{GL}_n(q)$, as it maps the stabiliser of a 1-space, consisting of the matrices of shape $\begin{pmatrix} \lambda & 0 \\ v & A \end{pmatrix}$, to the stabiliser of a hyperplane, consisting of the matrices of shape $\begin{pmatrix} \mu & w \\ 0 & B \end{pmatrix}$.

Let us sketch the proof that the outer automorphism group of $\mathrm{PSL}_n(q)$ is no bigger than this. We have to reconstruct the geometry from the abstract structure of the group (i.e. from its subgroups), and reconstruct the field from the abstract structure of the geometry (i.e. from its subspaces). The first step is accomplished by analysing the p-local subgroups, that is, the normalisers of p-subgroups. First observe that any p-local subgroup fixes a non-trivial proper subspace in the natural representation (that is, the action of $\mathrm{SL}_n(q)$ on $\mathbb{F}_q{}^n$), and therefore is contained in one of the maximal parabolic subgroups. Therefore any automorphism permutes the maximal parabolic subgroups amongst themselves. Now the stabiliser of a k-space is isomorphic to the stabiliser of an $(n-k)$-space, but not to the stabiliser of any other subspace. Therefore, multiplying by the duality automorphism if necessary, we may assume that our automorphism ϕ preserves the set of 1-spaces. Since any two distinct 1-spaces span a 2-space, or in group-theoretic terms the intersection of two 1-space stabilisers is contained in a unique 2-space stabiliser (but no $(n-2)$-space stabiliser, unless $n = 4$), the set of 2-spaces is also preserved. Similarly, by induction, ϕ preserves the set of k-spaces, for each k.

Hence by transitivity of $\mathrm{SL}_n(q)$ on maximal flags we may assume that ϕ preserves the standard maximal flag, and therefore (multiplying by a diagonal matrix if necessary) we may assume that ϕ fixes the standard basis. Next we have to prove that ϕ acts semilinearly. (A map ϕ on a vector space V is called *semilinear* if there is an automorphism σ of the underlying field such that $(\lambda u + v)^\phi = \lambda^\sigma u^\phi + v^\phi$ for all $u, v \in V$ and all scalars λ.) By induction it is enough to prove that ϕ acts semilinearly on the space spanned by two basis vectors, e and f, say (since the same field automorphism σ must be involved for every 2-space $\langle e, f' \rangle$, and therefore on the whole of V).

Now the stabiliser in $\mathrm{SL}_n(q)$ of the flag $0 < \langle e \rangle < \langle e, f \rangle$ acts on this 2-space as a group $E_q.C_{q-1}$ of lower-triangular matrices. The normal subgroup of order q determines the additive structure of the field, as it acts by $f \mapsto f + \lambda e$, and the composite of $f \mapsto f + \lambda e$ and $f \mapsto f + \mu e$ is $f \mapsto f + (\lambda + \mu)e$. The diagonal subgroup C_{q-1} determines the multiplicative structure, as it maps f to λf, and the composite of $f \mapsto \lambda f$ with $f \mapsto \mu f$ is $f \mapsto (\lambda\mu)f$. Therefore (multiplying by a suitable element of $E_q.C_{q-1}$) we may assume that ϕ acts on $\langle e, f \rangle$ as a field automorphism, as required. To sum up, we have proved

Theorem 3.2. *Let* $q = p^d$, *where* p *is prime.*

(i) If $n > 2$ *then* $\mathrm{Out}(\mathrm{PSL}_n(q)) \cong D_{2(n,q-1)} \times C_d$.
(ii) If $n = 2$ *then* $\mathrm{Out}(\mathrm{PSL}_2(q)) \cong C_{(2,q-1)} \times C_d$.

3.3.5 The projective line and some exceptional isomorphisms

There are many isomorphisms between the small linear groups and other groups. The most important are

$$\begin{aligned}
\mathrm{PSL}_2(2) &\cong S_3, \\
\mathrm{PSL}_2(3) &\cong A_4, \\
\mathrm{PSL}_2(4) \cong \mathrm{PSL}_2(5) &\cong A_5, \\
\mathrm{PSL}_2(7) &\cong \mathrm{PSL}_3(2), \\
\mathrm{PSL}_2(9) &\cong A_6, \\
\mathrm{PSL}_4(2) &\cong A_8.
\end{aligned} \tag{3.10}$$

We have already proved the first two of these in Section 3.3.1, and some of the others can be proved in much the same way. In this section we use the projective line, defined below, to prove the isomorphisms of $\mathrm{PSL}_2(4)$, $\mathrm{PSL}_2(5)$ and $\mathrm{PSL}_2(9)$ with alternating groups. The isomorphism $\mathrm{PSL}_4(2) \cong A_8$ can be proved by showing that both are isomorphic to $\Omega_6^+(2)$, using the Klein correspondence. [See Section 3.12 for a proof of this and of the isomorphism $\mathrm{PSL}_3(2) \cong \mathrm{PSL}_2(7)$.]

It is convenient to work in $\mathrm{PSL}_2(q)$ directly as a group of permutations of the 1-dimensional subspaces of $\mathbb{F}_q{}^2$, and to this end we label the 1-spaces by the ratio of the coordinates: that is $\langle (x, 1) \rangle$ is labelled x, and $\langle (1, 0) \rangle$ is labelled ∞. The set of 1-spaces is then identified with the set $\mathbb{F}_q \cup \{\infty\}$, called the *projective line* over $\mathbb{F}_q$, and denoted $\mathrm{PL}(q)$. The matrix $\begin{pmatrix} a & b \\ c & d \end{pmatrix} \in \mathrm{GL}_2(q)$ now acts on the projective line as $z \mapsto (az + c)/(bz + d)$, or, working in the traditional way with column vectors rather than row vectors,

$$\begin{pmatrix} a & b \\ c & d \end{pmatrix} : z \mapsto \frac{az+b}{cz+d}. \tag{3.11}$$

In this formula, ∞ obeys the usual rules: $1/0 = \infty$, $1/\infty = 0$, $\infty + z = \infty$, and (if $z \neq 0$) $\infty.z = \infty$. To avoid meaningless expressions such as ∞/∞ it is occasionally necessary to rewrite the formula:

$$\frac{az+b}{cz+d} = \frac{a+b/z}{c+d/z}. \tag{3.12}$$

If $z \mapsto z$ for all z in the projective line, then putting $z = 0$ gives $c = 0$ and putting $z = \infty$ gives $b = 0$, and putting $z = 1$ then gives $a = d$, so the matrix is a scalar. In other words, we get a faithful action of $\mathrm{PGL}_2(q)$ on the projective line. Notice that any two points of the projective line determine a basis of the 2-space, up to scalar multiplications of the two basis vectors separately. Given any change-of-basis matrix we can multiply by a diagonal matrix to make the determinant of the product 1. Thus $\mathrm{PSL}_2(q)$ is also 2-transitive on the points of the projective line. It is easy to see (Exercise 3.11) that $\mathrm{PSL}_2(q)$ is generated by the three maps $z \mapsto z + 1$, $z \mapsto \lambda^2 z$ and $z \mapsto -1/z$, where λ is a generator for the multiplicative group of $\mathbb{F}_q$.

The isomorphism $\mathrm{PSL}_2(4) \cong A_5$

Now $\mathrm{PSL}_2(4) \cong \mathrm{SL}_2(4)$ permutes the five points of $\mathrm{PL}(4)$ 2-transitively. The field $\mathbb{F}_4$ of order 4 may be defined as $\{0, 1, \omega, \overline{\omega}\}$ where $\overline{\omega} = \omega^2$ and $\omega^2 + \omega = 1$. Fixing the points 0 and ∞ in $\mathrm{PL}(4) = \{\infty, 0, 1, \omega, \overline{\omega}\}$ we still have the map $z \mapsto \overline{\omega}z/\omega = \omega z$ (defined by the matrix $\begin{pmatrix} \overline{\omega} & 0 \\ 0 & \omega \end{pmatrix}$), which acts as a 3-cycle on the remaining three points 1, ω, $\overline{\omega}$. Thus the action of $\mathrm{PSL}_2(4)$ contains at least a group A_5. But the orders of $\mathrm{PSL}_2(4)$ and A_5 are the same, and therefore the two groups are isomorphic.

The isomorphism $\mathrm{PSL}_2(5) \cong A_5$

The isomorphism $\mathrm{PSL}_2(5) \cong A_5$ may be shown by putting a projective line structure onto the set of six Sylow 5-subgroups of A_5. For example if we label $\langle(1, 2, 3, 4, 5)\rangle$ as ∞ and, reading modulo 5, label $\langle(t, t+1, t+2, t+4, t+3)\rangle$ as t for $t = 0, 1, 2, 3, 4$, then the generators $(1, 2, 3, 4, 5)$ and $(2, 3)(4, 5)$ of A_5 act on the line as $z \mapsto z + 1$ and $z \mapsto -1/z$ (see Exercise 3.12). Hence there is a homomorphism $\phi : A_5 \to \mathrm{PSL}_2(5)$, which is easily seen to be injective. Moreover $|A_5| = |\mathrm{PSL}_2(5)|$, so the two groups are isomorphic.

Notice that this isomorphism is intimately connected with the outer automorphism of S_6, as constructed in Section 2.4.2. Indeed, $\mathrm{PGL}_2(5)$ has order $(5^2 - 1)(5^2 - 5)/(5 - 1) = 120$ and permutes the six points of $\mathrm{PL}(5)$ transitively and faithfully. In particular, $\mathrm{PGL}_2(5)$ is a subgroup of index 6 in S_6, which therefore has an outer automorphism mapping the natural action to the action on the cosets of $\mathrm{PGL}_2(5)$. This simultaneously constructs the outer automorphism of S_6, and proves that the point stabilisers S_5 and $\mathrm{PGL}_2(5)$ are isomorphic.

The isomorphism $\mathrm{PSL}_2(9) \cong A_6$

The isomorphism $\mathrm{PSL}_2(9) \cong A_6$ is best shown by labelling the ten points of the projective line $\mathrm{PL}(9)$ with the ten partitions of six points into two subsets of size 3 (equivalently, the ten Sylow 3-subgroups of A_6). Take $\mathbb{F}_9 = \{0, \pm 1, \pm i, \pm 1 \pm i\}$, where $i^2 = -1$. Let the 3-cycle $(1,2,3)$ act on the points by $z \mapsto z + 1$ and let $(4,5,6)$ act by $z \mapsto z + i$. Then the point ∞ fixed by these two 3-cycles corresponds to the partition $(123 \mid 456)$, and we may choose the point 0 to correspond to the partition $(423 \mid 156)$, so that the rest of the correspondence is determined by the action of the 3-cycles above. We can now generate $\mathrm{PSL}_2(9)$ by adjoining the map $z \mapsto -1/z$, which we can check acts on the points in the same way as the permutation $(2,3)(1,4)$ (see Exercise 3.12). Thus we have a homomorphism from $\mathrm{PSL}_2(9)$ onto A_6, and since these two groups have the same order, they are isomorphic. Notice incidentally that an odd permutation of S_6 realises a field automorphism of $\mathbb{F}_9$: for example, the map $z \mapsto z^3$ corresponds to the transposition $(5,6)$. Thus $S_6 \cong \mathrm{P\Sigma L}_2(9)$, which is not isomorphic to $\mathrm{PGL}_2(9)$.

The actions of $\mathrm{PSL}_2(11)$ *on 11 points*

The action of $\mathrm{PSL}_2(q)$ on the $q+1$ points of the projective line is usually the smallest permutation action. However, we have seen that $\mathrm{PSL}_2(5) \cong A_5$ so has an action on 5 points. Similarly, $\mathrm{PSL}_2(7) \cong \mathrm{PSL}_3(2)$, so has an action on 7 points (indeed, it has two such actions: one on the seven 1-dimensional subspaces, and one on the seven 2-dimensional subspaces). In fact, the only other simple group $\mathrm{PSL}_2(q)$ which has an action on fewer than $q+1$ points is $\mathrm{PSL}_2(11)$, which has two distinct actions on 11 points. As this has ramifications later on (for example in the Mathieu groups, Sections 5.2, 5.3) we take time out to construct these actions here.

We follow Conway's construction [26], which is based on the work of Edge [56] and goes back ultimately to Galois. Consider the partition of the projective line $\mathrm{PL}(11)$ into six pairs $(\infty 0 \mid 12 \mid 36 \mid 48 \mid 5X \mid 79)$ where we write $X = 10$ to avoid confusion. It is easy to see that this partition has just 11 images under the subgroup 11:5 of $\mathrm{PSL}_2(11)$ generated by $z \mapsto z + 1$ and $z \mapsto 3z$. Now consider the action of $z \mapsto -1/z$ on these 11 partitions. Label them p_t, so that p_t is the partition in which ∞ is paired with t. A small calculation shows that $z \mapsto -1/z$ preserves this set of partitions, and acts as the permutation $(1,9)(2,6)(4,5)(7,8)$ on the p_t (see Exercise 3.12). Of course, the map $z \mapsto z + 1$ on $\mathrm{PL}(11)$ acts as $t \mapsto t + 1$, and similarly $z \mapsto 3z$ acts as $t \mapsto 3t$.

The other action on 11 points may be obtained by taking the image under $z \mapsto -z$ of the partition given above.

Projective planes

Analogous to the construction of a projective line from a 2-dimensional vector space is the construction of a *projective plane* from a 3-dimensional vector space. The projective plane consists of *points* (i.e. 1-dimensional subspaces of the vector space) and *lines* (i.e. 2-dimensional subspaces). If the underlying field is $\mathbb{F}_q$, then there are $q^2 + q + 1$ points and $q^2 + q + 1$ lines, with $q + 1$ points on each line, and $q + 1$ lines through each point.

For example if $q = 2$ there are seven points and seven lines. If the points are labelled by integers modulo 7, then the lines may be taken as the seven sets $\{t, t + 1, t + 3\}$. The automorphism group $\mathrm{PGL}_3(2) \cong \mathrm{PSL}_3(2)$ may then be generated by the permutations $t \mapsto t + 1$, $t \mapsto 2t$ and $(1, 2)(3, 6)$ of the points.

Similarly if $q = 3$ the thirteen points may be labelled by the integers modulo 13 (where for convenience we write $X = 10$, $E = 11$ and $T = 12$) in such a way that the lines are $\{t, t + 1, t + 3, t + 9\}$. Then the automorphism group $\mathrm{PGL}_3(3) \cong \mathrm{PSL}_3(3)$ is generated by the permutations $t \mapsto t + 1$, $t \mapsto 3t$ and $(1, 3)(2, 6)(8, E)(X, T)$ of the points.

3.3.6 Covering groups

It turns out that, in general, the special linear group $\mathrm{SL}_n(q)$ is the full covering group (see Section 2.7) of the simple group $\mathrm{PSL}_n(q)$. There are however a few exceptions. In fact, the exceptional part of the Schur multiplier of any classical or exceptional group is always a p-group, where p is the characteristic of the defining field of the group.

Most of the exceptional covers of groups $\mathrm{PSL}_n(q)$ arise as a result of the exceptional isomorphisms above. Thus A_5 has a double cover, and therefore so does $\mathrm{PSL}_2(4)$; similarly, A_6 has a triple cover, and therefore so does $\mathrm{PSL}_2(9)$. Both $\mathrm{PSL}_2(7)$ and A_8 have double covers, and therefore so do $\mathrm{PSL}_3(2)$ and $\mathrm{PSL}_4(2)$. The only other exceptional cover of a simple group $\mathrm{PSL}_n(q)$ is $4^2 \cdot \mathrm{PSL}_3(4)$. (See Section 3.12 for a sketch of a construction of this group.)

3.4 Bilinear, sesquilinear and quadratic forms

There are a number of useful inner products on real and complex vector spaces, and these inner products give rise to various bilinear, sesquilinear and quadratic forms. The situation is similar for vector spaces over finite fields, although the classification of the forms is rather different. In characteristic 0, the three interesting types of forms (generalised inner products) are

  (i) skew-symmetric bilinear,
 (ii) conjugate-symmetric sesquilinear, and
(iii) symmetric bilinear.

The corresponding generalised norms are called

 (iv) *Hermitian* forms (derived from (ii)) and
  (v) *quadratic* forms (derived from (iii)).

Over finite fields of odd characteristic we can use the same definitions, using a field automorphism of order 2 in place of complex conjugation (thus the field order must be a square in this case). But in characteristic 2, these definitions do not capture the interesting geometrical (and group-theoretical) phenomena. To remedy this, we replace skew-symmetric bilinear forms by alternating bilinear forms, and symmetric bilinear forms by quadratic forms. In both cases, the two concepts are equivalent if the characteristic of the field is not 2.

3.4.1 Definitions

First, a *bilinear form* on a vector space V over a field F is a map $f : V \times V \to F$ satisfying the laws $f(\lambda u + v, w) = \lambda f(u, w) + f(v, w)$ and $f(u, \lambda v + w) = \lambda f(u, v) + f(u, w)$. It is

$$\begin{aligned} symmetric \text{ if } f(u, v) &= f(v, u), \\ skew\text{-}symmetric \text{ or } anti\text{-}symmetric \text{ if } f(u, v) &= -f(v, u), \\ \text{and } alternating \text{ if } f(v, v) &= 0. \end{aligned} \tag{3.13}$$

Now an alternating bilinear form is always skew-symmetric, since

$$\begin{aligned} 0 &= f(u + v, u + v) \\ &= f(u, u) + f(u, v) + f(v, u) + f(v, v) \\ &= f(u, v) + f(v, u). \end{aligned} \tag{3.14}$$

Conversely, if the characteristic is not 2, then a skew-symmetric bilinear form is alternating, since $f(v, v) = -f(v, v)$. But if the characteristic is 2, then a bilinear form can be skew-symmetric without being alternating.

A *quadratic form* is a map Q from V to F satisfying

$$Q(\lambda u + v) = \lambda^2 Q(u) + \lambda f(u, v) + Q(v), \tag{3.15}$$

where f is a symmetric bilinear form. Thus a quadratic form always determines a symmetric bilinear form, called its *associated bilinear form*. And if the characteristic is not 2, the quadratic form can be recovered from the symmetric bilinear form as $Q(v) = \frac{1}{2} f(v, v)$. If the characteristic is 2, then the associated bilinear form is actually alternating, since

$$0 = Q(2v) = Q(v + v) = 2Q(v) + f(v, v) = f(v, v). \tag{3.16}$$

Warning. It is common in elementary textbooks to define $Q(v) = f(v, v)$, since this gives a closer analogy with norms and inner products in Euclidean spaces. However, the factor $\frac{1}{2}$ is essential for the correct generalisation to characteristic 2. This can lead to some confusion over the precise definitions of norms and inner products in the finite case.

Next we consider conjugate-symmetric sesquilinear forms. For this to make sense we need a field automorphism of order 2, to take the place of complex conjugation for $\mathbb{C}$. Thus the underlying field must have order q^2, for some $q = p^e$, and we write $\bar{x} = x^q$ for every element x of the field. Then a *conjugate-symmetric sesquilinear form* over a vector space V defined over $F = \mathbb{F}_{q^2}$ is a map $f : V \times V \to F$ satisfying

$$f(\lambda u + v, w) = \lambda f(u, w) + f(v, w),$$
$$f(w, v) = \overline{f(v, w)}. \tag{3.17}$$

Note that this implies $f(u, \lambda v + w) = \bar{\lambda} f(u, v) + f(u, w)$. The form H defined by $H(v) = f(v, v)$ is called a (or an) *Hermitian* (or Hermitean) form. Its defining property is that

$$H(\lambda u + v) = \lambda \bar{\lambda} H(u) + \lambda f(u, v) + \bar{\lambda} f(v, u) + H(v). \tag{3.18}$$

Any of the forms f considered in this section is determined by its values $f(e_i, e_j)$ on a basis $\{e_1, \ldots, e_n\}$. The matrix A whose (i, j)th entry is $f(e_i, e_j)$ is called the matrix of f (with respect to this ordered basis). It is easy to show that if $g : f_i \mapsto e_i$ is a base-change matrix, in the sense that $f_i = \sum_j g_{ij} e_j$, then the matrix of the form with respect to the new basis $\{f_1, \ldots, f_n\}$ is $gAg^\top$.

3.4.2 Vectors and subspaces

Many of the concepts, and much of the notation and nomenclature, are the same whichever type of bilinear or sesquilinear form f we have, although the quadratic forms are more complicated in characteristic 2. (See Section 3.4.7 for the modifications required in this case.) For example, we write $u \perp v$ to mean $f(u, v) = 0$ (which is equivalent to $f(v, u) = 0$ in each case), and say that u and v are *perpendicular* or *orthogonal* (with respect to the form f). We write $S^\perp = \{v \in V \mid v \perp s \text{ for all } s \in S\}$, for any subset S of V (and abbreviate $\{v\}^\perp$ to $v^\perp$). In many contexts $S^\perp$ is called the *orthogonal complement* of S, but since it is often not a complement in the vector-space sense, I prefer the term *perpendicular space* as being more accurate and less liable to be misunderstood.

A non-zero vector which is perpendicular to itself is called *isotropic*. More generally $f(v, v)$ is called the *norm* of v. (This is not the same as the usual definition over $\mathbb{C}$, where we take the square root of $f(v, v)$. Over finite fields, however, there is no sensible analogue of this square root.) The *radical* of f, written $\mathrm{rad}\, f$, is $V^\perp$, and f is *non-singular* if the radical is 0, and *singular* otherwise. We are usually (but not always) only interested in forms which are non-singular.

Given any subspace W of V, we can restrict the form f to W. In general this restriction, written $f|_W$, will be singular, and its radical is $W \cap W^\perp$. If $f|_W$ is non-singular, we say that W is *non-singular*, while if $f|_W$ is zero, we say W is *totally isotropic*. It is a straightforward exercise to show that if f is non-singular, and U is a subspace of V, then $(U^\perp)^\perp = U$ and

$$\dim(U) + \dim(U^\perp) = \dim(V), \qquad (3.19)$$

and hence if $U \cap U^\perp = 0$ then $V = U \oplus U^\perp$ (see Exercise 3.17).

3.4.3 Isometries and similarities

If f is a form on a vector space V, an *isometry* of f (or of V, if f is understood) is a linear map $g : V \to V$ which preserves the form, in the sense that $f(u^g, v^g) = f(u, v)$ for all $u, v \in V$. Similarly, an isometry of a quadratic form Q is a linear map g such that $Q(v^g) = Q(v)$ for all $v \in V$. We think of an isometry as preserving inner products and norms, and therefore preserving 'distances' and 'angles'. If we allow also changes of scale we obtain *similarities*, that is linear maps g such that $f(u^g, v^g) = \lambda_g f(u, v)$ for a scalar λ_g which depends on g but not on u or v. A similarity of a quadratic form Q is a linear map g such that $Q(v^g) = \lambda_g Q(v)$. Similarities preserve 'angles' but not necessarily 'distances'.

We obtain the finite classical groups from the groups of isometries of non-singular forms. In order to classify these groups, we need to classify the forms, which we do by choosing a basis for the space in such a way that the matrix of the form takes one of a small number of possible shapes. We consider the different types separately. In each case, we say two forms on V are *equivalent* if they become equal after a change of basis.

3.4.4 Classification of alternating bilinear forms

Given an alternating bilinear form f on a space V, we want to find a basis of V such that f looks as nice as possible. Our argument in this section applies to arbitrary fields, finite or infinite, of any characteristic. If there are any vectors u and v with $f(u, v) = \lambda \neq 0$, then choose the first two basis vectors $e_1 = u$ and $f_1 = \lambda^{-1}v$, say. Then with respect to the basis $\{e_1, f_1\}$ the form f satisfies

$$\begin{aligned} f(e_1, e_1) = f(f_1, f_1) &= 0, \\ f(e_1, f_1) = -f(f_1, e_1) &= 1. \end{aligned} \qquad (3.20)$$

Now restrict the form to $\{e_1, f_1\}^\perp$, and continue. Eventually we have chosen basis vectors $e_1, \ldots, e_m$ and $f_1, \ldots, f_m$, such that $f(u, v) = 0$ for all basis vectors u, v except $f(e_i, f_i) = -f(f_i, e_i) = 1$. Either we have a basis for the whole space, in which case f is non-singular and $\dim(V) = 2m$ is even, or else $f(u, v) = 0$ for all $u, v \in \{e_1, \ldots, f_m\}^\perp \neq 0$, in which case f is singular, and we can complete to a basis in any way we choose. Notice that in the latter case we have that $f(u, v) = 0$ for any $u \in \{e_1, \ldots, f_m\}^\perp$, and any $v \in V$. Usually (but not always) we shall be considering non-singular forms, in which case we have decomposed V as a perpendicular direct sum of m non-singular subspaces $\langle e_i, f_i \rangle$ of dimension 2, called *hyperbolic planes*. The basis

$\{e_1, \ldots, f_m\}$ is called a *symplectic basis*, and the form is called a *symplectic form*.

[The word 'symplectic' derives from the Greek cognate of 'complex', and in this context is said to have been coined by J. J. Sylvester, who had a particular affinity for words beginning with 'sy' (another example is the 'syntheme' introduced in Exercise 2.19). It is preferable, however, to emphasise the English pronunciation rather than the etymology, and realise that the symplectic groups are the *simplest* of the classical groups to understand.]

3.4.5 Classification of sesquilinear forms

To classify conjugate-symmetric sesquilinear forms, we again find a canonical basis for the space V with the form f. Recall that the underlying field is $\mathbb{F}_{q^2}$ which has an automorphism $x \mapsto \overline{x} = x^q$ of order 2. If there is a vector v with $f(v, v) \neq 0$, then $f(v, v) = \overline{f(v, v)}$ so $f(v, v)$ is in the fixed field $\mathbb{F}_q$ of the field automorphism $x \mapsto x^q$. Since the multiplicative group of the field is cyclic order $q^2 - 1 = (q+1)(q-1)$ there is a scalar $\lambda \in \mathbb{F}_{q^2}$ with $\lambda\overline{\lambda} = \lambda^{q+1} = f(v, v)$, so that $e_1 = \lambda^{-1}v$ satisfies $f(e_1, e_1) = 1$. Now restrict f to $(e_1)^\perp$, and carry on. If we find that $f(v, v) = 0$ for all vectors v in the space remaining, then for all v and w we have

$$
\begin{aligned}
0 &= f(v + \lambda w, v + \lambda w) \\
&= f(v, v) + \overline{\lambda}f(v, w) + \lambda f(w, v) + \lambda\overline{\lambda}f(w, w) \\
&= \overline{\lambda}f(v, w) + \lambda f(w, v).
\end{aligned}
\tag{3.21}
$$

Now we can choose two values of λ forming a basis for $\mathbb{F}_{q^2}$ over $\mathbb{F}_q$, say $\lambda_1 = 1$ and $\lambda_2 \neq \overline{\lambda_2}$, and solve the simultaneous equations to get $f(v, w) = f(w, v) = 0$, so that the form is identically 0. In particular, if the form is non-singular, then we have found an *orthonormal* basis for V, i.e. a basis of mutually perpendicular vectors each of norm 1.

3.4.6 Classification of symmetric bilinear forms

Suppose f is a symmetric bilinear form on a vector space V over a field F of odd characteristic p. Again we try to find a nice basis of V. If there are any vectors $u, v \in V$ with $f(u, v) \neq 0$ then there is a vector x (either u, or v or $u + v$) with $f(x, x) \neq 0$. If $f(x, x) = \lambda^2$, then writing $x' = \lambda^{-1}x$ we have $f(x', x') = 1$. Otherwise, we can choose our favourite non-square α in the field and scale so that $f(x', x') = \alpha$. (Here we use the finiteness of the field in an essential way.) Restricting the form now to $x'^\perp$ we continue until we find a perpendicular basis $x_1, \ldots, x_n$ such that for each i, $f(x_i, x_i) = 0, 1$ or α.

But if we have say $f(x_1, x_1) = f(x_2, x_2) = \alpha$, and $f(x_1, x_2) = 0$, then since the squares do not form a field we can choose λ and μ such that $\lambda^2 + \mu^2$ is a non-square, and by scaling appropriately we can choose $\lambda^2 + \mu^2 = \alpha^{-1}$. Then we find that $x_1' = \lambda x_1 + \mu x_2$ and $x_2' = \mu x_1 - \lambda x_2$ form an orthonormal basis of the 2-space spanned by x_1 and x_2, that is $f(x_1', x_1') = f(x_2', x_2') = 1$ and $f(x_1', x_2') = 0$. In this way, we can ensure that the matrix of the form is diagonal with all entries except at most one being 1 or 0.

In particular, we have shown that there are exactly two equivalence classes of non-singular symmetric bilinear forms under the action of the general linear group, in the case when F is a finite field of odd characteristic. Note that the finiteness of the field is essential: for example, over the polynomial ring $\mathbb{F}_p[t]$ there are infinitely many equivalence classes of quadratic forms, even in dimension 1.

If $n = 2$, the two forms may be taken as f_1 and f_2 given with respect to an orthogonal basis $\{x, y\}$ by $f_1(x, x) = f_1(y, y) = 1$, and $f_2(x, x) = 1$, $f_2(y, y) = \alpha \notin F^{\times 2}$. Now, if -1 is a square in F, say $-1 = i^2$, then $f_1(x + iy, x + iy) = 0$, while $f_2(x + \lambda y, x + \lambda y) = 1 + \lambda^2 \alpha$, which cannot be 0 (otherwise $\alpha = -\lambda^{-2} = (\lambda^{-1} i)^2$, which is a contradiction). On the other hand, if -1 is not a square, then $-\alpha$ is a square, say $-\alpha = \lambda^{-2}$ for some λ, so $f_2(x + \lambda y, x + \lambda y) = 0$, while $f_1(x + \lambda y, x + \lambda y)$ can never be 0. Now -1 is a square in $\mathbb{F}_q$ if and only if $q \equiv 1 \bmod 4$, so there is a non-zero isotropic vector for f_1 if and only if $q \equiv 1 \bmod 4$, and there is a non-zero isotropic vector for f_2 if and only if $q \equiv 3 \bmod 4$. We prefer the geometric distinction to the number-theoretic one, so we say the form is of *plus type* if there is an isotropic vector, and of *minus type* if there is not. Thus f_1 is of plus type and f_2 is of minus type if $q \equiv 1 \bmod 4$, and vice versa if $q \equiv 3 \bmod 4$.

More generally, a form in $2m$ dimensions is defined to be of *plus type* if there is a totally isotropic subspace of dimension m, and of *minus type* otherwise. A straightforward calculation shows that the form which has an orthonormal basis is of minus type just if $q \equiv 3 \bmod 4$ and m is odd. The maximal dimension of a totally isotropic subspace is often called the *Witt index* of the form. Thus the forms of plus type have Witt index m while those of minus type have Witt index $m - 1$.

3.4.7 Classification of quadratic forms in characteristic 2

First we need to extend some of our earlier definitions from Section 3.4.2. Suppose that Q is a quadratic form on V over a field $F = \mathbb{F}_q$ of characteristic 2, and that f is the associated bilinear form. The *radical of Q*, written $\mathrm{rad}\, Q$, is the subset of vectors $v \in \mathrm{rad}\, f$ such that $Q(v) = 0$. This is a subspace since $\mathrm{rad}\, f$ is a subspace and if $v, w \in \mathrm{rad}\, Q$ then

$$Q(v + \lambda w) = Q(v) + \lambda f(v, w) + \lambda^2 Q(w) = 0.$$

Indeed, if $v, w \in \mathrm{rad}\, f$ then $Q(v + \lambda w) = Q(v) + \lambda^2 Q(w)$, so Q restricts to a semilinear map from $\mathrm{rad}\, f$ to F, so $\mathrm{rad}\, Q$ has codimension 0 or 1 in $\mathrm{rad}\, f$.

The *norm* of a vector v is $Q(v)$, and v is called *isotropic* if it is a non-zero vector of norm 0. The form Q is called *non-singular* if $\mathrm{rad}\, Q = 0$, and *non-degenerate* (or *non-defective*) if $\mathrm{rad}\, f = 0$. Thus if Q is degenerate but non-singular then rad f has dimension 1, and $V/\mathrm{rad}\, f$ supports an alternating bilinear form induced by f. On the other hand, if $v_0 \in \mathrm{rad}\, f$ has $Q(v_0) = 1$, then $Q(v + \lambda v_0) = Q(v) + \lambda^2 Q(v_0)$, so every coset $v + \langle v_0 \rangle$ contains one vector of each possible norm.

It is not hard to show that every isometry of $V/\mathrm{rad}\, f$ can be lifted to a unique isometry of V, so the isometry groups of V (with the form Q) and $V/\mathrm{rad}\, f$ (with the form induced by f) are isomorphic. If we are only interested in the group theory rather than the geometry, therefore, we may, and do, restrict to the cases where rad $f = 0$, so dim V is even.

We pick a basis for the space in the same way as for alternating bilinear forms in Section 3.4.4, with the additional condition that we choose our basis vectors to be isotropic (i.e. $Q(v) = 0$) whenever possible. If $Q(e_i) = 0$, then $Q(f_i + \lambda e_i) = Q(f_i) + \lambda$, so replacing f_i by $f_i + Q(f_i)e_i$ we may assume $Q(f_i) = 0$. Moreover, if the dimension is at least 3 then there is always a pair of perpendicular vectors u, v say, and if u is not isotropic then set $\lambda = (Q(v)Q(u)^{-1})^{q/2}$ so that $\lambda^2 = Q(v)Q(u)^{-1}$ and therefore $Q(v + \lambda u) = 0$, so there is always a non-zero isotropic vector. To complete our basis, therefore, we only need to consider separately the case when the dimension is 2.

In this case, we may choose basis vectors v, w with $Q(v) = f(v, w) = 1$, and then for all λ we have $f(v, w + \lambda v) = 1$ and $Q(w + \lambda v) = Q(w) + \lambda^2 + \lambda$. Now for each μ the equation $\lambda^2 + \lambda = \mu$ has at most two solutions for λ, so there are at least $q/2$ distinct values for $\lambda^2 + \lambda$ as λ ranges over $\mathbb{F}_q$. So replacing w by $w + \lambda v$ we see that there are at most two possible quadratic forms, up to equivalence. Indeed, the equation $\lambda^2 + \lambda = 0$ has two solutions $\lambda = 0, 1$, so there are exactly two possible quadratic forms, up to equivalence. This argument also shows that there is a value of μ such that $x^2 + x + \mu = 0$ has no solutions, so $x^2 + x + \mu$ is an irreducible polynomial. Moreover, the two types of quadratic forms are represented by $Q(x, y) = xy$ and $Q(x, y) = x^2 + xy + \mu y^2$ where $x^2 + x + \mu$ is irreducible.

The first of these is called of *plus type*, as there are isotropic vectors, while the second is of *minus type* as there are not. More generally, in $2m$ dimensions, there is one form (called the *plus type*) which has isotropic m-spaces, and another (called the *minus type*) which does not.

3.4.8 Witt's Lemma

A key result which plays an important role in the study of the geometry of these spaces, and hence in the study of the subgroups of the classical groups, is Witt's Lemma (also known as Witt's Theorem). Essentially this says that the isometry groups of nonsingular forms are transitive on subspaces of any given isometry type. We prove here the cases where the forms are alternating bilinear, or conjugate-symmetric sesquilinear, or symmetric bilinear in odd characteristic. We shall not prove the corresponding result for quadratic forms in characteristic 2. More formally:

Theorem 3.3. *If (V, f) and (W, g) are isometric spaces, with f and g non-singular, and either alternating bilinear, or conjugate-symmetric sesquilinear, or symmetric bilinear in odd characteristic, then any isometry α between subspaces X of V and Y of W extends to an isometry of V with W.*

Proof. Suppose for a contradiction that Witt's Lemma is false, and pick a counterexample such that $\dim V$ is minimal, and X is as large as possible in V. We divide into two cases, according as X contains a non-trivial non-singular subspace U, or X is totally isotropic. In the first case $V = U \oplus U^\perp$, and the classification of non-singular forms in the previous sections shows that $U^\perp$ and $(U^\alpha)^\perp$ are isometric. Therefore, by induction, the restriction of α to $U^\perp \cap X$ extends to an isometry from $U^\perp$ to $(U^\alpha)^\perp$. Combining this with α on U gives the required isometry between V and W, extending α.

In the second case, pick $0 \neq x \in X$ and a complement Z to $\langle x \rangle$ in X, so that $X = \langle x \rangle \oplus Z$, and pick $x_1 \in Z^\perp \setminus X^\perp$. Scaling x_1 if necessary, we may assume $f(x, x_1) = 1$, and then replacing x_1 by $x_1 + \lambda x$ for suitable λ we may assume x_1 is isotropic. Similarly, $Y = \langle x^\alpha \rangle \oplus Z^\alpha$ and we pick an isotropic vector $y_1 \in (Z^\alpha)^\perp \setminus Y^\perp$ with $g(y, y_1) = 1$. Then we extend α to an isometry from $\langle X, x_1 \rangle$ to $\langle Y, y_1 \rangle$ by mapping x_1 to y_1. By induction, this map extends to an isometry from V to W, as required.

Witt's Lemma for orthogonal groups in characteristic 2 states that if (V, Q) and (W, R) are isometric spaces, where Q and R are non-degenerate quadratic forms, then any isometry between subspaces X of V and Y of W extends to an isometry of V and W. We leave the proof as an exercise (see Section 3.8 and Exercise 3.31). (There are more general versions of Witt's Lemma which apply to degenerate quadratic forms, or other singular forms, but they are more complicated to state and prove, and we shall not need them.)

3.5 Symplectic groups

The *symplectic group* $\mathrm{Sp}_{2m}(q)$ is the isometry group of a non-singular alternating bilinear form f on $V \cong \mathbb{F}_q^{2m}$, i.e. the subgroup of $\mathrm{GL}_{2m}(q)$ consisting of those elements g such that $f(u^g, v^g) = f(u, v)$ for all $u, v \in V$. Recall from Section 3.4.4 that V has a symplectic basis $\{e_1, \ldots, e_m, f_1, \ldots, f_m\}$ such that all basis vectors are perpendicular to each other except that $f(e_i, f_i) = 1$. To calculate the order of the symplectic group, we simply need to count the number of ways of choosing an (ordered) symplectic basis $e_1, \ldots, f_m$. Now e_1 can be any non-zero vector, so can be chosen in $q^{2m} - 1$ ways. Then $e_1{}^\perp$ has dimension $2m - 1$, so contains q^{2m-1} vectors. Thus there are $q^{2m} - q^{2m-1} = (q-1)q^{2m-1}$ vectors v with $f(u, v) \neq 0$. These come in sets of $q - 1$ scalar multiples, one with each possible value of $f(u, v)$, so there are just q^{2m-1} choices for f_1. Hence by induction the order of $\mathrm{Sp}_{2m}(q)$ is

$$|\mathrm{Sp}_{2m}(q)| = \prod_{i=1}^{m} (q^{2i} - 1) q^{2i-1} = q^{m^2} \prod_{i=1}^{m} (q^{2i} - 1). \qquad (3.22)$$

Observe that $f(\lambda u, \lambda v) = \lambda^2 f(u, v)$, which equals $f(u, v)$ if and only if $\lambda = \pm 1$. Therefore the only scalars in $\mathrm{Sp}_{2m}(q)$ are ± 1. The group $\mathrm{PSp}_{2m}(q)$ is defined to be the quotient of $\mathrm{Sp}_{2m}(q)$ by the subgroup (of order 1 or 2) of scalar matrices. It is usually simple, as we are about to see.

3.5.1 Symplectic transvections

Notice that $\mathrm{Sp}_2(q) = \mathrm{SL}_2(q)$. For if we write elements of $\mathrm{Sp}_2(q)$ with respect to a symplectic basis, then $\begin{pmatrix} a & b \\ c & d \end{pmatrix} \in \mathrm{Sp}_2(q)$ if and only if $f((a,b),(c,d)) = 1$, that is $ad - bc = 1$. In particular, every element of $\mathrm{Sp}_2(q)$ has determinant 1, and, from Section 3.3.2, $\mathrm{Sp}_2(q)$ is generated by transvections.

More generally, a *symplectic transvection* is a linear map

$$T_v(\lambda) : x \mapsto x + \lambda f(x, v)v, \qquad (3.23)$$

where f is a fixed symplectic (i.e. non-singular alternating bilinear) form on the space V, and $v \neq 0$ and $\lambda \neq 0$. We aim to show that the group S generated by symplectic transvections is the whole of $\mathrm{Sp}_{2m}(q)$. As well as feeding in to Iwasawa's Lemma, to prove simplicity of $\mathrm{PSp}_{2m}(q)$, this implies that (since the transvections have determinant 1) every element of $\mathrm{Sp}_{2m}(q)$ has determinant 1, so that $\mathrm{Sp}_{2m}(q) \leqslant \mathrm{SL}_{2m}(q)$. Our method is to prove that S acts transitively on the set of ordered symplectic bases. Since the stabiliser of an ordered basis is (obviously!) trivial, it will then follow immediately that $S = \mathrm{Sp}_{2m}(q)$.

So let v, w be two distinct non-zero vectors. If $f(v, w) = \lambda \neq 0$, then $T_{v-w}(\lambda^{-1}) : v \mapsto w$. Otherwise, pick x such that $f(v, x) \neq 0 \neq f(w, x)$: such an x exists because if not then by non-singularity there exist y and z with $f(v, y) = f(w, z) = 0$ and $f(v, z) \neq 0$ and $f(w, y) \neq 0$, whence a suitable linear combination of y and z has the required properties. Now we can map v to x and x to w, and deduce that S is transitive on non-zero vectors.

Similarly, suppose u is a fixed vector, and $f(u, v) = f(u, w) = 1$. If $f(v, w) = \lambda \neq 0$, then $T_{v-w}(\lambda^{-1}) : v \mapsto w$ and fixes u. Otherwise, let $x = u+v$, so that $f(u, x) = 1$ and $f(v, x) = f(w, x) = -1$, so we can map v to x and x to w while fixing u. Thus by induction S is transitive on ordered symplectic bases, as required.

3.5.2 Simplicity of $\mathrm{PSp}_{2m}(q)$

We usually disregard the case $m = 1$, because $\mathrm{Sp}_2(q) = \mathrm{SL}_2(q)$, as we saw in Section 3.5.1. The only other non-simple case is $\mathrm{Sp}_4(2) \cong S_6$. To see this isomorphism, let S_6 act on the 6-space $\mathbb{F}_2{}^6$ over $\mathbb{F}_2$ by permuting the coordinates. The subspace $U = \langle (1,1,1,1,1,1) \rangle$ of dimension 1 is fixed by S_6, as is the subspace W of dimension 5 consisting of vectors $x = (x_1, \ldots, x_6)$ satisfying $\sum_{i=1}^6 x_i = 0$. There is a natural alternating bilinear form f on W given by $f(x, y) = \sum_{i=1}^6 x_i y_i$, under which U is the radical of f.

Since $U < W$ we obtain (as in Section 3.4.7) an induced alternating bilinear form on the 4-space W/U and an induced action of S_6 on W/U preserving this form. Therefore there is a homomorphism from S_6 to $\mathrm{Sp}_4(2)$, and the image is certainly bigger than C_2, so the kernel of the homomorphism is trivial, and since the two groups have the same order they are isomorphic. (More on this isomorphism can be found in Section 4.2.1.)

To prove simplicity of the symplectic groups $\mathrm{PSp}_{2m}(q)$, for all $m > 2$, and for $m = 2$ and $q > 2$, we just need to verify the hypotheses of Iwasawa's Lemma (Theorem 3.1). We have already seen in Section 3.5.1 that the group $\mathrm{Sp}_{2m}(q)$ is generated by its symplectic tranvections. In the action of $\mathrm{Sp}_{2m}(q)$ on the 1-dimensional subspaces, we proved that the stabiliser of a point is transitive on the q^{2m-1} points which are not orthogonal to the fixed point. It is also transitive on the $(q^{2m-1} - 1)/(q - 1) - 1$ points which are orthogonal but not equal to it: for if v and w are both orthogonal to u, then either $f(v, w) = \lambda \neq 0$, in which case $T_{v-w}(\lambda^{-1}) : v \mapsto w$ while fixing u, or there exists a vector x with $f(v, x) \neq 0 \neq f(w, x)$ and we can map v via x to w while fixing u. Therefore the action is primitive, since the only possibilities for block sizes are now $1 + q^{2m-1}$ and $1 + (q^{2m-1} - 1)/(q - 1)$, neither of which divides number of points, $(q^{2m} - 1)/(q - 1)$.

It is obvious that the symplectic transvections $T_v(\lambda)$ for a fixed vector v form a normal abelian subgroup of stabiliser of the point $\langle v \rangle$, so the only remaining thing to check is that $\mathrm{Sp}_{2m}(q)$ is perfect. It is enough to check that the symplectic transvections are commutators. If $q > 3$, this is already true in $\mathrm{Sp}_2(q) \cong \mathrm{SL}_2(q)$, so we only need to check the two cases $\mathrm{Sp}_4(3)$ and $\mathrm{Sp}_6(2)$. This is left as an exercise (see Exercise 3.21).

3.5.3 Subgroups of symplectic groups

To construct groups B, N, T, U and W by analogy with the general linear groups (see Section 3.3.3), we take B to be the stabiliser of a maximal flag of the shape

$$0 < W_1 < \cdots < W_m = (W_m)^\perp < (W_{m-1})^\perp < \cdots < (W_1)^\perp < V.$$

We may as well take $W_k = \langle e_1, \ldots, e_k \rangle$, for simplicity, and order the basis as $e_1, \ldots, e_m, f_m, \ldots, f_1$ to show the structure of the flag. For all $i < j \leqslant m$ and all $\lambda \in \mathbb{F}_q$, define the maps $x_{ij}(\lambda)$ and $y_{ij}(\lambda)$ to fix all basis vectors e_k and f_k except

$$\begin{aligned}
x_{ij}(\lambda) : f_i &\mapsto f_i + \lambda f_j, \\
e_j &\mapsto e_j - \lambda e_i, \\
\text{and } y_{ij}(\lambda) : f_i &\mapsto f_i + \lambda e_j, \\
f_j &\mapsto f_j + \lambda e_i.
\end{aligned} \tag{3.24}$$

We then see that the unitriangular subgroup U is generated by the maps $x_{ij}(\lambda)$ and $y_{ij}(\lambda)$, together with the symplectic transvections $T_{e_i}(-\lambda) : f_i \mapsto f_i + \lambda e_i$, so that U has order q^{m^2} and is a Sylow p-subgroup.

The torus T is generated by diagonal elements $f_i \mapsto \lambda f_i, e_i \mapsto \lambda^{-1} e_i$, so is a direct product of m cyclic groups of order $q - 1$, and $B = UT$ as before. The normaliser N of this torus is generated modulo the torus by permutations of the subscripts $1, \ldots, m$, together with the element $e_1 \mapsto f_1 \mapsto -e_1$, and therefore the Weyl group N/T is isomorphic to the wreath product $C_2 \wr S_m$. As before, N is represented by monomial matrices, and the Weyl group is the quotient group of suitable permutations of the coordinate 1-spaces $\langle e_i \rangle$ and $\langle f_i \rangle$.

3.5.4 Subspaces of a symplectic space

Since the stabiliser of any subspace W of V must stabilise $W^\perp$ and $W \cap W^\perp$, we are generally only interested in the cases where $W \cap W^\perp = 0$ or $W \cap W^\perp = W$, so either W is a non-singular subspace, or W is totally isotropic.

The stabiliser of a non-singular subspace of dimension $2k$ preserves the decomposition $V = W \oplus W^\perp$, so is just $\mathrm{Sp}_{2k}(q) \times \mathrm{Sp}_{2m-2k}(q)$. This is usually a maximal subgroup of $\mathrm{Sp}_{2m}(q)$, unless $m = 2k$, in which case there is an element exchanging W and $W^\perp$, and extending the group to $\mathrm{Sp}_{2k}(q) \wr S_2$.

More generally, if $m = kl$ there is a subgroup $\mathrm{Sp}_{2k}(q) \wr S_l$ preserving a decomposition of V as a direct sum of mutually perpendicular non-singular spaces of dimension $2k$.

The stabiliser of an isotropic subspace W of dimension k preserves the flag $0 < W < W^\perp < V$, and there is an induced non-singular form on $W^\perp/W$. By Witt's lemma, we may choose our symplectic basis so that W is spanned by $e_1, \ldots, e_k$, and $W^\perp$ is spanned by $e_1, \ldots, e_m, f_{k+1}, \ldots, f_m$. We therefore see a basis of $W^\perp/W$ consisting of the images of $e_{k+1}, \ldots, e_m, f_{k+1}, \ldots, f_m$, and a quotient group $\mathrm{Sp}_{2m-2k}(q)$ acting on $W^\perp/W$. We also see a group $\mathrm{GL}_k(q)$ acting on W (and inducing the dual action on $V/W^\perp$), and a p-group of lower triangular matrices generated by elements (defined in (3.24)) $x_{ij}(\lambda)$ and $y_{ij}(\lambda)$ for all $i \leqslant k < j \leqslant m$. It can be shown that these elements $x_{ij}(\lambda)$, $y_{ij}(\lambda)$ generate a non-abelian group Q, such that $Z(Q) = \Phi(Q) = Q'$ is an elementary abelian group of order $q^{k(k+1)/2}$, and Q/Q' is an elementary abelian group of order $q^{2k(m-k)}$. (Here $\Phi(G)$ denotes the *Frattini subgroup*, i.e. the intersection of all the maximal subgroups of G. A p-group with the property that the centre, derived group and Frattini subgroup are equal is called a *special* group. We shall have more to say about them later.) The full stabiliser is therefore a group of shape $q^{k(k+1)/2}.q^{2k(m-k)}{:}(\mathrm{Sp}_{2m-2k}(q) \times \mathrm{GL}_k(q))$. In the case when $k = m$, the corresponding group has shape $q^{m(m+1)/2}{:}\mathrm{GL}_m(q)$. These stabilisers are the *maximal parabolic subgroups*.

Just as in Section 3.3.3 there is an interpretation of the Dynkin diagram in terms of these subgroups. In this case the Dynkin diagram is C_m and the kth node of the diagram corresponds to a totally isotropic k-dimensional subspace. Deleting this node from the diagram gives the Dynkin diagram of the Levi complement $\mathrm{Sp}_{2m-2k}(q) \times \mathrm{GL}_k(q)$ of the subspace stabiliser.

Given a maximal isotropic subspace $W = W^\perp$, of dimension m, we can choose a complement U which is also totally isotropic. For example, if $W = \langle e_1, \ldots, e_m \rangle$ we may take $U = \langle f_1, \ldots, f_m \rangle$. The stabiliser of the direct sum decomposition $V = W \oplus U$ is $\mathrm{GL}_m(q).2$, in which the elements swapping W and U induce the duality automorphism on $\mathrm{GL}_m(q)$ (see Section 3.3.4).

See section 3.10 below for more subgroups.

3.5.5 Covers and automorphisms

The generic covers are described by $\mathrm{Sp}_{2n}(q) = 2{\cdot}\mathrm{PSp}_{2n}(q)$ for q odd, while for q even $\mathrm{Sp}_{2n}(q) \cong \mathrm{PSp}_{2n}(q)$ and there is usually no proper cover.

It turns out that there are just two exceptional covers of finite symplectic groups, including the exceptional double cover of $\mathrm{Sp}_4(2) \cong S_6$, which we constructed earlier (see Section 2.7.2). [The alert reader may object that we constructed *two* double covers of S_n, in one of which the transpositions lift to involutions, while in the other they lift to elements of order 4. Such a reader is encouraged to prove, by using the outer automorphism of S_6, or otherwise, that the two double covers of S_6 are isomorphic: see Exercise 2.38.] The other is $2{\cdot}\mathrm{Sp}_6(2)$, which may be seen for example inside the exceptional cover $2^2{\cdot}\Omega_8^+(2)$ of the orthogonal group $\Omega_8^+(2)$, which we shall construct later, in Section 3.12.

The diagonal automorphisms are induced by similarities of the form, that is elements g of the general linear group which satisfy $f(u^g, v^g) = \lambda_g f(u, v)$ for all $u, v \in V$, and a scalar λ_g depending only on g and not on u and v. As scalar multiplication by λ multiplies f by λ^2, the only scalar multiples which can correspond to non-trivial outer automorphisms are those by $F^\times / F^{\times 2}$, where $F^{\times 2} = \{x^2 \mid x \in F^\times\}$. This quotient group of scalars has order 2 if q is odd, and order 1 if q is a power of 2. We obtain a group $\mathrm{PGSp}_{2m}(q) = \mathrm{PSp}_{2m}(q).2$ for all odd q.

The field automorphisms are obtained by applying an automorphism of the underlying field to all the matrix entries, when the matrices are written with respect to the standard symplectic basis $\{e_1, \ldots, f_m\}$.

The graph automorphisms are harder to describe: they only exist in dimension 4, and only for fields of characteristic 2 (cf. the outer automorphism of $S_6 \cong \mathrm{Sp}_4(2)$ described in Section 2.4.2). They are best constructed using the Klein correspondence (see the end of Section 3.7) and the natural isomorphism $\mathrm{Sp}_4(2^e) \cong \Omega_5(2^e)$ (see Section 3.8). As their main application is as the crucial ingredient in the construction of the Suzuki groups (see Section 4.2), we postpone discussion of these automorphisms until Chapter 4.

3.5.6 The generalised quadrangle

As we saw in Section 3.5.3, the Weyl group of $\mathrm{Sp}_4(q)$ is $C_2 \wr C_2 \cong D_8$. It turns out that when the Weyl group of a group of Lie type is dihedral of order $2n$, so acts as the group of symmetries of an n-gon, the group itself acts on a so-called *generalised n-gon*. We briefly describe the generalised quadrangle for $Sp_4(q)$.

The points are the (isotropic) 1-spaces, and the lines are the totally isotropic 2-spaces. We say a point and a line are *incident* if the corresponding 1-space is contained in the corresponding 2-space. We know that there are $(q^4 - 1)/(q - 1) = q^3 + q^2 + q + 1$ points, and easily compute that there is the same number of lines. Each line is incident with $q + 1$ points, and therefore each point is incident with $q + 1$ lines. [So putting $q = 1$ would give us back the ordinary quadrangle, with four points and four lines, each incident with two of the other type.]

Often the generalised quadrangle is described in terms of its point–line incidence graph: this is the bipartite graph with $q^3 + q^2 + q + 1$ vertices corresponding to points, and $q^3 + q^2 + q + 1$ vertices corresponding to lines, and edges joining all incident pairs. If we fix a point, there are $q + 1$ lines incident with it, and $q(q + 1)$ further points incident with one of those lines. There are then $q^2(q + 1)$ more lines incident with one of these further points. That leaves just q^3 points, and we summarise the structure of the graph as follows, where figures below the nodes denote the number of vertices, and figures above the edges denote the number of edges incident with each vertex.

$$
\begin{array}{ccccc}
\overset{q+1}{\bullet} \!\!\!\!\! & \overset{1\quad q}{\circ} & \overset{1\quad q}{\circ} & \overset{1\quad q\quad q+1}{\bullet} \\
1 & q+1 & q(q+1) & q^2(q+1) & q^3
\end{array}
$$

Other examples of generalised n-gons are the generalised quadrangles for the unitary groups $U_4(q)$ and $U_5(q)$ (see Section 3.6.4), the generalised hexagons for $G_2(q)$ (see Section 4.3.8) and $^3D_4(q)$ (see Section 4.6.4), and the generalised octagon for $^2F_4(q)$ (see Section 4.9.4). Moreover, a projective plane (see Section 3.3.5) may also be regarded as a generalised triangle.

3.6 Unitary groups

We obtain the unitary groups in much the same way, starting from a non-singular conjugate-symmetric sesquilinear form instead of a non-singular alternating bilinear form. The *(general) unitary group* $GU_n(q)$ is defined as the isometry group of a non-singular conjugate-symmetric sesquilinear form f, i.e. the subgroup of $GL_n(q^2)$ consisting of the elements g which preserve the form, in the sense that $f(u^g, v^g) = f(u, v)$ for all $u, v \in V$. To calculate its order, we need to count the number of vectors of norm 1, and use induction. Let z_n denote the number of vectors of norm 0, and y_n denote the number of vectors of norm 1. Then the total number of vectors in the space is $q^{2n} = 1 + z_n + (q - 1)y_n$, and we calculate $z_{n+1} = z_n + (q^2 - 1)y_n$ so $z_{n+1} = (q^{2n} - 1)(q + 1) - qz_n$. Since $z_0 = z_1 = 0$ we may solve the recurrence relation to get $z_n = (q^n - (-1)^n)(q^{n-1} + (-1)^n)$, and therefore $y_n = q^{n-1}(q^n - (-1)^n)$.

Now an arbitrary element of $\mathrm{GU}_n(q)$ may be specified by picking an orthonormal basis one vector at a time, so the order of $\mathrm{GU}_n(q)$ is

$$|\mathrm{GU}_n(q)| = \prod_{i=1}^{n} q^{i-1}(q^i - (-1)^i)$$

$$= q^{n(n-1)/2} \prod_{i=1}^{n} (q^i - (-1)^i). \tag{3.25}$$

Writing elements g of $\mathrm{GU}_n(q)$ with respect to an orthononormal basis, the rows of g are orthonormal vectors. This fact can be expressed by the equation $g\overline{g}^\top = I_n$. In particular if $\det(g) = \lambda$ then $\lambda\overline{\lambda} = \lambda^{q+1} = 1$. Now as λ is in a cyclic group of order $q^2 - 1 = (q+1)(q-1)$ this equation says that λ is in the unique subgroup of order $q + 1$. In particular, $\mathrm{GU}_n(q)$ has a subgroup $\mathrm{SU}_n(q)$ of index $q + 1$ consisting of all the elements with determinant 1.

Similarly, the scalars in $\mathrm{GU}_n(q)$ are those λ for which $\lambda\overline{\lambda} = \lambda^{q+1} = 1$. There are exactly $q + 1$ such λ in $\mathbb{F}_{q^2}$. In particular, $\mathrm{GU}_n(q)$ has a central subgroup Z of order $q + 1$. The quotient $\mathrm{PGU}_n(q) = \mathrm{GU}_n(q)/Z$ is called the projective unitary group.

The scalars of determinant 1 are those λ for which both $\lambda^n = 1$ and $\lambda^{q+1} = 1$. The number of such scalars is precisely the greatest common divisor $d = (n, q+1)$. The quotient $\mathrm{PSU}_n(q)$ of $\mathrm{SU}_n(q)$ by the scalars it contains is usually a simple group. The exceptions are those explained by the isomorphisms $\mathrm{PSU}_2(q) \cong \mathrm{PSL}_2(q)$, together with the group $\mathrm{PSU}_3(2)$, which is a soluble group of order 72 (see Exercises 3.23 and 3.24).

To see that $\mathrm{SU}_2(q) \cong \mathrm{SL}_2(q)$ we take the natural module for $\mathrm{SU}_2(q)$ over $\mathbb{F}_{q^2}$, and find a 2-dimensional $\mathbb{F}_q$-subspace which is invariant under the action of the group. We first pick an element $\mu \in \mathbb{F}_{q^2}$ with $\mu\overline{\mu} = \mu^{1+q} = -1 \in \mathbb{F}_q$, and then take all vectors of the form $(\lambda, \mu\overline{\lambda})$ where $\lambda \in \mathbb{F}_{q^2}$. This is clearly a 2-dimensional $\mathbb{F}_q$-subspace, and we can check it is invariant under an arbitrary element $\begin{pmatrix} \alpha & \beta \\ -\overline{\beta} & \overline{\alpha} \end{pmatrix}$ of $\mathrm{SU}_2(q)$. [For this matrix maps the typical vector $(\lambda, \mu\overline{\lambda})$ to $(\alpha\lambda - \mu\overline{\beta}\lambda, \beta\lambda + \mu\overline{\alpha}\lambda)$ and

$$\mu(\overline{\alpha\lambda - \mu\overline{\beta}\lambda}) = \mu\overline{\alpha}\overline{\lambda} - \mu\overline{\mu}\beta\lambda = \beta\lambda + \mu\overline{\alpha}\overline{\lambda}$$

since $\mu\overline{\mu} = -1$.] The kernel of this action is obviously trivial, and $|\mathrm{SU}_2(q)| = |\mathrm{SL}_2(q)|$, so the groups are isomorphic.

Warning. Here as elsewhere notation varies within the literature. In particular, a notation such as $\mathrm{U}_n(q)$ or $U(n, q)$ is sometimes used for the general unitary group which I have called $\mathrm{GU}_n(q)$, and sometimes for $\mathrm{PSU}_n(q)$.

3.6.1 Simplicity of unitary groups

We merely sketch the proof of simplicity of the unitary groups $\mathrm{PSU}_n(q)$ here, as it is very similar to the case of the symplectic groups. Since $\mathrm{PSU}_2(q) \cong$

$\mathrm{PSL}_2(q)$ we may assume $n > 2$. The proof uses Iwasawa's Lemma applied to the permutation representation on the isotropic 1-spaces, and the unitary transvections as generators. These transvections are defined in the same way as for the symplectic groups, by the rule $T_v(\lambda) : x \mapsto x + \lambda f(x, v)v$, where v is isotropic, so $f(v, v) = 0$. A straightforward calculation shows that $T_v(\lambda)$ is an isometry if and only if $\lambda = 0$ or $\lambda^{q-1} = -1$; thus we define the *unitary transvections* to be the maps $T_v(\lambda)$ with $\lambda^{q-1} = -1$ (see Exercise 3.22).

The unitary transvections for fixed v form an abelian normal subgroup of the stabiliser of $\langle v \rangle$, since $T_v(\lambda)T_v(\mu) = T_v(\lambda + \mu)$, and $\lambda^q = -\lambda$, $\mu^q = -\mu$, so $(\lambda + \mu)^q = \lambda^q + \mu^q = -\lambda - \mu$. It is straightforward to show that the special unitary group acts primitively on the set of isotropic 1-spaces. Also $\mathrm{SU}_n(q)$ can be shown to be generated by the unitary transvections, except for the case $\mathrm{SU}_3(2)$. Explicit calculation shows that every unitary transvection is a commutator of elements of $\mathrm{SU}_n(q)$, provided $n > 3$, or $n = 3$ and $q > 2$. Hence Iwasawa's Lemma shows $\mathrm{PSU}_n(q)$ is simple in all the remaining cases.

3.6.2 Subgroups of unitary groups

First we change basis so that we can see the 'BN-pair' (i.e. the subgroups B, N and related subgroups $T = B \cap N$, $W = N/T$ and U such that $B = UT$) clearly. If V_1 is a non-singular unitary 2-space, and q is odd, then we can find a symplectic basis of V_1 by taking $e_1 = (\alpha, \beta)$ and $f_1 = (\alpha, -\beta)$, where $\alpha\overline{\alpha} + \beta\overline{\beta} = 0$ (i.e. $(\alpha/\beta)^{q+1} = -1$) and $\alpha\overline{\alpha} - \beta\overline{\beta} = 1$. (Such α and β exist because $x\overline{x} = \lambda$ has solutions for all $\lambda \in \mathbb{F}_q$. If q is even we may take instead $e_1 = (1, 1)$ and $f_1 = (\omega, \omega^2)$, where $\omega^2 + \omega + 1 = 0$ in $\mathbb{F}_4$.) Write V as a perpendicular direct sum $V_1 \oplus \cdots \oplus V_m$ of non-singular 2-spaces (together with a 1-space W in odd dimensions), and pick a symplectic basis $\{e_i, f_i\}$ for V_i as above. Choosing $w \in W$ of norm 1 if necessary, we order our basis $e_1, \ldots, e_m, (w,)f_m, \ldots, f_1$.

With respect to this ordered basis there is a Sylow p-subgroup U acting as lower unitriangular matrices, and a torus T acting as diagonal matrices. Now if $e_i \mapsto \lambda e_i$ then $f_i \mapsto \mu f_i$, where $\mu = \overline{\lambda^{-1}}$, so the torus is a direct product of m cyclic groups of order $q^2 - 1$ in both $\mathrm{SU}_{2m}(q)$ and $\mathrm{SU}_{2m+1}(q)$. In both cases the normaliser N of the torus acts on the set $\{\langle e_i \rangle, \langle f_i \rangle\}$ of 1-dimensional subspaces as a permutation group of shape $C_2 \wr S_m$, with the wreathing group S_m permuting the subscripts $1, \ldots, m$, and the ith factor of the base group interchanging $\langle e_i \rangle$ with $\langle f_i \rangle$.

The maximal parabolic subgroups are again the stabilisers of totally isotropic subspaces, such as $W_k = \langle e_1, \ldots, e_k \rangle$, and have shape

$$q^{k(2n-3k)} {:} (\mathrm{GL}_k(q^2) \times \mathrm{SU}_{n-2k}(q)),$$

where the normal subgroup Q denoted by $q^{k(2n-3k)}$ is a p-group with $Z(Q) = Q' = \Phi(Q)$ of order q^{k^2}. In particular, Q is a *special* group as defined in Section 3.5.4.

The stabilisers in $\mathrm{GU}_n(q)$ of non-singular subspaces of dimension k have the shape $\mathrm{GU}_k(q) \times \mathrm{GU}_{n-k}(q)$. If $n = km$ there are imprimitive groups of shape $\mathrm{GU}_k(q) \wr S_m$, preserving a decomposition of the space as a perpendicular direct sum of non-singular k-spaces. If $n = 2m$ there is also a group $\mathrm{GL}_m(q^2).2$ preserving a disjoint pair of maximal isotropic subspaces (of dimension m).

3.6.3 Outer automorphisms

Here there is not much to say: the diagonal automorphism group has order $d = (n, q+1)$ and comes from $\mathrm{PGU}_n(q)$. In other words it is $\mathrm{PGU}_n(q)/\mathrm{PSU}_n(q) \cong C_{(n,q+1)}$. The field automorphisms form a cyclic group of order $2e$ where $q = p^e$, since we are working over the field of order q^2. There are no graph automorphisms.

Various groups may be defined by adjoining field automorphisms to the various versions of the unitary groups. Since a field automorphism is defined by its action on matrix entries, it is heavily dependent on the basis chosen. Since we have defined the standard basis for the unitary groups to be an orthonormal basis, we shall use this also for our definition of the standard field automorphisms. We then obtain groups $\Gamma\mathrm{U}_n(q)$ and $\Sigma\mathrm{U}_n(q)$, by adjoining the standard field automorphisms to $\mathrm{GU}_n(q)$ and $\mathrm{SU}_n(q)$ respectively. The quotients of these groups by the scalar matrices they contain are denoted $\mathrm{P}\Gamma\mathrm{U}_n(q)$ and $\mathrm{P}\Sigma\mathrm{U}_n(q)$.

Warning. In the literature one sometimes finds definitions of these groups without specifing the basis. Such groups are not always well-defined, even up to isomorphism! Even if the basis is specified, it may not be the same as ours. In particular, a field automorphism defined with respect to a symplectic basis is likely to result in a definition of $\Sigma\mathrm{U}_n(q)$ and $\mathrm{P}\Sigma\mathrm{U}_n(q)$ incompatible with ours.

3.6.4 Generalised quadrangles

Both $\mathrm{PSU}_4(q)$ and $\mathrm{PSU}_5(q)$ have Weyl group D_8 so give rise to generalised quadrangles as described in Section 3.5.6 for $\mathrm{Sp}_4(q)$. Again, the *points* are the isotropic 1-spaces and the *lines* are the isotropic 2-spaces, with incidence being given by inclusion. We have already shown that the number of points is

$$(q^4 - 1)(q^3 + 1)/(q^2 - 1) = (q^2 + 1)(q^3 + 1)$$

for $\mathrm{PSU}_4(q)$ and

$$(q^4 - 1)(q^5 + 1)/(q^2 - 1) = (q^2 + 1)(q^5 + 1)$$

for $\mathrm{PSU}_5(q)$. A similar argument shows that for $\mathrm{PSU}_4(q)$ each point is in $q + 1$ lines, and for $\mathrm{PSU}_5(q)$ each point is in $q^3 + 1$ lines, while in both cases each line contains $q^2 + 1$ points. Therefore the number of lines is repectively $(q+1)(q^3+1)$ and $(q^3+1)(q^5+1)$. Drawing black circles for points and white circles for lines we have the following pictures for $\mathrm{PSU}_4(q)$:

$$\overset{q+1 \quad\; 1\;\; q^2 \qquad\; 1\;\; q \qquad\; 1\;\; q^2 \;\; q+1}{\underset{1 \qquad\quad q+1 \qquad q^2(q+1)\;\; q^3(q+1) \qquad q^5}{\bullet \!-\!-\!\circ\!-\!-\!\bullet\!-\!-\!\circ\!-\!-\!\bullet}}$$

$$\overset{q^2+1 \;\; 1\;\; q \qquad\; 1\;\; q^2 \qquad\; 1\;\; q \;\; q^2+1}{\underset{1 \qquad\quad q^2+1 \qquad q(q^2+1)\;\; q^3(q^2+1) \qquad q^4}{\circ \!-\!-\!\bullet\!-\!-\!\circ\!-\!-\!\bullet\!-\!-\!\circ}}$$

and for $\mathrm{PSU}_5(q)$:

$$\overset{q^3+1 \;\; 1\;\; q^2 \qquad\; 1\;\; q^3 \qquad\; 1\;\; q^2 \;\; q^3+1}{\underset{1 \qquad\quad q^3+1 \qquad q^2(q^3+1)\;\; q^5(q^3+1) \qquad q^7}{\bullet \!-\!-\!\circ\!-\!-\!\bullet\!-\!-\!\circ\!-\!-\!\bullet}}$$

$$\overset{q^2+1 \;\; 1\;\; q^3 \qquad\; 1\;\; q^2 \qquad\; 1\;\; q^3 \;\; q^2+1}{\underset{1 \qquad\quad q^2+1 \qquad q^3(q^2+1)\;\; q^5(q^2+1) \qquad q^8}{\circ \!-\!-\!\bullet\!-\!-\!\circ\!-\!-\!\bullet\!-\!-\!\circ}}$$

3.6.5 Exceptional behaviour

There is one exceptional isomorphism, namely $\mathrm{PSU}_4(2) \cong \mathrm{PSp}_4(3)$, which is best treated as an isomorphism of orthogonal groups $\mathrm{P}\Omega_6^-(2) \cong \mathrm{P}\Omega_5(3)$ (see Section 3.12 below). This is related to the fact that the automorphism group, $\mathrm{PGO}_6^-(2) \cong \mathrm{PGO}_5(3)$, is the Weyl group of type E_6. Since $\mathrm{PSp}_4(3)$ has a double cover $\mathrm{Sp}_4(3)$, it follows that $\mathrm{PSU}_4(2)$ has an exceptional cover $2{\cdot}\mathrm{PSU}_4(2)$. There are also two more exceptional covers $3^2{\cdot}\mathrm{PSU}_4(3)$ and $2^2{\cdot}\mathrm{PSU}_6(2)$. Some of these are important for the sporadic groups (see Chapter 5). Since $\mathrm{PSU}_4(3) \cong \mathrm{P}\Omega_6^-(3)$, we shall discuss this case with the other orthogonal groups in Section 3.12. The existence of $2{\cdot}\mathrm{PSU}_6(2)$, from which the existence of $2^2{\cdot}\mathrm{PSU}_6(2)$ follows easily, is proved in Section 5.7.1 in the course of the construction of the sporadic group Fi_{22}.

3.7 Orthogonal groups in odd characteristic

Recall from Section 3.4.6 that, up to equivalence, there are exactly two non-singular symmetric bilinear forms f on a vector space V over a finite field F of odd order. The (general) orthogonal group $\mathrm{GO}(V, f)$ is defined as the group of linear maps g satisfying $f(u^g, v^g) = f(u, v)$ for all $u, v \in V$. Clearly $\mathrm{GO}(V, \alpha f) = \mathrm{GO}(V, f)$ for any scalar α. If n is even then f and αf are always equivalent forms. On the other hand, if n is odd, and $\alpha \in F$ is a non-square, then f and αf are inequivalent, so there is only one orthogonal group (up to isomorphism) in this case, and we write it as $\mathrm{GO}(V)$ or $\mathrm{GO}_n(q)$ without ambiguity. If n is even, however, we have two thoroughly different orthogonal groups (they do not even have the same order, as we shall see). If f is of plus type we write $\mathrm{GO}_{2m}^+(q)$ for $\mathrm{GO}(V, f)$, while if f is of minus type we write $\mathrm{GO}_{2m}^-(q)$.

3.7.1 Determinants and spinor norms

Now in any of these orthogonal groups G the elements have determinant ± 1. For if M is the matrix of the form, and $g \in G$, then $gMg^\top = M$ so $\det(g) = \det(M(g^\top)^{-1}M^{-1}) = (\det g)^{-1}$.

The elements of determinant 1 form a subgroup of index 2, called the *special orthogonal group* $SO_n(q)$. The only scalars in the orthogonal groups are ± 1, and -1 is in the special orthogonal group if and only if the dimension is even. The corresponding quotient groups are the *projective orthogonal* groups $PGO_n(q)$ and *projective special orthogonal* groups $PSO_n(q)$. In contrast to the other three families of classical groups, however, these groups are not in general simple. The special orthogonal group has itself a subgroup of index 2, defined as the kernel of another invariant, called the *spinor norm*. Thus there is in most cases (the exceptions are the groups $PSO_{2m}^\varepsilon(q)$ where $q^m + \varepsilon \equiv 0 \bmod 4$, as we show at the end of this section) a further subgroup of index 2 in $PSO_n(q)$.

This new invariant works in much the same way as the concept of even and odd permutations. We first write our arbitrary element of the special orthogonal group as a product of reflections. [It is not completely obvious that this can always be done, but this is not important for us as we shall not use this fact in the formal definition of the spinor norm in Section 3.9.2, nor in the proof of its basic properties.] Since a reflection may be defined by the property that it negates a certain 1-space $\langle v \rangle$ and fixes all vectors orthogonal to v, it may be defined by the formula

$$r_v : x \mapsto x - 2\frac{f(x,v)}{f(v,v)}v. \tag{3.26}$$

Since reflections have determinant -1, this product contains an even number of reflections. Now there are two types of reflections: those which negate a vector of norm 1, and those which negate a vector of norm a non-square α. So there is a subgroup of the special orthogonal group (of index 1 or 2) consisting of those elements which are a product of a set of reflections, consisting of an even number of each type (these are called the elements of spinor norm 1). To show that this subgroup has index 2, it suffices to show that there is an element which cannot be written in this way (these are called the elements of spinor norm -1), or to show that the identity element cannot be written as such a product, with an odd number of reflections of each type. This is surprisingly difficult to prove, and we postpone this until Section 3.9.2.

The kernel of the spinor norm map is denoted $\Omega(V, f)$ or $\Omega_n^\pm(q)$ as appropriate, and the quotients $\Omega(V, f)/\{\pm 1\}$ by $P\Omega(V, f)$ etc. The groups $P\Omega_n^\pm(q)$ for q odd are always simple if $n \geqslant 5$.

Note that in even dimensions -1 has spinor norm 1 if and only if there is an orthonormal basis, that is if and only if $q^m \equiv \varepsilon \bmod 4$, where the orthogonal group is $GO_{2m}^\varepsilon(q)$. In particular, if this condition does not hold then $SO_{2m}^\varepsilon(q) = 2 \times \Omega_{2m}^\varepsilon(q)$ and $PSO_{2m}^\varepsilon(q) = P\Omega_{2m}^\varepsilon(q)$.

3.7.2 Orders of orthogonal groups

To calculate the orders of these groups we first prove (by induction) a formula for the number of isotropic vectors in an orthogonal space. Then we show that the stabiliser of an isotropic vector in $GO_n^\varepsilon(q)$ is $q^{n-2}{:}GO_{n-2}^\varepsilon(q)$. Thus we obtain (by induction again) a formula for the order of the full orthogonal group.

In fact, this formula, and the argument used to derive it, works also in characteristic 2, provided the dimension is even. However, the formulae for the order of $SO_n(q)$ and the generically simple group $P\Omega_n(q)$ differ slightly between odd and even characteristic (see Section 3.8.2 below).

For the base case we need to know the orders of the 1- and 2-dimensional orthogonal groups. It is obvious that $GO_1(q) \cong C_2$ for odd q. For q odd and $n = 2$ we can choose an orthogonal basis such that the quadratic form is $x^2 + \lambda y^2$, with either $\lambda = 1$ or λ a fixed non-square. For the plus type, there are just two solutions of $(x/y)^2 + \lambda = 0$, so (up to multiplication by scalars) just two isotropic vectors. In both plus type and minus type the stabiliser in $GO_2^\varepsilon(q)$ of a non-isotropic vector v has order 2 (consisting of the reflection in $v^\perp$), and therefore (since by Witt's Lemma the orthogonal group acts transitively on the vectors of any given norm) the number of vectors of norm 1 is equal to the number of vectors of norm α (a fixed non-square). In particular, there are up to sign just $\frac{1}{2}(q+1)$ vectors of norm 1 in the case $GO_2^-(q)$, and $\frac{1}{2}(q-1)$ in the case $GO_2^+(q)$. Therefore $|GO_2^+(q)| = 2(q-1)$ and $|GO_2^-(q)| = 2(q+1)$. In fact (see Exercise 3.28) these are dihedral groups, since they may be generated by two reflections, in vectors with suitable norms and inner product.

Now we are ready for the first induction, which is really three separate inductions corresponding to the three separate base cases. Let z_m denote the number of (non-zero!) isotropic vectors in an orthogonal space of dimension $2m$ or $2m + 1$. Our inductive hypothesis is that

$$
\begin{aligned}
z_m &= q^{2m} - 1 \text{ in dimension } 2m + 1, \\
z_m &= (q^m - 1)(q^{m-1} + 1) \text{ for a space of plus type,} \\
\text{and } z_m &= (q^m + 1)(q^{m-1} - 1) \text{ for a space of minus type.} \quad (3.27)
\end{aligned}
$$

Note that these formulae give $z_0 = 0$ for a 1-space, $z_1 = 2(q-1)$ for a plus-type 2-space, and $z_1 = 0$ for a minus-type 2-space, so the induction starts.

For the inductive step, we split the $(n+2)$-space V as $V = U \oplus W$, where U is a 2-space of plus type, and W is an orthogonal space of dimension n, of the same type as the original space V. Now every isotropic vector is of the form $u + w$, where $u \in U$ and $w \in W$. Either u and w both have norm 0 (but are not both the zero vector), or u has norm $\lambda \neq 0$ and w has norm $-\lambda$. Since U contains $2q - 1$ vectors of norm 0 (including the zero vector), and $q - 1$ vectors of every non-zero norm, we count

$$
\begin{aligned}
z_{m+1} &= (2q - 1)(1 + z_m) + (q - 1)(q^n - 1 - z_m) - 1 \\
&= qz_m + (q - 1)(q^n + 1), \quad (3.28)
\end{aligned}
$$

and it is a simple matter to complete the proof by induction in each of the three cases.

Next we determine the stabiliser of an isotropic vector v_0. Certainly this is contained in the stabiliser of the flag $0 < \langle v_0 \rangle < v_0^\perp < V$. Fixing v_0 implies that the quotient $V/v_0^\perp$ is also fixed. The possible actions on $v_0^\perp/\langle v_0 \rangle$ form a group $\mathrm{GO}_{n-2}(q)$. By choosing a basis $\{v_0, w_1, \ldots, w_{n-2}, v_1\}$ such that $w_i \perp v_0$, we see that the maps f_i defined by

$$
\begin{aligned}
f_i : w_i &\mapsto w_i + v_0, \\
v_1 &\mapsto v_1 - w_i, \\
w_j &\mapsto w_j \ (j \neq i)
\end{aligned}
\tag{3.29}
$$

generate the kernel of this action. Hence the stabiliser of v_0 has order $q^{n-2}|\mathrm{GO}_{n-2}(q)|$. Moreover, we can choose v_1 orthogonal to $w_1, \ldots, w_{n-2}$, so that V is the orthogonal direct sum of $\langle v_0, v_1 \rangle$ and $\langle w_1, \ldots, w_{n-2} \rangle$. Therefore if n is even then the orthogonal space $\langle w_1, \ldots, w_{n-2} \rangle$ has the same type as V.

Finally, since $\mathrm{GO}_1(q) \cong C_2$ we have

$$
\begin{aligned}
|\mathrm{GO}_{2m+1}(q)| &= 2 \prod_{k=1}^{m} ((q^{2k} - 1)q^{2k-1}) \\
&= 2q^{m^2}(q^2 - 1)(q^4 - 1)\cdots(q^{2m} - 1).
\end{aligned}
\tag{3.30}
$$

Similarly, since $\mathrm{GO}_2^+(q)$ has order $2(q-1)$ we have

$$
\begin{aligned}
|\mathrm{GO}_{2m}^+(q)| &= 2(q-1) \prod_{k=2}^{m} ((q^k - 1)(q^{k-1} + 1)q^{2k-2}) \\
&= 2q^{m(m-1)}(q^2 - 1)(q^4 - 1)\cdots(q^{2m-2} - 1)(q^m - 1),
\end{aligned}
\tag{3.31}
$$

and also $\mathrm{GO}_2^-(q)$ has order $2(q+1)$ so

$$
|\mathrm{GO}_{2m}^-(q)| = 2q^{m(m-1)}(q^2 - 1)(q^4 - 1)\cdots(q^{2m-2} - 1)(q^m + 1).
\tag{3.32}
$$

In odd characteristic, the special orthogonal group has index 2 in the general orthogonal group, so we get corresponding formulae for their orders by deleting the factor 2.

3.7.3 Simplicity of $\mathrm{P\Omega}_n(q)$

As in the earlier cases we can prove simplicity of the groups $\mathrm{P\Omega}_n^\varepsilon(q)$ for q odd and $n \geqslant 5$ using Iwasawa's Lemma. However, there are no orthogonal transvections (except in characteristic 2, and even there they do not lie in $\Omega_n(q)$), which makes the proof a little more complicated. Instead, provided $n \geqslant 5$, there are elements called *(long) root elements* (also known as Siegel transformations, or Eichler transformations), which are defined for pairs (u, v) of perpendicular isotropic vectors by the formula

$$T_{u,v}(\lambda) : x \mapsto x + \lambda f(x,u)v - \lambda f(x,v)u. \tag{3.33}$$

It is straightforward to show that the root elements $T_{u,v}(\lambda)$ for fixed u generate an abelian normal subgroup of the stabiliser of $\langle u \rangle$, and that $\Omega_n^\varepsilon(q)$ acts primitively on the set of isotropic 1-dimensional subspaces. It is less easy to show that $\Omega_n^\varepsilon(q)$ is generated by root elements. Explicit calculations show that every root element is a commutator of root elements, and hence $\Omega_n^\varepsilon(q)$ is perfect. Then we can apply Iwasawa's Lemma as before.

An alternative proof was suggested to me by John Bray. This uses a generalisation of Iwasawa's Lemma to show that the only proper non-trivial normal subgroup of the special orthogonal group $\mathrm{PSO}_n^\varepsilon(q)$ is $\mathrm{P}\Omega_n^\varepsilon(q)$.

The generalisation of Iwasawa's Lemma which we use here is the following:

Theorem 3.4. *If the finite group G acts faithfully and primitively on a set Ω, and A is a normal subgroup of the point stabiliser H, such that the G-conjugates of A generate G, then any proper quotient of G is isomorphic to a quotient of A.*

Proof. Suppose that K is a non-trivial normal subgroup of G. Then K is not contained in H, since the action of G is faithful, so $G = HK$ by maximality of H, since the action is primitive. Therefore, by the same argument as in Theorem 3.1, $G = AK$, whence $G/K = AK/K \cong A/(A \cap K)$.

We apply this to the group $G = \mathrm{PSO}_{2m+1}(q)$ with q odd, and $m \geqslant 1$, acting on vectors of a suitable norm. This norm is chosen so that G contains (the images modulo scalars of) the reflections in these vectors. Moreover, G is generated by these reflections—again, this is not obvious, but must be checked by elementary but tedious calculations—so by the theorem, any proper non-trivial quotient of G is of order 2. Thus $\mathrm{P}\Omega_{2m+1}(q)$ is the only proper non-trivial normal subgroup of $\mathrm{PSO}_{2m+1}(q)$.

Now consider the orthogonal groups in even dimension. In this case, $\mathrm{GO}_{2m}^\varepsilon(q) \cong \Omega_{2m}^\varepsilon(q).2^2$ has three subgroups of index 2, one of which is $\mathrm{SO}_{2m}^\varepsilon(q)$. Another contains (and is generated by) the reflections in the vectors of norm 1, while the third contains (and is generated by) the reflections in the vectors of norm a non-square. Thus the subgroup of $\mathrm{GO}_{2m}^\varepsilon(q)$ generated by reflections in vectors of a given norm is a subgroup of index 2. Let G be the image of such a subgroup modulo scalars. The same argument as above shows that any proper quotient of G has order 2, provided $m \geqslant 2$.

It follows from Corollary 2.9 that in each case $\mathrm{P}\Omega_n^\varepsilon(q)$ is characteristically simple. But it is easy to see that it is non-abelian provided $n \geqslant 3$, so it is either simple, or $\mathrm{P}\Omega_n^\varepsilon(q) \cong T \times T$ for a simple subgroup T. Moreover, in the latter case T acts regularly on the set of 1-spaces containing vectors of some fixed norm. Now the stabiliser of such a vector in $\mathrm{GO}_n(q)$ is a group of type $\mathrm{GO}_{n-1}(q)$, and by Witt's Lemma $\mathrm{GO}_n(q)$ is transitive on vectors of any given norm, so the number of them is given by the order formulae in Section 3.7.2. We get $q^m(q^m + 1)$ or $q^m(q^m - 1)$ in $\mathrm{GO}_{2m+1}(q)$, according as the stabiliser is $\mathrm{GO}_{2m}^+(q)$ or $\mathrm{GO}_{2m}^-(q)$, and $q^{m-1}(q^m - 1)$ in $\mathrm{GO}_{2m}^+(q)$, and $q^{m-1}(q^m + 1)$ in $\mathrm{GO}_{2m}^-(q)$. Hence we know the number of such 1-spaces, and the order of the group, and we can easily check that this only happens in the case $\mathrm{P}\Omega_4^+(q)$. In particular, $\mathrm{P}\Omega_n^\varepsilon(q)$ is simple for q odd and $n \geqslant 5$. Conversely, we shall show later, in Section 3.11.1, that $\mathrm{P}\Omega_4^+(q) \cong \mathrm{PSL}_2(q) \times \mathrm{PSL}_2(q)$.

3.7.4 Subgroups of orthogonal groups

We continue to discuss the case when the characteristic is odd, although the characteristic 2 case is very similar.

To exhibit the structure of the 'BN-pair' we choose a different basis from the orthogonal one used to classify the forms. By analogy with the symplectic groups we choose a basis $e_1, \ldots, e_m$ for a maximal isotropic subspace W for $\mathrm{GO}_{2m}^+(q)$, $\mathrm{GO}_{2m+1}(q)$ or $\mathrm{GO}_{2m+2}^-(q)$. Extend this arbitrarily to a basis of $W^\perp$, and then adjoin $f_m, \ldots, f_1$ such that $f(e_i, f_i) = 1$, and $f(e_i, f_j) = 0$ otherwise, and the f_j are orthogonal to the chosen basis vectors for $W^\perp$ outside W. Let U be the group of lower unitriangular matrices, T the group of diagonal matrices (except in the case of $\mathrm{GO}_{2m+2}^-(q)$ where we allow the cyclic subgroup C_{q+1} of $\mathrm{GO}_2^-(q)$ acting on the space spanned by the two extra basis vectors), and then put $B = UT$, $N = N(T)$ and $W = N/T$ as before.

The details of the calculations are omitted as they are very similar to earlier cases. We find that U is a Sylow p-subgroup, and B is its normaliser. The torus T is $(C_{q-1})^m$ in the case of $\Omega_{2m}^+(q)$ or $\Omega_{2m+1}(q)$, and $(C_{q-1})^m \times C_{q+1}$ in $\Omega_{2m+2}^-(q)$. The normaliser of the torus acts by permuting the subscripts $1, \ldots, m$ and swapping $\langle e_i \rangle$ with $\langle f_i \rangle$ (in the case $\Omega_{2m}^+(q)$ we must swap an even number of these pairs). Thus $N/T \cong C_2 \wr S_m$ in $\Omega_{2m+1}(q)$ and $\Omega_{2m+2}^-(q)$, and a subgroup of index 2 therein in $\Omega_{2m}^+(q)$.

The maximal parabolic subgroups are again the stabilisers of totally isotropic subspaces. Specifically, the stabiliser of an isotropic k-space in $\mathrm{GO}_n^\varepsilon(q)$ is a group of shape $q^{k(k-1)/2+k(n-2k)}{:}(\mathrm{GL}_k(q) \times \mathrm{GO}_{n-2k}^\varepsilon(q))$, in which the normal p-subgroup is a special group with centre of order $q^{k(k-1)/2}$. In the case when $\varepsilon = +$ and $n = 2k$ the stabiliser is a group $q^{k(k-1)/2}{:}\mathrm{GL}_k(q)$ all of whose elements have determinant 1. It follows that $\mathrm{SO}_{2k}^+(q)$ has two orbits on totally isotropic k-subspaces, fused in $\mathrm{GO}_{2k}^+(q)$. It can be shown that two of these subspaces, U and W, are in the same $\mathrm{SO}_{2k}^+(q)$-orbit if and only if $U \cap W$ has even codimension in each of U and W. (The *codimension* of $U \cap W$ in U is by definition $\dim U - \dim U \cap W$.)

The types of totally isotropic subspaces again correspond with the nodes of the Dynkin diagram. In the case of $\mathrm{GO}_{2m}^+(q)$ the diagram is D_m, and the $m-2$ nodes along the stem of the diagram correspond to the spaces of dimension $1, 2, \ldots, m-2$, while the last two nodes correspond to the two different types of m-dimensional totally isotropic spaces. In the case of $\mathrm{GO}_{2m+1}(q)$ the diagram is B_m.

The stabilisers in $\mathrm{GO}_n(q)$ of non-singular subspaces are of the form $\mathrm{GO}_k(q) \times \mathrm{GO}_{n-k}(q)$, but we need to look more closely to determine the signs

of these orthogonal groups in the cases when it makes a difference. It is easi-
est to turn the question round and ask, for each group $\mathrm{GO}_k(q) \times \mathrm{GO}_{n-k}(q)$,
whether it fixes a quadratic form of plus type or minus type or both. Clearly
if $n = 2m + 1$ is odd we have both $\mathrm{GO}_{2k}^\pm(q) \times \mathrm{GO}_{n-2k}(q) < \mathrm{GO}_n(q)$. Next
consider $\mathrm{GO}_{2k+1}(q) \times \mathrm{GO}_{2m-2k-1}(q)$. By multiplying the form on the first
$(2k + 1)$-space by a non-square, the whole form changes from plus type to
minus type or vice versa, so

$$\mathrm{GO}_{2k+1}(q) \times \mathrm{GO}_{2m-2k-1}(q) < \mathrm{GO}_{2m}^\pm(q). \tag{3.34}$$

Using the symplectic bases defined at the beginning of this section, we see
that

$$\mathrm{GO}_{2k}^+(q) \times \mathrm{GO}_{2m-2k}^+(q) < \mathrm{GO}_{2m}^+(q),$$
$$\mathrm{GO}_{2k}^-(q) \times \mathrm{GO}_{2m-2k}^+(q) < \mathrm{GO}_{2m}^-(q), \tag{3.35}$$

so the only remaining case to consider is $\mathrm{GO}_{2k}^-(q) \times \mathrm{GO}_{2m-2k}^-(q)$. We have
already seen that if α is a non-square then the quadratic form $\alpha x^2 + \alpha y^2$ is
equivalent to $x^2 + y^2$. Therefore $\mathrm{GO}_{2k}^-(q) \times \mathrm{GO}_{2m-2k}^-(q) < \mathrm{GO}_{2m}^+(q)$.

If $n = km$ is odd there is an imprimitive subgroup $\mathrm{GO}_k(q) \wr S_m$ in $\mathrm{GO}_n(q)$.
Similarly, if $n = 2km$ there is a subgroup $\mathrm{GO}_{2k}^+(q) \wr S_m$ in $\mathrm{GO}_{2km}^+(q)$, and then
$\mathrm{GO}_{2k}^-(q) \wr S_m$ is in $\mathrm{GO}_{2km}^+(q)$ if m is even and in $\mathrm{GO}_{2km}^-(q)$ if m is odd. Also if
$n = 2km$ and k is odd then $\mathrm{GO}_k(q) \wr S_{2m}$ lies in whichever of $\mathrm{GO}_{2km}^\pm(q)$ admits
an orthonormal basis, that is $\mathrm{GO}_{2km}^-(q)$ if $km(q-1)/2$ is odd, and $\mathrm{GO}_{2km}^+(q)$
otherwise. In $\mathrm{GO}_{2m}^+(q)$ it is possible to find two disjoint maximal isotropic
subspaces U, W, and then the stabiliser of the decomposition $V = U \oplus W$ is
a subgroup $\mathrm{GL}_m(q).2$.

3.7.5 Outer automorphisms

For $n \neq 8$ there are no outer automorphisms besides the ones induced by
the full orthogonal group, those induced by similarities (which multiply the
form by scalars) and the field automorphisms. If the defining quadratic form
(or symmetric bilinear form) is invariant under automorphisms of the field,
then the field automorphisms of the group act just by applying the field automorphisms to the matrix entries. Note that, as for the symplectic groups, a
scalar λ multiplies the form by λ^2, so if n is even then the similarities induce
an outer automorphism group of order 2. If n is odd then all similarities are
scalar multiples of isometries. For $n = 8$ there is an additional automorphism
of $\mathrm{P\Omega}_8^+(q)$, called triality, which is described in Section 4.7.

The structures of these outer automorphism groups are not immediately
obvious. If $q = p^e$ is odd then for n odd the outer automorphism group is
$C_2 \times C_e$. For $n = 2m$ and $q^m \equiv -\varepsilon \bmod 4$ it is $C_2 \times C_{2e}$, while for $n = 2m$
and $q^m \equiv \varepsilon \bmod 4$ it is $D_8 \times C_e$, except when $m = 4$ and $\varepsilon = +$, in which case
it is $S_4 \times C_e$.

3.8 Orthogonal groups in characteristic 2

In characteristic 2, everything is different. The quadratic form has a different definition, the canonical forms are different, there are no reflections, the determinant tells us nothing, and there is no spinor norm. Nevertheless, in even dimensions, the formulae for the group orders still hold, and there is still a mysterious subgroup of index 2, although to define it we need a new invariant, which is called the *quasideterminant* or *pseudodeterminant* (it is more closely analogous to the determinant than to the spinor norm). Indeed, the structure of the orthogonal groups in characteristic 2 is simpler than in odd characteristic, since the determinants are all 1 and there are no non-trivial scalars in the orthogonal group (for if $Q(\lambda v) = Q(v) \neq 0$ then $\lambda^2 = 1$ so $\lambda = 1$). Recall from Section 3.4.7 that in characteristic 2 we have $\mathrm{GO}_{2m+1}(q) \cong \mathrm{Sp}_{2m}(q)$ and therefore we do not need to consider the odd-dimensional case.

3.8.1 The quasideterminant and the structure of the groups

The elements which in characteristic 2 play the role played by the reflections in odd characteristic are the *orthogonal transvections* (some people even call them reflections, as the formula defining them is obtained from the formula (3.26) by replacing $\frac{1}{2}f(v,v)$ by $Q(v)$). For each vector v of norm 1 the corresponding orthogonal transvection t_v is defined by

$$t_v : w \mapsto w + f(w,v)v. \tag{3.36}$$

Clearly this is a linear map, and it preserves the quadratic form since

$$Q(w + f(w,v)v) = Q(w) + f(w,v)^2 + f(w,v)^2 Q(v) = Q(w). \tag{3.37}$$

Now the orthogonal group can be generated by these transvections (at least if the dimension is 6 or more: the proof is left as an exercise) and the quasideterminant of an element x is defined to be 1 or -1 according as x can be written as a product of an even or an odd number of orthogonal transvections. In order to prove that this is well-defined we show that the transvections act as odd permutations of the set of maximal isotropic subspaces (in the case $\mathrm{GO}_{2m}^+(q)$). I am grateful to Bill Kantor for supplying this elegant argument.

First consider the case $\mathrm{GO}_2^+(q)$. Here there are just two isotropic 1-spaces, since if x is isotropic and y is scaled so that $f(x,y) = 1$, then $Q(\lambda x + y) = Q(y) + \lambda$ so is zero for exactly one value of λ. Choosing y to be isotropic, then, the vectors of norm 1 are exactly the vectors $\lambda x + \lambda^{-1} y$, and every orthogonal transvection $t_{\lambda x + \lambda^{-1} y}$ swaps $\langle x \rangle$ with $\langle y \rangle$.

More generally, we need to look at maximal isotropic subspaces in the $2m$-space for $\mathrm{GO}_{2m}^+(q)$. Suppose we have a maximal isotropic subspace U, and a vector v of norm 1. Note that $v^\perp$ has codimension 1, and does not contain U since $v \notin U = U^\perp$. Therefore $v^\perp \cap U$ has codimension 1 in U, and we may choose a basis $u_1, \ldots, u_m$ for U so that $u_1, \ldots, u_{m-1}$ span $v^\perp \cap U$ and

$f(u_m, v) = 1$. Then t_v fixes $u_1, \ldots, u_{m-1}$ and maps u_m to $u_m + v \notin U$, so t_v does not fix U. Hence the transvections act fixed-point-freely on the set of totally isotropic subspaces of dimension m.

Now we prove by induction on m that the number of such subspaces is $\prod_{i=0}^{m-1}(q^i + 1)$. Just as in the odd characteristic case (Section 3.7.2), we see that the number of isotropic vectors is $(q^m - 1)(q^{m-1} + 1)$. Since $\mathrm{GO}_{2m}^+(q)$ acts transitively on the isotropic vectors (this follows from our classification of the quadratic forms, as the form looks the same whichever isotropic vector we take as e_1), we may choose the first isotropic vector to be e_1. Then we see that the stabiliser of e_1 has the shape $q^{2m-2}{:}\mathrm{GO}_{2m-2}^+(q)$. By induction, since we count each m-dimensional totally isotropic subspace $q^m - 1$ times in this way, the number of them is

$$\left((q^m - 1)(q^{m-1} + 1) \prod_{i=0}^{m-2}(q^i + 1) \right) \bigg/ (q^m - 1) = \prod_{i=0}^{m-1}(q^i + 1), \quad (3.38)$$

which is twice an odd number. Therefore t_v acts as an odd permutation on the set of m-dimensional totally isotropic subspaces.

We can therefore define the quasideterminant of an element of $\mathrm{GO}_{2m}^+(q)$ to be the sign of the permutation describing its action on this set. The kernel of the quasideterminant map is a subgroup of index 2 in $\mathrm{GO}_{2m}^+(q)$, which we denote $\Omega_{2m}^+(q)$. (This subgroup is sometimes denoted $\mathrm{SO}_{2m}^+(q)$, but this notation can be confusing, so is not recommended.)

For $\mathrm{GO}_{2m}^-(q)$ this argument does not go through directly, since the maximal isotropic subspaces have dimension $m - 1$ and it is possible that $U \leqslant v^\perp$. However, if we extend the field to $\mathbb{F}_{q^2}$, we obtain maximal isotropic subspaces of dimension m, and can apply the preceding argument. (Incidentally, this shows that $\mathrm{GO}_{2m}^-(q) < \mathrm{GO}_{2m}^+(q^2)$.) In fact it is possible to extend this argument to show that the transvections in $\mathrm{GO}_{2m}^+(q)$ interchange two families of maximal isotropic subspaces: two such subspaces U and W are in the same family if and only if $U \cap W$ has even codimension in each of them. Another useful fact is that an element x in $\mathrm{GO}_{2m}^\varepsilon(q)$ is in $\Omega_{2m}^\varepsilon(q)$ if and only if the rank of $1 + x$ (as a $2m \times 2m$ matrix) is even.

3.8.2 Properties of orthogonal groups in characteristic 2

We have already seen in Section 3.7 how to calculate the orders of the orthogonal groups by induction. In characteristic 2 the base cases of the induction are again $\mathrm{GO}_2^+(q) \cong D_{2(q-1)}$ and $\mathrm{GO}_2^-(q) \cong D_{2(q+1)}$ (see Exercise 3.28). Therefore we have

$$|\mathrm{GO}_{2m}^+(q)| = 2q^{m(m-1)}(q^2 - 1)(q^4 - 1)\cdots(q^{2m-2} - 1)(q^m - 1) \text{ and}$$
$$|\mathrm{GO}_{2m}^-(q)| = 2q^{m(m-1)}(q^2 - 1)(q^4 - 1)\cdots(q^{2m-2} - 1)(q^m + 1). \quad (3.39)$$

The generically simple groups $\Omega_{2m}^\varepsilon(q)$ have index 2 in $\mathrm{GO}_{2m}^\varepsilon(q)$. They are simple for all $m \geqslant 3$ and all q. Simplicity can be proved analogously to the proof in odd characteristic (see Section 3.7.3), either using the root elements

$$T_{u,v}(\lambda) : x \mapsto x + \lambda f(x, u)v - \lambda f(x, v)u$$

where $\langle u, v \rangle$ is a totally isotropic 2-space, or using the orthogonal transvections and the argument following Theorem 3.4 (see also Exercise 3.33).

Unless $m = 4$ and $\varepsilon = +$, the full automorphism group of $\Omega_{2m}^{\varepsilon}(q)$ in characteristic 2 is obtained from $\mathrm{GO}_{2m}^{\varepsilon}(q)$ by adjoining field automorphisms. If $m = 4$ and $\varepsilon = +$, there is an additional 'triality' automorphism of $\mathrm{P}\Omega_8^+(q)$, which is defined and described in some detail in Section 4.7. The Schur multiplier is trivial except for the cases $\Omega_6^+(2) \cong \mathrm{PSL}_4(2) \cong A_8$ and $\Omega_6^-(2) \cong \mathrm{PSU}_4(2) \cong \mathrm{PSp}_4(3)$, both of which, as already noted, have proper double covers, and $\Omega_8^+(2)$ which has a cover $2^2 \cdot \Omega_8^+(2)$ which can be constructed from the Weyl group of type E_8 (see Section 3.12.4), using either the spin group (see Section 3.9.2) or the triality automorphism (see Section 4.7).

The subgroups of the orthogonal groups in characteristic 2 are very similar to the subgroups of orthogonal groups in odd characteristic. The main difference is that the orthogonal groups in odd dimensions usually do not arise, except that $\Omega_{2m-1}(q) \cong \mathrm{Sp}_{2m-2}(q)$ is a maximal subgroup of both groups $\Omega_{2m}^{\varepsilon}(q)$. See Section 3.10, especially Section 3.10.10, for details.

3.9 Clifford algebras and spin groups

For all the classical groups there is a 'duality' between the generic part of the Schur multiplier and the diagonal part of the outer automorphism group. In the case of $\mathrm{PSL}_n(q)$, the centre of $\mathrm{SL}_n(q)$ consists of scalars λ such that $\lambda^n = 1$, giving a cyclic group of order $d = (n, q - 1)$. Dually, the determinant map maps the scalars to the subset $\{\lambda^n\} = \{\lambda^d\}$ of the non-zero elements of the field, so induces a map from $\mathrm{PGL}_n(q)$ onto $F^\times / \{\lambda^d\}$, which is again a cyclic group of order d. Similarly for $\mathrm{PSU}_n(q)$, replacing $q - 1$ by $q + 1$. The symplectic groups have centre and diagonal automorphism group each of order 2.

In the case of the orthogonal groups, in odd characteristic and even dimension there is a centre of order 2, dual to the determinant ± 1, but what is dual to the spinor norm? To answer this question we have to construct a double cover of the orthogonal group, called the *spin group*. The usual way to do this is via the so-called Clifford algebra. Indeed, we can (and do, below) use the Clifford algebra to define the spinor norm, or rather to prove that it is well-defined, and hence prove that $\Omega_n(q)$ has index 2 in $\mathrm{SO}_n(q)$ for q odd. (A variant of this construction also works in characteristic 2, although there is no subgroup of index 2, or double cover, to construct in this case.)

For the purposes of this section, an *algebra* is a vector space A over a field F, together with a multiplication $A \times A \to A$ which is F-bilinear, such that A is a ring (with or without a 1) with respect to this multiplication and the vector space addition. In particular, multiplication in A is associative. Later on, we shall drop this associativity condition, when discussing octonion algebras, Jordan algebras, Lie algebras, and so on.

3.9.1 The Clifford algebra

Let V be a vector space over a field $F = \mathbb{F}_q$ of odd order, and let f be a non-singular symmetric bilinear form on V, with $Q(v) = \frac{1}{2}f(v,v)$. Pick a basis $\{e_1, \ldots, e_n\}$ of V. Construct a vector space of dimension 2^n with basis vectors labelled by formal products of $e_1, \ldots, e_n$ (including the empty product, representing an identity 1), subject to the relations $e_i e_i = Q(e_i)$ and $e_i e_j + e_j e_i = f(e_i, e_j)$. Extending the product bilinearly, this implies that for any vector $v \in V = \langle e_1, \ldots, e_n \rangle$ we have $vv = Q(v)$ and therefore $vw + wv = f(v,w)$ for all $v, w \in V$.

This algebra is called the *Clifford algebra*, written $C(V, f)$, or $C(V, Q)$, or $C(V)$ if f is understood. It splits as the vector space direct sum of the *even part* C_0, spanned by the even products $e_{i_1} \cdots e_{i_{2m}}$ for $m \leqslant n/2$, and the *odd part* C_1, spanned by the odd products $e_{i_1} \cdots e_{i_{2m+1}}$ for $m < n/2$, in such a way that $C_i C_j \subseteq C_{i+j}$, with subscripts read modulo 2. Assume from now on that the basis $\{e_1, \ldots, e_n\}$ of V is orthogonal for f, i.e. $f(e_i, e_j) = 0$ whenever $i \neq j$. The quadratic form Q on V can be extended to the whole of $C(V)$ by defining its values on the given basis by

$$Q(e_{i_1} \cdots e_{i_k}) = Q(e_{i_1}) \cdots Q(e_{i_k}), \tag{3.40}$$

and defining these basis vectors to be perpendicular (with respect to the associated bilinear form). It can be shown that $Q(vw) = Q(v)Q(w)$ for any $v, w \in V$, and more generally that $Q(xy) = Q(x)Q(y)$ for any $x, y \in C(V)$ (see Exercise 3.35).

The orthogonal group G acts on $C(V)$ as a group of automorphisms, induced from its action on V. (For those who know some representation theory, the action of G on the Clifford algebra is as on the exterior algebra

$$\Lambda(V) = 1 \oplus V \oplus \Lambda^2 V \oplus \Lambda^3 V \oplus \cdots \oplus \Lambda^n V,$$

where $\Lambda^k V$ is the kth skew-symmetric (or exterior) power of V, spanned by the $\binom{n}{k}$ vectors $e_{i_1} e_{i_2} \cdots e_{i_k}$ with $i_1 < \cdots < i_k$.) Indeed, this action can be expressed by conjugation within the algebra, since for any non-singular vector $v \in V$ the map

$$\begin{aligned}
w \mapsto v^{-1}wv \\
= Q(v)^{-1}v(f(v,w) - vw) \\
= -w + \frac{2f(v,w)}{f(v,v)}v
\end{aligned} \tag{3.41}$$

is just minus the reflection in v. These maps generate $GO(V, f)$ if n is even, or $SO(V, f)$ if n is odd.

3.9.2 The Clifford group and the spin group

The *Clifford group* is usually defined as the subgroup of invertible elements of $C(V)$ which preserve V under conjugation. Certainly the Clifford group contains all $v \in V$ with $Q(v) \neq 0$. However, this group is a little too big for our purposes, especially when V has odd dimension. The *even Clifford group* is defined to be the intersection of the Clifford group with the even part C_0 of the Clifford algebra $C(V)$. There is now an obvious group homomorphism from the even Clifford group onto the special orthogonal group. Moreover, Q is a group homomorphism from the Clifford group to the scalars, with the property that $Q(\lambda) = \lambda^2$ for every scalar λ. Therefore there is a homomorphism from the even Clifford group modulo scalars, to the group of order 2. This factors through the special orthogonal group, so that Q induces the spinor norm. Finally we have shown that the special orthogonal group has a subgroup of index 2, as promised in Section 3.7.1.

The corresponding *spin group* is defined as the subgroup of the even Clifford group consisting of elements x with $Q(x) = 1$. There is an obvious group homomorphism from the spin group onto the orthogonal group $\Omega_n(q)$, and we want to determine the kernel of this homomorphism. Now x acts trivially on the Clifford algebra (or equivalently, on V) if and only if x is in the centre of the Clifford algebra. If n is even, the centre has a basis $\{1\}$, while if n is odd it has a basis $\{1, e_1 e_2 \cdots e_n\}$. Hence the intersection of the centre with C_0 is just the set of scalar multiples of the identity. These elements have $Q(\lambda 1) = \lambda^2$, so the only scalars in the spin group are ± 1. Therefore the kernel of the natural map from the spin group to the orthogonal group has order 2. It is easy to find elements of the spin group which square to -1, and hence the spin group is a proper double cover of the orthogonal group. We write $\mathrm{Spin}_n^\varepsilon(q)$ for this group of shape $2 \cdot \Omega_n^\varepsilon(q)$.

If n is odd, or if $n = 2m$ and $q^m \equiv -\varepsilon \bmod 4$, then $\Omega_n^\varepsilon(q)$ is already simple, and the spin group has the structure $2 \cdot \Omega_n^\varepsilon(q)$. If $n = 2m$ and $q^m \equiv \varepsilon \bmod 4$, then $\Omega_n^\varepsilon(q)$ has a centre of order 2, and the spin group has the structure $4 \cdot \mathrm{P}\Omega_n^\varepsilon(q)$ if m is odd, and the structure $2^2 \cdot \mathrm{P}\Omega_n^\varepsilon(q)$ (necessarily with $\varepsilon = +$) if m is even. Notice that in the case $n = 3$, the even part of the Clifford algebra has dimension 4, and is isomorphic to the quaternion algebra (see Section 4.3.1).

3.9.3 The spin representation

The Clifford algebra $C(V)$ has been defined abstractly as an (associative) algebra of dimension 2^n, where $\dim V = n$. It is possible to realise the Clifford algebra more concretely as an algebra of matrices, with the algebra product being matrix multiplication. Here we only treat the quadratic forms of even dimension and plus type: the others can be obtained as subalgebras of larger Clifford algebras, since $\mathrm{GO}_{2m}^-(q) < \mathrm{GO}_{2m+1}(q) < \mathrm{GO}_{2m+2}^+(q)$. If $W = \{0\}$, then $C(W)$ consists of the 1×1 identity matrix. Otherwise, decompose V as a direct sum $V = W \oplus \langle e, f \rangle$, where $W = \langle e, f \rangle^\perp$ and $Q(e) = Q(f) = 0$, $Q(e + f) = 1$. By induction $C(W)$ is written as a matrix algebra, and we write $C(V)$ as an algebra of 2×2 matrices over $C(W)$ by mapping $w \in W$

to $\begin{pmatrix} w & 0 \\ 0 & -w \end{pmatrix}$, and e to $\begin{pmatrix} 0 & 1 \\ 0 & 0 \end{pmatrix}$ and f to $\begin{pmatrix} 0 & 0 \\ 1 & 0 \end{pmatrix}$. It is straightforward to check that these matrices satisfy the required relations $we+ew = wf+fw = 0$, $e^2 = f^2 = 0$, $ef + fe = 1$. For example the 4-dimensional Clifford algebra $C(V)$, where $V = \langle e, f \rangle$, is spanned as a vector space by $1, e, f, ef$, which are represented respectively by the matrices

$$\begin{pmatrix} 1 & 0 \\ 0 & 1 \end{pmatrix}, \begin{pmatrix} 0 & 1 \\ 0 & 0 \end{pmatrix}, \begin{pmatrix} 0 & 0 \\ 1 & 0 \end{pmatrix}, \begin{pmatrix} 1 & 0 \\ 0 & 0 \end{pmatrix}.$$

Thus in the case when V has dimension $2m$ and the form is of plus type, we have written $C(V)$ as an algebra of $2^m \times 2^m$ matrices. Abstractly this is the full matrix algebra, as it has dimension 2^{2m}, but the important ingredient which makes it a Clifford algebra is the embedding of V inside it. In particular, $C(V)$ acts (faithfully) on a 2^m-dimensional space—this is the so-called *spin representation*. Clearly we may restrict this representation to the Clifford group, the even Clifford group, and the spin group. It is easy to see that on restriction to the even Clifford group the representation breaks up as the direct sum of two representations of degree 2^{m-1}. Indeed, just index the rows and columns by binary numbers and pick just those numbers which have an even number of bits to make one of the summands. In particular we obtain a (faithful) representation of the spin group, of dimension 2^{m-1}.

In the case $n = 8$, we have $2^{m-1} = 2^3 = 8$, and so the spin group also has a representation of degree 8. Therefore there is a corresponding homomorphism from the spin group to the orthogonal group on this new 8-space. Under this homomorphism, -1 maps to -1. But in the *natural* quotient map from the spin group to the orthogonal group, -1 maps to $+1$. Therefore we have constructed an *outer* automorphism of the spin group, and also of $\mathrm{P\Omega}_8^+(q)$, which is known as *triality*. We shall discuss this in more detail in Chapter 4, where we shall give an alternative description of the spin group $2{\cdot}\Omega_8^+(q)$ in terms of octonions.

3.10 Maximal subgroups of classical groups

We wish to classify maximal subgroups of classical groups along the lines of the O'Nan–Scott theorem (see Sections 2.5 and 2.6) for symmetric and alternating groups. For classical groups over $\mathbb{C}$, such a result was obtained by Dynkin in 1952 [55]. The first part of his argument works also for finite fields, and a more detailed version of the theorem in the finite case was published by Aschbacher in 1984 [5]. Nevertheless, Aschbacher's Theorem falls far short of the degree of explicitness of the O'Nan–Scott Theorem. An early attempt I made [171] to put the flesh on the bones gave an almost correct result for the symplectic groups. But it was not until 1990 that Kleidman and Liebeck [108] provided (for all classical groups) the level of detail which is required to write down explicit lists of maximal subgroups in particular cases.

First we describe the types of subgroups which arise in this classification. A subgroup H of $\mathrm{GL}(V)$ is said to be *reducible* if there is a subspace $0 < W < V$ which is invariant under H, and *irreducible* otherwise. Similarly, H is called *imprimitive* if there is a non-trivial direct sum decomposition of V which is invariant under H, and *primitive* otherwise. The maximal reducible subgroups of classical groups are obviously the full stabilisers of certain subspaces. If there is a form on the space we may assume the subspace is either non-singular or totally isotropic. Any imprimitive subgroup preserves a decomposition of the space as a direct sum of subspaces of the same dimension. If there is a form then either the subspaces are non-singular or there are precisely two of them. The other types of subgroups either arise from tensor products or extraspecial groups, or are almost simple (modulo scalars). We describe these next.

3.10.1 Tensor products

We first need to define the concept of a *tensor product* of vector spaces. If U is a vector space with basis $\{u_1, \ldots, u_k\}$ and W is a vector space with basis $\{w_1, \ldots, w_m\}$, over the same field F, we define the *tensor product space* $V = U \otimes W$ to be the vector space of dimension $n = km$ with basis $\{v_1, \ldots, v_{km}\}$, where we write $v_{i+k(j-1)} = u_i \otimes w_j$ to exhibit the connection with U and W. If $A = (a_{ij})$ is a $k \times k$ matrix acting on U, and $B = (b_{ij})$ is an $m \times m$ matrix acting on W, then we get an action on $U \otimes W$ by sending $u_i \otimes w_j$ to $u_i A \otimes w_j B$, interpreted as

$$u_i A \otimes w_j B = \sum_{r=1}^{k} \sum_{s=1}^{m} a_{ir} b_{js} (u_r \otimes w_s).$$

The corresponding $n \times n$ matrix with entries $a_{ir} b_{js}$ (with rows indexed by $i + k(j - 1)$ and columns indexed by $r + k(s - 1)$) is called the *Kronecker product* of A and B, and written $A \otimes B$.

If we take all possibilities for A in $\mathrm{GL}_k(q)$ and B in $\mathrm{GL}_m(q)$ respectively, we get an action of $\mathrm{GL}_k(q) \times \mathrm{GL}_m(q)$ on $U \otimes W$. However, this is not a faithful action. For the scalar matrices in both $\mathrm{GL}_k(q)$ and $\mathrm{GL}_m(q)$ act as scalars on $U \otimes W$ (more precisely, $(\lambda I_k) \otimes I_m = \lambda I_{km} = I_k \otimes (\lambda I_m)$), so that the kernel of the action consists of the elements $(\lambda I_k, \lambda^{-1} I_m)$ of $\mathrm{GL}_k(q) \times \mathrm{GL}_m(q)$. The quotient of $\mathrm{GL}_k(q) \times \mathrm{GL}_m(q)$ by this kernel of order $q - 1$ is an example of a central product. In general a *central product* $G \circ H$ of two groups G and H is a quotient of $G \times H$ by a subgroup of the centre. Usually the subgroup we quotient by is, as in this case, a diagonal subgroup of $Z(G) \times Z(H)$.

Thus we have $\mathrm{GL}_k(q) \circ \mathrm{GL}_m(q)$ as a subgroup of $\mathrm{GL}_{km}(q)$. In this case it is clearer to work modulo scalars, in the sense that $\mathrm{PGL}_k(q) \times \mathrm{PGL}_m(q) < \mathrm{PGL}_{km}(q)$. This subgroup is usually maximal, unless $k = m$ in which case we can identify U with W, and there is a map taking $u_i \otimes u_j$ to $u_j \otimes u_i$ which acts on $U \otimes U$ and extends the group to $\mathrm{PGL}_k(q) \wr S_2$.

Since we can take the tensor product of two spaces, we can take the tensor product of several, say $V = V_1 \otimes V_2 \otimes \cdots \otimes V_m$. If all the V_i are isomorphic, say $\dim V = k$, then $n = \dim V = k^m$, and we have an embedding of $\mathrm{PGL}_k(q) \wr S_m$ in $\mathrm{PGL}_n(q)$. These groups correspond to the primitive wreath products in the symmetric groups.

3.10.2 Extraspecial groups

Recall from Section 3.5.4 that a p-group G is called *special* if $Z(G) = G' = \Phi(G)$, where $\Phi(G)$ is the Frattini subgroup. A special group is called *extraspecial* if also $|Z(G)| = p$. For any group G, the commutator map from $G/Z(G) \times G/Z(G)$ to G' satisfies $[h, g] = [g, h]^{-1}$, and if G is special then $G/Z(G) = G/\Phi(G)$ is elementary abelian, so is a vector space over $\mathbb{F}_p$. Moreover, in this case $[g, hk] = [g, k][g, h]^k = [g, h][g, k]$, so if G is extraspecial then this commutator map is a skew-symmetric bilinear form (written multiplicatively). Indeed, it is alternating since $[g, g] = 1$. Moreover, by the assumption $Z(G) = \Phi(G)$, it follows that for every $g \notin \Phi(G)$ there is an $h \in G$ with $[g, h] \neq 1$. In other words, this alternating bilinear form is non-singular.

For every g in an extraspecial p-group G, we have that $g^p \in Z(G)$, and therefore

$$
\begin{aligned}
(gh)^p &= g(hgh^{-1}g^{-1})g(h^2gh^{-2}g^{-1})g\cdots(h^{p-1}gh^{-p+1}g^{-1})gh^p \\
&= g^p[h^{-1},g^{-1}]^{g^{p-1}}[h^{-2},g^{-1}]^{g^{p-2}}\cdots[h^{-p+1},g^{-1}]^g h^p \\
&= g^p[h^{-1},g^{-1}][h^{-2},g^{-1}]\cdots[h^{-p+1},g^{-1}]h^p \\
&= g^p h^p[h,g]\cdots[h,g]^{p-1} \\
&= g^p h^p[h,g]^{p(p-1)/2}.
\end{aligned}
\tag{3.42}
$$

Now if $p = 2$ this reduces to $(gh)^2 = g^2 h^2[h, g]$, which is just the multiplicative version of the definition of a quadratic form in (3.15), so the squaring map $G/Z(G) \to Z(G)$ is a quadratic form. On the other hand if p is odd it reduces to $(gh)^p = g^p h^p$, so either all elements have order p, or the elements of order 1 or p form a characteristic subgroup of index p.

The classification of non-singular (quadratic or alternating bilinear) forms in Sections 3.4.4, 3.4.7, 3.4.6 implies that there are exactly two isomorphism types of extraspecial p-groups of each order. We write 2^{1+2m}_ε for the extraspecial group of order 2^{1+2m} whose associated quadratic form is of type ε. For p odd we write p^{1+2m}_+ for the extraspecial group of exponent p, and p^{1+2m}_- for the one of exponent p^2.

It is easy to see that $D_8 \cong 2^{1+2}_+$ and $Q_8 \cong 2^{1+2}_-$. Taking central products of these in such a way that all the central involutions are identified gives us constructions of $2^{1+2m}_+ = D_8 \circ \cdots \circ D_8$ and $2^{1+2m}_- = D_8 \circ \cdots \circ D_8 \circ Q_8$. The 2-dimensional representations of D_8 and Q_8 (which exist over any field of odd characteristic) can therefore be tensored together to get representations of 2^{1+2m}_ε of dimension 2^m.

Similarly for p odd we obtain a p-dimensional representation of p^{1+2}_+ by taking an element cycling the p coordinates, and a diagonal element diag$(1, \alpha, \ldots, \alpha^{p-1})$, where α is an element of order p in the field. (This can be modified for p_-^{1+2} by replacing the coordinate permutation

$$e_1 \mapsto e_2 \mapsto \cdots \mapsto e_p \mapsto e_1$$

by

$$e_1 \mapsto e_2 \mapsto \cdots \mapsto e_p \mapsto \alpha e_1,$$

but we shall not be using this case.) A representation of degree p^m of p_+^{1+2m} is then obtained by tensoring together m copies of these matrices. (Note that we need elements of order p in the field, so the order of the field must be congruent to 1 modulo p.)

What has all this got to do with maximal subgroups of classical groups? A faithful representation of G of degree n over F is nothing more than an embedding of G into $\mathrm{GL}_n(F)$. Thus we have constructed subgroups of $\mathrm{GL}_n(F)$ isomorphic to p_ε^{1+2m}, where $n = p^m$. The subgroups we are after are the normalisers in $\mathrm{GL}_n(F)$ (or in other classical groups) of these extraspecial groups. In the rest of this section we consider in more detail the structure of these normalisers.

The given representations of extraspecial groups extend to representations of $2_\varepsilon^{1+2m}\mathrm{GO}_{2m}^\varepsilon(2)$ for $p = 2$, or $p_+^{1+2m}{:}\mathrm{Sp}_{2m}(p)$ for p odd. To see this we need a little representation theory, specifically the fact that the extraspecial group G has a unique representation of degree p^m such that a given generator for its centre acts as a given scalar. It follows that any automorphism of $G \cong p^{1+2m}$ which centralises the centre $Z(G)$ can be realised inside the general linear group in dimension p^m over any field of order $r \equiv 1 \bmod p$. But every isometry of the form on $G/Z(G)$ lifts to p^{2m} automorphisms of G, so we obtain in this way an extension of G by the isometry group of the form. If p is odd this extension splits since the involution centraliser is $C_p \times \mathrm{Sp}_{2m}(p)$, while if $p = 2$ it is almost always non-split. If the field F contains 4th roots of 1, i.e. square roots of -1, then there is a 2-dimensional representation of $4 \circ D_8 \cong 4 \circ Q_8$. Therefore there is a representation of $4 \circ 2^{1+2m}$ in 2^m dimensions, and working modulo $\{\pm 1\}$ we get a quadratic form on a space of dimension $2m + 1$ over $\mathbb{F}_2$, with isometry group $\mathrm{GO}_{2m+1}(2) \cong \mathrm{Sp}_{2m}(2)$. Thus we get a representation of $4 \circ 2^{1+2m}\mathrm{Sp}_{2m}(2)$ in 2^m dimensions over F.

All these groups of extraspecial type are in $\mathrm{SL}_{p^m}(r)$, where r is a prime power congruent to 1 modulo p (or modulo 4, in the last case). In some cases, they also fix forms and so are in smaller classical groups. Thus D_8 fixes a quadratic form (of plus type if and only if $r \equiv 1 \bmod 4$) and Q_8 fixes a symplectic form, so 2_+^{1+2m} fixes a quadratic form and 2_-^{1+2m} fixes a symplectic form. Therefore, for r odd and $m \geqslant 2$,

$$\begin{aligned}
2_+^{1+2m}\Omega_{2m}^+(2) &< \mathrm{SO}_{2m}^+(r), \\
2_-^{1+2m}\Omega_{2m}^-(2) &< \mathrm{Sp}_{2m}(r).
\end{aligned} \tag{3.43}$$

Similarly, p_+^{1+2} fixes a unitary form over $\mathbb{F}_{r^2}$ if and only if p divides $r+1$, and so the same is true of p_+^{1+2m}. This gives $p_+^{1+2m}:\mathrm{Sp}_{2m}(p) < \mathrm{SU}_{p^m}(r)$ provided $p|(r+1)$. Similarly, the groups $4 \circ 2^{1+2m}$ are unitary whenever $4|(r+1)$.

3.10.3 The Aschbacher–Dynkin theorem for linear groups

It is relatively easy to show that every subgroup of $\mathrm{PGL}_n(q)$ which does not contain $\mathrm{PSL}_n(q)$ is either contained in a maximal subgroup of one of the types we have seen above (namely the stabilisers of subspaces, the imprimitive groups, the groups constructed from tensor product decompositions of the underlying vector space, and the groups of extraspecial type) or is of *almost simple type*, which means that its intersection with $\mathrm{PSL}_n(q)$ is almost simple (recall that a group G is almost simple if $S \leqslant G \leqslant \mathrm{Aut}S$ for some non-abelian simple group S). This is a special case of Aschbacher's theorem [5], but the proof we sketch is essentially due to Dynkin [55]. The proof requires a little (modular) representation theory but is otherwise elementary.

Theorem 3.5. *Any subgroup of $\mathrm{GL}_n(q)$ not containing $\mathrm{SL}_n(q)$ is contained in one of the following subgroups:*

  (i) a reducible group $q^{km}:(\mathrm{GL}_k(q) \times \mathrm{GL}_m(q))$, the stabiliser of a k-space, where $k+m=n$;

  (ii) an imprimitive group $\mathrm{GL}_k(q) \wr S_m$, the stabiliser of a direct sum decomposition into m spaces of dimension k, where $km=n$;

  (iii) a simple tensor product $\mathrm{GL}_k(q) \circ \mathrm{GL}_m(q)$, the stabiliser of a tensor product decomposition $F^k \otimes F^m$, where $km=n$ and $F = \mathbb{F}_q$;

  (iv) a wreathed tensor product, the preimage of $\mathrm{PGL}_k(q) \wr S_m$, the stabiliser of a tensor product decomposition $F^k \otimes \cdots \otimes F^k$, where $k^m = n$;

  (v) the preimage of $p^{2k}:\mathrm{Sp}_{2k}(p)$, where $n = p^k$ (or $2^{2k}\cdot\mathrm{GO}_{2k}^\varepsilon(2)$ if $p=2$ and $q \equiv 3 \bmod 4$);

  (vi) the preimage of an almost simple group, acting irreducibly.

Proof. Given any subgroup H of $G = \mathrm{PGL}_n(q)$ not containing $\mathrm{PSL}_n(q)$, let $\widetilde{H}$ denote its preimage in $\widetilde{G} = \mathrm{GL}_n(q)$. The *socle* of H, written $\mathrm{soc}\,H$, is the product of all the minimal normal subgroups of H. Writing $N = \mathrm{soc}\,H$, we are interested in the representation ρ of $\widetilde{N}$ on the underlying n-dimensional vector space V. If ρ is not completely reducible (a representation is *completely reducible* if it is a direct sum of irreducibles), then there is a unique largest subspace W of V such that $\rho|_W$ is completely reducible. Therefore $\widetilde{H}$ fixes W (case (i)).

If ρ is completely reducible but not homogeneous (a representation is *homogeneous* if it is a direct sum of isomorphic irreducibles) then $\widetilde{H}$ preserves the decomposition of V as a direct sum of its homogeneous components, so $\widetilde{H}$ is either reducible (case (i) again) or imprimitive (case (ii)).

If ρ is completely reducible and homogeneous, but not irreducible, then $\widetilde{N} \circ C_{\widetilde{G}}(\widetilde{N})$ acts as a tensor product (case (iii)). Similarly, if H has more than one minimal normal subgroup, then $\widetilde{N}$ acts as a tensor product (case (iii) again).

So we have reduced to the case that N is the unique minimal normal subgroup of H. This may be either abelian, in which case it lifts to an extraspecial group (case (v)), or non-abelian simple (case (vi)), or non-abelian non-simple, in which case the representation of $\widetilde{N}$ is again a tensor product (case (iv)). This completes the proof of this easy version of the Aschbacher–Dynkin Theorem.

It is possible then to look more closely at the subgroups of almost simple type. Some are 'really' written over a smaller field, so are contained in a subgroup $\mathrm{PGL}_n(q_0)$ of $\mathrm{PGL}_n(q)$, where $q = q_0^e$ and e is prime. Some are 'really' of smaller dimension over some extension field, so are contained in a subgroup $\mathrm{P\Gamma L}_{n/k}(q^k)$ for some prime k. Some are other classical groups in their natural representations. And the more one knows about the representations of the quasisimple groups, the more one can extend or refine this list.

Aschbacher's 1984 version of the list of maximal subgroups comprises nine types, as follows:

  (i) subspace stabilisers,
 (ii) imprimitive wreath products,
(iii) simple tensors,
 (iv) wreathed tensors,
  (v) extraspecial type,
 (vi) subfield groups,
(vii) extension field groups,
(viii) classical type,
 (ix) other almost simple groups.

There is a version of this theorem for each of the classical groups, in which case more details can be given of many of these subgroups.

3.10.4 The Aschbacher–Dynkin theorem for classical groups

In order to understand how the nine types of subgroups of the linear groups behave in the presence of forms of various types, we need to look at the behaviour of the forms under the operations of tensor products, and restriction and extension of fields. (In Section 3.5.4 we looked at the subspaces in the case of the symplectic groups, and saw that we can restrict attention to non-singular subspaces and totally isotropic subspaces. It is clear that the same applies in the case of unitary and orthogonal groups.) Without going into too much detail at this stage, we can incorporate the forms into the Aschbacher–Dynkin theorem as follows. In this version, the natural classical groups are denoted $\widetilde{G}$, and the corresponding projective groups by G. Thus for example we might have $\widetilde{G} = \mathrm{Sp}_{2n}(q)$ and $G = \mathrm{PSp}_{2n}(q)$.

Theorem 3.6. *If G_0 is a finite simple classical group, $G_0 \leqslant G \leqslant \mathrm{Aut}(G_0)$, and G does not involve the triality automorphism of $\mathrm{P\Omega}_8^+(q)$ or the graph automorphism of $\mathrm{PSp}_4(2^a)$, and M is a maximal subgroup of G, not containing G_0, then either M stabilises one of the following structures on the natural module for $\widetilde{G}$:*

(i) a non-singular subspace;
(ii) a totally isotropic subspace;
(iii) a partition into at least two isometric non-singular subspaces;
(iv) a partition into two totally isotropic subspaces;
(v) a partition into non-singular subspaces defined over an extension field of prime degree;
(vi) a decomposition as a tensor product of two non-isometric spaces;
(vii) a decomposition as a tensor product of at least two isometric spaces;
(viii) a proper subfield, of prime degree;

or one of the following holds:

(ix) M is a classical group of the same dimension and with the same field of definition as G;
(x) M is an automorphism group of a simple group S, the representation of $\widetilde{S}$ being irreducible and not writable over any proper subfield, where $\widetilde{S}$ is the preimage of S in $\widetilde{G}$;
(xi) M is an automorphism group of an extraspecial group r^{1+2m} with r dividing d, where d is the order of the generic part of the Schur multiplier of G_0; or of $C_4 \circ 2^{1+2m}$ when the generic part of the Schur multiplier has C_4 as a quotient.

This form of the theorem is stated and proved in [171]. Its proof does not require any more work than we have done already. However, if we want to provide more detail of the structures of the corresponding subgroups, and to decide which ones are in fact maximal, then we need to do a lot more work. The book of Kleidman and Liebeck [108] is then essential reading, as Aschbacher's paper does not give the details required. We begin by studying the ways in which the forms interact with the geometry.

3.10.5 Tensor products of spaces with forms

The general idea is that given a form f on U and a form h on W, we define $f \otimes h$ on $U \otimes W$ via

$$(f \otimes h)(u_1 \otimes w_1, u_2 \otimes w_2) = f(u_1, u_2)h(w_1, w_2), \qquad (3.44)$$

where u_i range over a basis of U and w_i range over a basis of W, and extending bilinearly if f and h are bilinear, and sesquilinearly if f and h are sesquilinear. Now the linearity of f and h implies that

$$(f \otimes h)(u_1' \otimes w_1', u_2' \otimes w_2') = f(u_1', u_2')h(w_1', w_2')$$

for all vectors $u'_i \in U$ and $w'_i \in W$. Hence $f \otimes h$ is invariant under the isometry groups of f and h. Indeed, we can extend to similarities of f and h provided f is multiplied by a scalar λ and h is multiplied by the inverse scalar λ^{-1}.

Clearly $f \otimes h$ is a non-singular bilinear form if and only if both f and h arc non-singular bilinear forms, and similarly for sesquilinear forms. In the sesquilinear case, we are only interested in the forms which are conjugate-symmetric. For $f \otimes h$ to be conjugate-symmetric both f and h must be conjugate-symmetric. In the bilinear case we can mix-and-match symmetry and skew-symmetry: an even number of the forms f, h and $f \otimes h$ are skew-symmetric. Thus, if the characteristic of the underlying field is odd, we have the inclusions:

$$
\begin{aligned}
\mathrm{PSU}_k(q) \times \mathrm{PSU}_l(q) &< \mathrm{PSU}_{kl}(q), \\
\mathrm{PSp}_{2k}(q) \times \mathrm{PSp}_{2l}(q) &< \mathrm{P\Omega}_{4kl}(q), \\
\mathrm{PSp}_{2k}(q) \times \mathrm{P\Omega}_l(q) &< \mathrm{PSp}_{2kl}(q), \text{ and} \\
\mathrm{P\Omega}_k(q) \times \mathrm{P\Omega}_l(q) &< \mathrm{P\Omega}_{kl}(q),
\end{aligned}
\tag{3.45}
$$

where in the second and fourth cases the types of the orthogonal groups remain to be determined. In the third case, all types of orthogonal group occur.

In both cases PSp_{2k} and $\mathrm{P\Omega}^+_{2k}$ there is a totally isotropic subspace in U of dimension k, and tensoring this with W gives a totally isotropic subspace of $U \otimes W$ of exactly half the dimension, so in these cases the bilinear form $f \otimes h$ is of plus type, and we get

$$
\begin{aligned}
\mathrm{PSp}_{2k}(q) \times \mathrm{PSp}_{2l}(q) &< \mathrm{P\Omega}^+_{4kl}(q) \text{ and} \\
\mathrm{P\Omega}^+_{2k}(q) \times \mathrm{P\Omega}^{\pm}_l(q) &< \mathrm{P\Omega}^+_{2kl}(q).
\end{aligned}
\tag{3.46}
$$

In the case $\mathrm{P\Omega}^-_{2k} \otimes \mathrm{P\Omega}_{2l+1}$ choose a maximal isotropic subspace X of W, so that $U \otimes X$ is an isotropic subspace of dimension $2kl$ in $U \otimes W$. Now choose $x \in X^\perp \setminus X$, and Y a maximal isotropic subspace of U, so that $Y \otimes \langle x \rangle$ has dimension $k - 1$. Moreover, the space $Z = (U \otimes X) + (Y \otimes \langle x \rangle)$ is an isotropic subspace of dimension $2kl + k - 1$. Therefore $Z^\perp / Z$ has dimension 2, and is spanned by $u \otimes x$ and $v \otimes x$, where u, v span $Y^\perp$ modulo Y. In particular, $Z^\perp / Z$ has minus type, and therefore (by Witt's lemma) so does $f \otimes h$. Thus

$$
\mathrm{P\Omega}^-_{2k}(q) \times \mathrm{P\Omega}_{2l+1}(q) < \mathrm{P\Omega}^-_{2k(2l+1)}(q).
\tag{3.47}
$$

Similarly in the case $\mathrm{P\Omega}^-_{2k} \otimes \mathrm{P\Omega}^-_{2l}$ we may choose a maximal isotropic subspace X of dimension $l - 1$ in W, so that $U \otimes X$ is isotropic of dimension $2k(l - 1)$ in $U \otimes W$. Now choosing a complement $\langle x_1, x_2 \rangle$ to X in $X^\perp$ and a maximal isotropic subspace Y of dimension $k - 1$ in U gives an isotropic subspace Z of dimension $2k(l - 1) + 2(k - 1) = 2kl - 2$, and we reduce to considering a 4-space $Z^\perp / Z$ of type $\mathrm{P\Omega}^-_2 \otimes \mathrm{P\Omega}^-_2$. Since the characteristic is odd, we may take orthogonal bases for both tensor factors, and deduce that the product form is diagonal with entries $(1, \lambda, \lambda, \lambda^2)$ for some λ. But this form is equivalent to the one with entries $(1, 1, 1, 1)$, which is of plus type. Therefore

$$P\Omega_{2k}^-(q) \times P\Omega_{2l}^-(q) < P\Omega_{4kl}^+(q). \tag{3.48}$$

In characteristic 2 things are (for a change) much easier, since both orthogonal groups $P\Omega_{2k}^{\pm}(q)$ are contained in $PSp_{2k}(q)$. We therefore just need to verify that $PSp_{2k}(q) \times PSp_{2l}(q) < P\Omega_{4kl}^+(q)$ in this case also. The quadratic form is constructed by defining $Q(u \otimes w) = 0$ for all basis vectors u of U and w of W, and it is easy to verify that this implies $Q(x \otimes y) = 0$ for all $x \in U$ and $y \in W$, and hence Q is invariant under the group $Sp_{2k}(q) \circ Sp_{2l}(q)$. It is obvious that Q is of plus type.

Wreathing iterated tensor products gives rise to some more inclusions:

$$PSU_k(q) \wr S_l < PSU_{k^l}(q) \text{ for all } q,$$
$$PSp_{2k}(q) \wr S_l < PSp_{(2k)^l}(q) \text{ for } ql \text{ odd, and}$$
$$PSp_{2k}(q) \wr S_l < P\Omega_{(2k)^l}^+(q) \text{ for } ql \text{ even}. \tag{3.49}$$

For q odd we also have

$$P\Omega_k(q) \wr S_l < P\Omega_{k^l}(q) \text{ for } k \text{ odd, and}$$
$$P\Omega_{2k}^{\varepsilon_1}(q) \wr S_l < P\Omega_{(2k)^l}^{\varepsilon_2}(q), \tag{3.50}$$

where $\varepsilon_2 = -$ if l is odd and $\varepsilon_1 = -$, and $\varepsilon_2 = +$ otherwise.

The full stabilisers of these tensor product decompositions are usually maximal subgroups of the appropriate classical group. However, the details are messy and there are some exceptions. See the book by Kleidman and Liebeck [108].

3.10.6 Extending the field on spaces with forms

Next we examine what happens when we extend the field. Thus we have a form f (or Q) on a vector space V over $\mathbb{F}_q$, and embed V in a vector space V^* over $\mathbb{F}_{q^k}$ by extending the scalars. Concretely this can be done by choosing a basis $\{e_1, \ldots, e_n\}$ of V and defining $V^* = \{\sum_{i=1}^n \lambda_i e_i \mid \lambda_i \in \mathbb{F}_{q^k}\}$, or more abstractly by $V^* = \mathbb{F}_{q^k} \otimes_{\mathbb{F}_q} V$. Now we want to know how f can extend to a form f^* on V^*. We shall show that there are various possibilities, giving rise to the following embeddings of the corresponding isometry groups.

$$Sp_{2m}(q) < Sp_{2m}(q^k) \text{ for all } k,$$
$$Sp_{2m}(q) < SU_{2m}(q),$$
$$\Omega_n^\varepsilon(q) < SU_n(q) \text{ for all } \varepsilon,$$
$$SU_n(q) < SU_n(q^k) \text{ for } k \text{ odd},$$
$$\Omega_n^{\varepsilon_1}(q) < \Omega_n^{\varepsilon_2}(q^k) \text{ for } \varepsilon_2 = \varepsilon_1{}^k. \tag{3.51}$$

In the case of symmetric and skew-symmetric forms (and alternating and quadratic forms in characteristic 2), it is clear that the form can extend to a form of the same type over the larger field, ignoring the signs of the orthogonal groups for the moment. In the case of sesquilinear forms, this can happen provided the field automorphism extends to the larger field, i.e. if k is odd.

Now consider the signs of the quadratic forms: it is easy to see that $\mathrm{GO}_2^\pm(q) < \mathrm{GO}_2^+(q^2)$, since every quadratic polynomial over $\mathbb{F}_q$ has a root in $\mathbb{F}_{q^2}$. Similarly if k is odd then $\mathrm{GO}_2^-(q) < \mathrm{GO}_2^-(q^k)$, since the irreducible polynomial $x^2 - \alpha$ or $x^2 + x + \mu$ used to define $\mathrm{GO}_2^-(q)$ remains irreducible over $\mathbb{F}_{q^k}$. Therefore

$$
\begin{aligned}
\mathrm{GO}_{2n}^-(q) &< \mathrm{GO}_{2n}^+(q^2), \\
\mathrm{GO}_{2n}^+(q) &< \mathrm{GO}_{2n}^+(q^k) \text{ for all } k, \\
\mathrm{GO}_{2n}^-(q) &< \mathrm{GO}_{2n}^-(q^k) \text{ for } k \text{ odd.}
\end{aligned}
\tag{3.52}
$$

It is also possible, however, that the form f^* is of a different type from the original form f. Specifically, if f is bilinear and k is even, then f^* can be sesquilinear. In the case when f is symmetric, such an f^* can be defined by extending f sesquilinearly. In other words,

$$
f^*\left(\sum \lambda_i e_i, \sum \mu_j e_j\right) = \sum_{i=1}^n \sum_{j=1}^n \lambda_i \overline{\mu_j} f(e_i, e_j)
\tag{3.53}
$$

for all $\lambda_i, \mu_j \in \mathbb{F}_{q^k}$. In the case when f is skew-symmetric we modify this construction by first multiplying the form by β, where $\beta \in \mathbb{F}_{q^k}$ satisfies $\overline{\beta} = -\beta$. Then we extend the form βf sesquilinearly—this is possible because for $u, v \in V$ we have $\beta f(u, v) = -\beta f(v, u) = \overline{\beta} f(v, u)$ so that βf is already conjugate-symmetric on V. The case when f is alternating, over a field of characteristic 2, is a special case of both of these cases.

Warning. In the literature, the larger group is usually regarded as the primary one, and therefore the subgroup is called a 'subfield' subgroup. Conversely, what we shall describe in the next section as 'restricting' the field is usually thought of as the 'extension field' case.

3.10.7 Restricting the field on spaces with forms

Finally, we examine what happens when we forget some of the field structure, and identify $\mathbb{F}_{q^k}$ with a k-dimensional vector space over $\mathbb{F}_q$. Here we start with a vector space V of dimension m over $\mathbb{F}_{q^k}$, endowed with a form f (or Q) with values in $\mathbb{F}_{q^k}$. We may assume k is prime, and then use recursion to build the general case. We need a form f^* (or Q^*) on the same set of vectors, with values only in $\mathbb{F}_q$. The general method is to compose f (or Q) with a suitable $\mathbb{F}_q$-linear map $\mathbb{F}_{q^k} \to \mathbb{F}_q$. In fact, any non-zero $\mathbb{F}_q$-linear map will do: they are all of the form $x \mapsto \mathrm{Tr}(\lambda x)$, where $0 \neq \lambda \in \mathbb{F}_{q^k}$ and the *trace* map, $\mathrm{Tr} : \mathbb{F}_{q^k} \to \mathbb{F}_q$ is defined by

$$
\mathrm{Tr} : x \mapsto x + x^q + \cdots + x^{q^{k-1}}.
\tag{3.54}
$$

This is the same as the trace of 'multiplication by x', considered as an $\mathbb{F}_q$-linear map on the k-dimensional vector space $\mathbb{F}_{q^k} \cong \mathbb{F}_q{}^k$.

Therefore f^* is symmetric if f is symmetric, and f^* is skew-symmetric if f is skew-symmetric. On the other hand if f is conjugate-symmetric, then f^* may be either conjugate-symmetric (if k is odd), or symmetric (f^* is essentially the 'real part' $\frac{1}{2}(f + \overline{f})$ of f in the case $k = 2$ and q odd), or skew-symmetric (f^* is the 'imaginary part' $\frac{1}{2}(f - \overline{f})$ of f, also in the case $k = 2$ and q odd).

This gives us the following inclusions of isometry groups, apart from the signs on the orthogonal groups:

$$
\begin{aligned}
\mathrm{Sp}_{2m}(q^k) &< \mathrm{Sp}_{2mk}(q), \\
\Omega_m^\varepsilon(q^k) &< \Omega_{mk}^\varepsilon(q) \text{ for } m \text{ even}, \\
\Omega_m(q^2) &< \Omega_{2m}^\varepsilon(q) \text{ for } m, q \text{ odd}, \\
\Omega_m(q^k) &< \Omega_{mk}(q) \text{ for } m, k, q \text{ odd}, \\
\mathrm{SU}_m(q^k) &< \mathrm{SU}_{mk}(q) \text{ for } k \text{ odd}, \\
\mathrm{SU}_m(q) &< \Omega_{2m}^\varepsilon(q) \text{ for } \varepsilon = (-1)^m, \\
\mathrm{SU}_m(q) &< \mathrm{Sp}_{2m}(q).
\end{aligned}
\tag{3.55}
$$

To show that $\mathrm{SU}_m(q) < \Omega_{2m}^\varepsilon(q)$, where $\varepsilon = (-1)^m$, we pick an orthonormal basis for the unitary form and show that this decomposes the orthogonal space as a perpendicular direct sum of minus-type 2-spaces. Thus it is sufficient to prove the case $m = 1$. If q is odd, pick $\alpha \in \mathbb{F}_{q^2}$ such that $\overline{\alpha} = \alpha^q = -\alpha$, so that with respect to the $\mathbb{F}_q$-basis $\{1, \alpha\}$ the form $f^*(x, y) = x\overline{y} + \overline{x}y$ has matrix $\mathrm{diag}(2, -2\alpha^2)$, where α^2 is a non-square in $\mathbb{F}_q$. It is now easy to check that f^* is of minus type, as required. A similar argument works in characteristic 2.

It remains to consider the embeddings of orthogonal groups in each other. If m and k are both odd, there is nothing left to prove, so consider first the case m odd, and $k = 2$ (and we may assume q is odd). As above it is sufficient to consider the case $m = 1$, where we may take $f(x, y) = \lambda xy$ for some fixed $\lambda \in \mathbb{F}_{q^2}$. We show that for different choices of λ, we obtain f^* to be either plus type or minus type. For f^* is of plus type if and only if there is a non-zero solution x to $\mathrm{Tr}(\lambda x^2) = 0$. This is equivalent to $\lambda x^2 + \lambda^q x^{2q} = 0$, or $\lambda^{q-1} x^{2(q-1)} = -1$. We can certainly choose λ so that $\lambda^{q-1} = -1$, so that $x = 1$ is a solution. On the other hand, if $\lambda = \alpha\beta$, where $\alpha^{q-1} = -1$ and β has order $q^2 - 1$, then $-\lambda^{q-1}$ has order $q + 1$, whereas $x^{2(q-1)}$ has order dividing $(q+1)/2$, so there is no solution.

Finally we consider the case m even. Pick a totally isotropic $(m/2 - 1)$-space W for f, and observe that (by Witt's Lemma) it is sufficient to determine the signs of the forms on $W^\perp/W$. In other words we may assume that $m = 2$, and we want to prove that $\Omega_2^\varepsilon(q^k) < \Omega_{2k}^\varepsilon(q)$. Suppose that q is odd. We first show that the determinant of (the matrix of) f is a square in $\mathbb{F}_{q^k}$ if and only if the determinant of (the matrix of) f^* is a square in $\mathbb{F}_q$. We use the *determinant* map $\det : \mathbb{F}_{q^k} \to \mathbb{F}_q$ defined by

$$
\det : x \mapsto x.x^q.\cdots.x^{q^{k-1}} = x^{(q^k-1)/(q-1)}
\tag{3.56}
$$

which is the same as the usual determinant of 'multiplication by x' regarded as an $\mathbb{F}_q$-linear map on $\mathbb{F}_{q^k}$. Thus if f is diagonalised as $\mathrm{diag}(1,\alpha)$, then $\det f = \alpha$ and $\det(f^*) = \det \alpha = \alpha^{(q^k-1)/(q-1)}$, proving our claim.

Now f is of plus type if and only if either

(i) $\det f$ is a square and $q^k \equiv 1 \bmod 4$, or
(ii) $\det f$ is a non-square and $q^k \equiv 3 \bmod 4$.

In other words, either

(i) $\det f^*$ is a square in $\mathbb{F}_q$, and either $k = 2$ or $q \equiv 1 \bmod 4$; or
(ii) $\det f^*$ is a non-square in $\mathbb{F}_q$, and k is odd and $q \equiv 3 \bmod 4$.

But this is exactly the condition for f^* to be of plus type. Similar arguments work in characteristic 2 to show that $\Omega_2^\varepsilon(2^k) < \Omega_{2k}^\varepsilon(2)$.

3.10.8 Maximal subgroups of symplectic groups

We are now in a position to describe in more detail the structures of the subgroups of the classical groups appearing in Theorem 3.6. As usual, the symplectic groups are the easiest case to work with. In characteristic 2 we have the following result proved in [171]. This is more explicit than the theorems in [5]. Since [171] remains unpublished, and [108] is far more comprehensive, it seems reasonable to call Theorems 3.7, 3.8, 3.9, 3.10, 3.11, 3.12 collectively the Kleidman–Liebeck Theorems, although in fact Kleidman and Liebeck prove much more general results.

Theorem 3.7. *If M is a maximal subgroup of $\mathrm{Sp}_{2n}(q)$, with $q = 2^a$, then either M is an automorphism group of a simple group S, and the representation of S is absolutely irreducible, symplectic but not orthogonal, and not writable over any proper subfield of $\mathbb{F}_q$, or M is one of the following groups:*

(i) $q^{k(k+1)/2}.q^{2km}{:}(\mathrm{GL}_k(q) \times \mathrm{Sp}_{2m}(q))$, with $0 < k \leqslant k + m = n$;
(ii) $\mathrm{Sp}_{2k}(q) \times \mathrm{Sp}_{2m}(q)$, with $0 < k < m < k + m = n$;
(iii) $\mathrm{Sp}_{2k}(q) \wr S_m$, with $0 < k < km = n$, and if $q = 2$ then $k \neq 1$;
(iv) $\mathrm{Sp}_{2k}(q^m).m$, where $km = n$ and m is prime;
(v) $\mathrm{Sp}_{2n}(q_0)$, where $q = q_0{}^b$ and b is prime;
(vi) $\mathrm{GO}_{2n}^+(q)$;
(vii) $\mathrm{GO}_{2n}^-(q)$;
(viii) $G_2(q)$, if $n = 3$;
(ix) $\mathrm{Sz}(q)$, if $n = 2$ and a is odd.

Note that the last two cases are not necessary in the statement of the theorem. They are included merely to draw attention to these particular subgroups.

From the same source, we have in odd characteristic:

Theorem 3.8. *If M is a maximal subgroup of $\mathrm{PSp}_{2n}(q)$, with q odd, then either M is an automorphism group of a simple group S, and the representation of $\widetilde{S}$ (the preimage of S in $\mathrm{Sp}_{2n}(q)$) is absolutely irreducible, symplectic, and not writable over any proper subfield of $\mathbb{F}_q$, or M is one of the following groups:*

   (i) $q^{k(k+1)/2}.q^{2km}{:}(\mathrm{GL}_k(q) \circ \mathrm{Sp}_{2m}(q))$, *with* $0 < k \leqslant k+m = n$;

   (ii) $\mathrm{Sp}_{2k}(q) \circ \mathrm{Sp}_{2m}(q)$, *with* $0 < k < m < k+m = n$;

   (iii) $2^{m-1}{\cdot}(\mathrm{PSp}_{2k}(q) \wr S_m)$, *with* $0 < k < km = n$;

   (iv) $\mathrm{GL}_n(q).2/\{\pm 1\}$, *with* $n \geqslant 3$;

   (v) $\mathrm{GU}_n(q).2/\{\pm 1\}$;

   (vi) $\mathrm{PSp}_{2k}(q^m).m$, *with* $km = n$ *and* m *prime;*

   (vii) $(\mathrm{PSp}_{2k}(q) \times \mathrm{PGO}_m^\varepsilon(q)).2$, *where* $km = n$ *and* $m \geqslant 2$, *excluding the cases* $(k,m,\varepsilon) = (1,4,+)$ *and* $(m,q) = (2,3)$, $(3,3)$ *and* $(m,q,\varepsilon) = (2,5,+)$;

  (viii) $(\mathrm{PSp}_{2k}(q).2 \wr S_m)\frac{1}{2}$, *where* $(2k)^m = 2n$ *and* m *is odd;*

   (ix) $\mathrm{PSp}_{2n}(q_0)$, *where* $q = q_0{}^b$ *and* b *is odd;*

   (x) $\mathrm{PSp}_{2n}(q_0).2$, *where* $q = q_0{}^2$;

   (xi) $2^{2m}\Omega_{2m}^-(2)$, *if* q *is prime,* $n = 2^{m-1}$, *and* $q \equiv \pm 3 \bmod 8$;

  (xii) $2^{2m}\mathrm{GO}_{2m}^-(2)$, *if* q *is prime,* $n = 2^{m-1}$, *and* $q \equiv \pm 1 \bmod 8$.

In the above theorem, the notation $G\frac{1}{2}$ is used to denote a subgroup of index 2 in G.

3.10.9 Maximal subgroups of unitary groups

The following result is distilled from [108]:

Theorem 3.9. *If M is any subgroup of $\mathrm{GU}_n(q)$ not containing $\mathrm{SU}_n(q)$, then either M is almost simple modulo scalars, so that M is the normaliser of a quasisimple group S, and the representation of S is absolutely irreducible, unitary, and not writable over any proper subfield of $\mathbb{F}_{q^2}$, or M is contained in one of the following subgroups:*

   (i) $\mathrm{GU}_k(q) \times \mathrm{GU}_{n-k}(q)$, *the stabiliser of a non-singular k-space;*

   (ii) $q^{k^2}.q^{2k(n-k)}.(\mathrm{GL}_k(q^2) \times \mathrm{GU}_{n-2k}(q))$, *the stabiliser of a totally isotropic k-space, with $k \leqslant n/2$;*

  (iii) $\mathrm{GU}_k(q) \wr S_m$, *the stabiliser of a decomposition as a perpendicular direct sum of m non-singular k-spaces, with $n = km$;*

   (iv) $\mathrm{GL}_{n/2}(q^2).2$, *the stabiliser of a decomposition as a direct sum of two totally isotropic $n/2$-spaces, if n is even;*

   (v) $\mathrm{GU}_k(q) \circ \mathrm{GU}_m(q)$, *the stabiliser of a tensor product decomposition, with $n = km$;*

   (vi) $C_{q+1}.(\mathrm{PGU}_k(q) \wr S_m)$, *the stabiliser of a decomposition as a tensor product of m spaces of dimension k, with $n = k^m$;*

  (vii) $\mathrm{GU}_m(q^k).k$, *the stabiliser of a vector space structure over an extension field of order q^{2k}, with $n = mk$ and k an odd prime;*

(viii) $C_{q+1}.\mathrm{PGU}_n(q_0)$, the stabiliser of a subfield of order q_0, where $q = q_0{}^k$ and k is an odd prime;

(ix) $C_{q+1} \circ \mathrm{GO}_n^\varepsilon(q)$, for any ε, if q is odd;

(x) $C_{q+1} \circ \mathrm{Sp}_n(q)$, if n is even;

(xi) $C_{q+1} \circ r^{1+2m}\mathrm{Sp}_{2m}(r)$, with $n = r^m$, and either r is an odd prime dividing $q + 1$, or $r = 2$ and $q \equiv 3 \bmod 4$.

3.10.10 Maximal subgroups of orthogonal groups

We follow the example of Kleidman and Liebeck, and consider separately the three cases

(i) odd dimension,
(ii) even dimension, minus type, and
(iii) even dimension, plus type.

The structures of the stabilisers of subspaces, and of partitions into non-singular subspaces, are discussed in Section 3.7.4. The rest of the subgroups listed here are discussed in Sections 3.10.5 to 3.10.7.

Theorem 3.10. *If q is odd, $n \geqslant 2$ and M is any subgroup of $\mathrm{GO}_{2n+1}(q)$ not containing $\Omega_{2n+1}(q)$ then either M is almost simple modulo $\{\pm 1\}$, so that M is the normaliser of a simple group S, and the representation of S is absolutely irreducible, orthogonal, and not writable over any proper subfield of $\mathbb{F}_q$, or M is contained in one of the following subgroups:*

(i) $q^{k(k-1)/2}.q^{k(2n-2k+1)}.(\mathrm{GL}_k(q) \times \mathrm{GO}_{2n-2k+1}(q))$, the stabiliser of a totally isotropic k-space, with $1 \leqslant k \leqslant n$;

(ii) $\mathrm{GO}_{2k}^\varepsilon(q) \times \mathrm{GO}_{2(n-k)+1}(q)$, the stabiliser of a non-singular subspace of dimension $2k$ and type ε $(= +$ or $-)$;

(iii) $\mathrm{GO}_k(q) \wr S_m$, with $km = 2n + 1$ and $m > 1$;

(iv) $2 \times \mathrm{SO}_k(q) \times \mathrm{SO}_m(q)$, with $km = 2n + 1$ and $1 < k < 2n + 1$;

(v) $2 \times \mathrm{SO}_k(q) \wr S_m$, with $k^m = 2n + 1$ and $m > 1$;

(vi) $\mathrm{GO}_m(q^k).k$ with $km = 2n + 1$ and k prime;

(vii) $\mathrm{GO}_{2n+1}(q_0)$, with $q_0{}^k = q$ and k prime.

Notice that the second type of subgroups in this list appears at first glance to contradict Witt's Lemma (Theorem 3.4.8), as there are *two* orbits of the orthogonal group on non-singular subspaces of each odd dimension. However, these spaces are not isometric, since the forms on the two spaces differ by multiplication by a non-square, as described in Section 3.7.4.

Theorem 3.11. *If $n \geqslant 3$ and M is any subgroup of $\mathrm{GO}_{2n}^-(q)$ not containing $\Omega_{2n}^-(q)$ then either M is almost simple modulo $\{\pm 1\}$, so that M is the normaliser of a quasisimple group S, and the representation of S is absolutely irreducible, orthogonal, and not writable over any proper subfield of $\mathbb{F}_q$, or M is contained in one of the following subgroups:*

(i) $q^{k(k-1)/2}.q^{2k(n-k)}.(\mathrm{GL}_k(q) \times \mathrm{GO}_{2(n-k)}^-(q))$, *the stabiliser of a totally isotropic subspace of dimension* k, $1 \leqslant k < n$;

(ii) $\mathrm{GO}_{2k}^+(q) \times \mathrm{GO}_{2m}^-(q)$, *with* $k + m = n$ *and* $0 < k < n$;

(iii) $\mathrm{GO}_{2m+1}(q) \times \mathrm{GO}_{2k+1}(q)$ *if* q *is odd, with* $k + m + 1 = n$, *and either* $k < m$ *or* $k = m$ *and* $q \equiv 1 \bmod 4$;

(iv) $\mathrm{Sp}_{2n-2}(q)$ *if* q *is even*;

(v) $\mathrm{GO}_{2k}^-(q) \wr S_m$ *with* $km = n$ *and* m *odd*;

(vi) $\mathrm{GO}_k(q) \wr S_m$ *with* $km = 2n$, k *and* n *both odd, and* $q \equiv 3 \bmod 4$;

(vii) $\mathrm{SO}_m(q) \times \mathrm{GO}_{2k}^-(q)$, *with* $km = n$, *and* m *and* q *both odd*;

(viii) $\mathrm{GO}_{2n}^-(q_0)$ *with* $q_0^{\,k} = q$ *and* k *an odd prime*;

(ix) $\mathrm{GO}_{2m}^-(q^k).k$ *with* $km = n$ *and* k *prime*;

(x) $\mathrm{GO}_n(q^2).2$ *if* n *and* q *are both odd*;

(xi) $\mathrm{GU}_n(q).2$, *if* n *is odd*.

Theorem 3.12. *If* $n \geqslant 3$ *and* M *is any subgroup of* $\mathrm{GO}_{2n}^+(q)$ *not containing* $\Omega_{2n}^+(q)$ *then either* M *is almost simple modulo* $\{\pm 1\}$, *so that* M *is the normaliser of a quasisimple group* S, *and the representation of* S *is absolutely irreducible, orthogonal, and not writable over any proper subfield of* $\mathbb{F}_q$, *or* M *is contained in one of the following subgroups:*

(i) $q^{k(k-1)/2}.q^{2k(n-k)}.(\mathrm{GL}_k(q) \times \mathrm{GO}_{2(n-k)}^+(q))$, *the stabiliser of a totally isotropic subspace of dimension* k, *with* $1 \leqslant k < n - 1$ *or* $k = n$;

(ii) $\mathrm{GO}_{2k}^+(q) \times \mathrm{GO}_{2m}^+(q)$, *with* $k + m = n$ *and* $0 < k < n$;

(iii) $\mathrm{GO}_{2k}^-(q) \times \mathrm{GO}_{2m}^-(q)$, *with* $k + m = n$ *and* $0 < k < n$;

(iv) $\mathrm{GO}_{2m+1}(q) \times \mathrm{GO}_{2k+1}(q)$ *if* q *is odd, with* $k + m + 1 = n$, *and either* $k < m$ *or* $k = m$ *and* $q \equiv 3 \bmod 4$;

(v) $\mathrm{Sp}_{2n-2}(q)$ *if* q *is even*;

(vi) $\mathrm{GO}_{2k}^+(q) \wr S_m$ *with* $km = n$;

(vii) $\mathrm{GO}_{2k}^-(q) \wr S_m$ *with* $km = n$ *and* m *even*;

(viii) $\mathrm{GO}_k(q) \wr S_m$ *with* $km = 2n$, k *and* q *both odd, and either* n *is even or* $q \equiv 1 \bmod 4$;

(ix) $\mathrm{GL}_n(q).2$;

(x) $\mathrm{GU}_n(q).2$, *if* n *is even*.

(xi) $\mathrm{GO}_{2m}^+(q^k).k$ *with* $km = n$ *and* k *prime*;

(xii) $\mathrm{GO}_n(q^2).2$ *if* n *and* q *are both odd*;

(xiii) $(\mathrm{Sp}_{2m}(q) \circ \mathrm{Sp}_{2k}(q)).2$, *with* $n = 2km$ *and* $k \neq m$;

(xiv) $\mathrm{SO}_m(q) \times \mathrm{GO}_{2k}^+(q)$, *with* $km = n$, *and* m *and* q *both odd*;

(xv) $\mathrm{GO}_{2k}^{\varepsilon_1}(q) \circ \mathrm{GO}_{2m}^{\varepsilon_2}(q)$, *if* q *is odd, with* $2km = n$;

(xvi) $2.(\mathrm{PSp}_{2m}(q).2 \wr S_k)\frac{1}{2}$ *where* $2n = (2m)^k$ *and either* k *or* q *is even*;

(xvii) $2.(\mathrm{PGO}_{2m}^\varepsilon(q).2 \wr S_k)\frac{1}{2}$, *where* $2n = (2m)^k$, q *is odd*;

(xviii) $\mathrm{GO}_{2n}^+(q_0)$ *with* $q_0^{\,k} = q$ *and* k *prime*;

(xix) $\mathrm{GO}_{2n}^-(q_0)$ *with* $q_0^2 = q$;

(xx) $2_+^{1+2m}\Omega_{2m}^+(2)$, *with* $n = 2^{m-1}$, *if* q *is prime*, $q \equiv \pm 3 \bmod 8$;

(xxi) $2_+^{1+2m}\mathrm{GO}_{2m}^+(2)$, *with* $n = 2^{m-1}$, *if* q *is prime*, $q \equiv \pm 1 \bmod 8$.

3.11 Generic isomorphisms

We have already proved the generic isomorphisms $\mathrm{PSp}_2(q) \cong \mathrm{PSL}_2(q)$ in Section 3.5.1, and $\mathrm{PSU}_2(q) \cong \mathrm{PSL}_2(q)$ in Section 3.6. Here we consider the orthogonal groups, many of which are isomorphic to other classical groups. Indeed, all those in dimension up to 6 have other names as follows:

$$\begin{aligned}
\mathrm{P}\Omega_3(q) &\cong \mathrm{PSL}_2(q), \\
\mathrm{P}\Omega_4^+(q) &\cong \mathrm{PSL}_2(q) \times \mathrm{PSL}_2(q), \\
\mathrm{P}\Omega_4^-(q) &\cong \mathrm{PSL}_2(q^2), \\
\mathrm{P}\Omega_5(q) &\cong \mathrm{PSp}_4(q), \\
\mathrm{P}\Omega_6^+(q) &\cong \mathrm{PSL}_4(q), \\
\mathrm{P}\Omega_6^-(q) &\cong \mathrm{PSU}_4(q).
\end{aligned} \tag{3.57}$$

A holistic view of these isomorphisms may be achieved by a thorough understanding of triality (see Section 4.7), but for the moment we adopt a more concrete, case-by-case, representation-theoretic approach. We construct certain representations of the groups on the right-hand side, using tensor products and symmetric and skew-symmetric squares. We then show that these representations support quadratic forms of the appropriate types, whence the isomorphism follows from the fact that the two groups have the same order.

3.11.1 Low-dimensional orthogonal groups

The first three cases in (3.57) all concern groups of type $\mathrm{PSL}_2(q) \cong \mathrm{PSp}_2(q)$, and can be treated similarly. Take two 2-dimensional spaces V_1 and V_2, with bases $\{e_1, f_1\}$ and $\{e_2, f_2\}$, say, and put a symmetric bilinear form h on $V_1 \otimes V_2$, as in Section 3.10.5, by defining $h(x \otimes y, z \otimes t) = b_1(x, z) b_2(y, t)$, where b_i is the alternating bilinear form on V_i defined by $b_i(e_i, f_i) = 1$. First let $V_1 = V_2 = V$ be the natural module for $\mathrm{SL}_2(q)$, with q odd, and let W be the submodule of $V \otimes V$ spanned by $e_1 \otimes e_1$, $e_1 \otimes f_1 + f_1 \otimes e_1$ and $f_1 \otimes f_1$. (Thus W is the *symmetric square* of V.) The above form h restricts to W as the form with matrix $\begin{pmatrix} 0 & 0 & 1 \\ 0 & -2 & 0 \\ 1 & 0 & 0 \end{pmatrix}$, and it is easy to check that this form is invariant under the generators $e_1 \mapsto e_1 + \lambda f_1$ and $e_1 \leftrightarrow f_1$ of $\mathrm{SL}_2(q)$, which act respectively as the matrices

$$\begin{pmatrix} 1 & \lambda & \lambda^2 \\ 0 & 1 & 2\lambda \\ 0 & 0 & 1 \end{pmatrix} \quad \text{and} \quad \begin{pmatrix} 0 & 0 & 1 \\ 0 & -1 & 0 \\ 1 & 0 & 0 \end{pmatrix}.$$

Since -1 in $\mathrm{SL}_2(q)$ acts trivially, we have an embedding of $\mathrm{PSL}_2(q)$ in $\mathrm{GO}_3(q)$, and by comparing orders we have $\mathrm{PSL}_2(q) \cong \Omega_3(q)$, for q odd.

In the second case, let V_1 and V_2 be natural modules for two distinct copies of $\mathrm{Sp}_2(q) \cong \mathrm{SL}_2(q)$, so that $V_1 \otimes V_2$ becomes a module for $\mathrm{SL}_2(q) \times \mathrm{SL}_2(q)$, with $(-1, -1)$ acting trivially. In characteristic 2, define the quadratic form Q with h as associated bilinear form by

$$Q(e_1 \otimes e_2) = Q(e_1 \otimes f_2) = Q(f_1 \otimes e_2) = Q(f_1 \otimes f_2) = 0.$$

It is easy to see that the quadratic form is of plus type, as $e_1 \otimes e_2$ and $e_1 \otimes f_2$ span a totally isotropic 2-space. To check that the form is invariant under the group it suffices (by symmetry) to check invariance under the element $e_1 \mapsto e_1 + \lambda f_1$. This is left as an exercise. It follows that $\mathrm{SL}_2(q) \circ \mathrm{SL}_2(q)$ embeds in $\mathrm{GO}_4^+(q)$, and $\mathrm{PSL}_2(q) \times \mathrm{PSL}_2(q) \cong \mathrm{P\Omega}_4^+(q)$.

In the third case, take V_1 to be the natural module for $\mathrm{SL}_2(q^2)$, and V_2 to be its image under the field automorphism $x \mapsto x^q = \overline{x}$. Thus the element $e_1 \mapsto e_1 + \lambda f_1$ acts as $e_2 \mapsto e_2 + \overline{\lambda} f_2$. Now take the $\mathbb{F}_q$-space spanned by $e_1 \otimes e_2$, $f_1 \otimes f_2$, and $\mu e_1 \otimes f_2 + \overline{\mu} f_1 \otimes e_2$, for $\mu \in \mathbb{F}_{q^2}$, and observe that this has dimension 4 over $\mathbb{F}_q$. Again in characteristic 2 define Q to be 0 on the simple tensors. Moreover, the first two basis vectors are non-orthogonal isotropic vectors, so span a 2-space of plus type, whose orthogonal complement consists of the remaining listed vectors. These last satisfy

$$Q(\mu e_1 \otimes f_2 + \overline{\mu} f_1 \otimes e_2) = -\mu\overline{\mu} \neq 0 \tag{3.58}$$

(for any $\mu \neq 0$), so form a 2-space of minus type. It is easy to check that the generators $e_i \mapsto f_i \mapsto -e_i$ and $e_i \mapsto e_i + \lambda_i f_i$ of $\mathrm{SL}_2(q^2)$, where $\lambda_2 = (\lambda_1)^q$, preserve this subspace. Therefore by order considerations $\mathrm{P\Omega}_4^-(q) \cong \mathrm{PSL}_2(q^2)$.

3.11.2 The Klein correspondence

The last three isomorphisms in the list (3.57) all concern groups acting on 4-spaces, and can be treated similarly. I shall call these three isomorphisms collectively the *Klein correspondence*, although this name is sometimes restricted to the case $\mathrm{P\Omega}_6^+(q) \cong \mathrm{PSL}_4(q)$. They can be proved by putting a rather different quadratic form on the skew-symmetric square of the natural 4-dimensional module. If $\{e_1, \ldots, e_4\}$ is a basis for the natural module V of $\mathrm{SL}_4(q)$, define $e_{ij} = e_i \otimes e_j - e_j \otimes e_i = -e_{ji}$. (We shall sometimes write $e_i \wedge e_j$ instead of e_{ij}.) Then it is easy to check that $\{e_{ij} \mid i < j\}$ forms a basis for a 6-dimensional subspace of the tensor product $V \otimes V$, invariant under the induced action of $\mathrm{SL}_4(q)$. This space is called the *skew-symmetric square* (or *exterior square*) of V, and is denoted $\Lambda^2(V)$. It supports an inner product (i.e. a non-singular symmetric bilinear form) g defined by

$$g(e_{ij}, e_{kl}) = 1 \text{ if } i, j, k, l \text{ is an even permutation of } 1, 2, 3, 4,$$
$$g(e_{ij}, e_{kl}) = -1 \text{ if it is an odd permutation, and}$$
$$g(e_{ij}, e_{kl}) = 0 \text{ otherwise.} \tag{3.59}$$

To check this is invariant under the action of the group it is enough by symmetry to check the map $e_1 \mapsto e_1 + \lambda e_2$, which acts on $\Lambda^2(V)$ as

$$e_{13} \mapsto e_{13} + \lambda e_{23},$$
$$e_{14} \mapsto e_{14} + \lambda e_{24},$$
$$e_{ij} \mapsto e_{ij} \text{ otherwise.} \tag{3.60}$$

All inner products are obviously unchanged by this map, except $g(e_{13}, e_{14})$ which goes to $g(e_{13} + \lambda e_{23}, e_{14} + \lambda e_{24}) = \lambda(g(e_{23}, e_{14}) + g(e_{13}, e_{24})) = 0$. Finally we see that the space spanned by e_{12}, e_{13}, e_{23} is an isotropic subspace of dimension 3, so this quadratic form is of plus type, and since the two groups have the same order it follows that $\mathrm{P\Omega}_6^+(q) \cong \mathrm{PSL}_4(q)$, for q odd. In characteristic 2 we define the quadratic form Q with associated bilinear form g by $Q(e_{ij}) = 0$, and then the same argument proves that $\mathrm{P\Omega}_6^+(q) \cong \mathrm{PSL}_4(q)$ also for q even.

Restricting to the subgroup $\mathrm{Sp}_4(q)$ of $\mathrm{SL}_4(q)$ in the case q odd, we might as well choose $e_3 = f_2$ and $e_4 = f_1$ to get a symplectic basis. Then the symplectic form on V is a bilinear map f, which gives rise to a linear map f^* on the skew-symmetric square via $f^*(v \otimes w - w \otimes v) = f(v, w)$. We have to check that f^* is well-defined, using the fact that f is bilinear and skew-symmetric. Moreover, the kernel of f^* is a 5-dimensional subspace, specifically the perpendicular space to $e_{14} + e_{23}$, spanned by $e_{12}, e_{13}, e_{24}, e_{34}$ and $e_{14} - e_{23}$. Again by comparing orders we obtain $\mathrm{PSp}_4(q) \cong \mathrm{P\Omega}_5(q)$, for q odd.

Finally we work inside the Klein correspondence $\mathrm{PSL}_4(q^2) \cong \mathrm{P\Omega}_6^+(q^2)$ to show that $\mathrm{PSU}_4(q) \cong \mathrm{P\Omega}_6^-(q)$. This time pick $e_1, \ldots, e_4$ to be an orthonormal basis with respect to the sesquilinear form, and consider the 6-space over $\mathbb{F}_q$ spanned by the vectors $\lambda e_{ij} + \overline{\lambda} e_{kl}$, where (i, j, k, l) is any even permutation of $(1, 2, 3, 4)$ and $\overline{\lambda} = \lambda^q$ for $\lambda \in \mathbb{F}_{q^2}$. We calculate the natural inner product g on the 6-space by

$$g(\lambda e_{ij} + \overline{\lambda} e_{kl}, \mu e_{ij} + \overline{\mu} e_{kl}) = \lambda \overline{\mu} + \overline{\lambda} \mu$$
$$= \lambda \mu^q + \lambda^q \mu \in \mathbb{F}_q, \tag{3.61}$$

and for $\lambda = \mu$ this is always non-zero, so for each choice of i, j, k, l the 2-space $\{\lambda e_{ij} + \overline{\lambda} e_{kl} \mid \lambda \in \mathbb{F}_{q^2}\}$ supports a quadratic form of minus type over $\mathbb{F}_q$, for q odd. The corresponding result in characteristic 2 is

$$Q(\lambda e_{ij} + \overline{\lambda} e_{kl}) = \lambda \overline{\lambda} = \lambda^{q+1} \neq 0. \tag{3.62}$$

Thus on the whole 6-space we have a quadratic form of minus type. It is obvious that the 6-space is invariant under even permutations of $e_1, \ldots, e_4$, and it is easy to check invariance under the elements $g_{\alpha\beta}$ defined by

$$e_1 \mapsto \alpha e_1 + \beta e_2,$$
$$e_2 \mapsto -\overline{\beta} e_1 + \overline{\alpha} e_2 \tag{3.63}$$

(for any α and β satisfying $\alpha\overline{\alpha} + \beta\overline{\beta} = \alpha^{1+q} + \beta^{1+q} = 1$), which are sufficient to generate the full group $\mathrm{SU}_4(q)$. Again by comparing orders we obtain $\mathrm{PSU}_4(q) \cong \mathrm{P\Omega}_6^-(q)$, for all q.

3.12 Exceptional covers and isomorphisms

The rest of this chapter is concerned with what might be termed 'sporadic' behaviour of classical groups, that is, special properties of individual groups that are different from the generic properties of the families of groups. Most, if not all, of this material is important for the study of sporadic simple groups, and some of it will also be used in Chapter 4 for constructing exceptional groups of Lie type.

We shall sketch constructions of some of the exceptional covers, and prove all the remaining isomorphisms among classical groups. We have already proved $\mathrm{PSL}_2(4) \cong \mathrm{PSL}_2(5) \cong A_5$ and $\mathrm{PSL}_2(9) \cong A_6$ in Section 3.3.5 and $\mathrm{Sp}_4(2) \cong S_6$ in Section 3.5.2, so we shall begin with the isomorphisms $A_8 \cong \mathrm{PSL}_4(2)$ and $\mathrm{PSL}_2(7) \cong \mathrm{PSL}_3(2)$. Then we use the 'hexacode' and the construction of the spin groups to construct $3^2 {\cdot} \mathrm{PSU}_4(3)$ and $4^2 {\cdot} \mathrm{PSL}_3(4)$. We use the Weyl groups of types E_6, E_7 and E_8 to construct the exceptional covers $2 {\cdot} \mathrm{PSU}_4(2)$, $2 {\cdot} \mathrm{Sp}_6(2)$ and $2^2 {\cdot} \Omega_8^+(2)$, and (using E_6) to prove that $\mathrm{PSp}_4(3) \cong \mathrm{PSU}_4(2)$.

The other exceptional covers $2^2 {\cdot} \mathrm{PSU}_6(2)$ and $3 {\cdot} \mathrm{P\Omega}_7(3)$ are unfortunately out of our reach at this stage, although they are of crucial importance when it comes to studying the sporadic Fischer groups (see Section 5.7). Indeed, our construction of Fi_{22} in that section contains a proof of the existence of $2^2 {\cdot} \mathrm{PSU}_6(2)$, although one might prefer a more direct proof.

3.12.1 Isomorphisms using the Klein correspondence

To show that $A_8 \cong \Omega_6^+(2)$ (and hence by the Klein correspondence $A_8 \cong \mathrm{PSL}_4(2)$), consider the permutation representation of A_8 written over the field of order 2. There is a natural inner product $\sum_{i=1}^8 x_i y_i$ on this space, and the vector $v_0 = (1,1,1,1,1,1,1,1)$ and its orthogonal complement are fixed by the action of A_8. Thus there is an induced action of A_8 (indeed S_8) on the 6-space $v_0^\perp / \langle v_0 \rangle$. This space not only inherits the inner product, but also supports a quadratic form under which the vectors with 2 (or 6) non-zero coordinates have norm 1, while those with 4 non-zero coordinates have norm 0. This quadratic form is non-singular of plus type as the images of the vectors $(1,1,1,1,0,0,0,0)$, $(1,1,0,0,1,1,0,0)$, and $(1,0,1,0,1,0,1,0)$ span a totally isotropic 3-space. Therefore S_8 is a subgroup of $\mathrm{GO}_6^+(2)$ and by comparing orders we have $S_8 \cong \mathrm{GO}_6^+(2)$ and $A_8 \cong \Omega_6^+(2)$. (An alternative proof of the isomorphism $A_8 \cong \mathrm{PSL}_4(2)$ is given in Section 5.2.6, using the Mathieu group M_{24}.)

The isomorphism $\mathrm{PSL}_2(7) \cong \mathrm{PSL}_3(2)$ can be seen in the same setting, by labelling the coordinates with the 8 points $\infty, 0, \ldots, 6$ of the projective line over $\mathbb{F}_7$. Any such labelling will do. For example if we label the coordinates in the order ∞, 0, 1, 3, 2, 6, 4, 5 then the above totally isotropic 3-space is invariant under the action of $\mathrm{PSL}_2(7)$ permuting the coordinates. Alternatively, you may prefer the order ∞, 0, 1, 2, 3, 4, 5, 6, in which case $\mathrm{PSL}_2(7)$ acts on the 3-space spanned by $(1, 1, 1, 0, 1, 0, 0, 0)$ (modulo complementation) and its images under rotations of the last 7 coordinates. It is then easy to check that the resulting homomorphism $PSL_2(7) \to PSL_3(2)$ is onto, and since the groups have the same order they are isomorphic.

3.12.2 Covering groups of $PSU_4(3)$

Next we sketch a construction of $3 \cdot PSU_4(3)$, from which we obtain $3^2 \cdot PSU_4(3)$ by applying an automorphism of $PSU_4(3)$ which does not fix the triple cover. Indeed, we first construct $6 \cdot PSU_4(3)$, and then take the quotient by the central involution to obtain $3 \cdot PSU_4(3)$. Recall the construction in Sections 2.7.3 and 2.7.4 of $3 \cdot A_6$ and $3 \cdot A_7$ as groups of symmetries of sets of vectors of shape $(2, 0, 0, 0, 0, 0)$ and $(0, 0, 1, 1, 1, 1)$ and $(0, 1, 0, 1, \omega, \overline{\omega})$. Now extend this set of vectors by allowing sign changes on the 6 coordinates independently. Thus the group of monomial symmetries of this enlarged set Ω of vectors is a group of shape $2^6 : 3 \cdot A_6 \cong 6 \cdot (2^5 : A_6)$. There are two orbits of this group on Ω: up to scalars there are 6 vectors of shape $(2, 0, 0, 0, 0, 0)$ and 120 of shape $(0, 0, 1, 1, 1, 1)$. We also see a group $3 \cdot A_7$, and in fact the groups $6.2^5 A_6$ and $3 \cdot A_7$ together generate $6 \cdot PSU_4(3).2$. However, we do not use this, and instead show directly that the full group of symmetries of Ω is $6 \cdot PSU_4(3).2$.

The (complex) reflection in a vector v is defined by the formula

$$x \mapsto x - 2 \frac{x.v}{v.v} v, \tag{3.64}$$

in which $x.v$ denotes the complex inner product $\sum x_i \overline{v_i}$. [The complex reflections defined here have order 2. Often a more general definition of a complex reflection of order n is used, which is given by the formula

$$x \mapsto x - (1 - \lambda) \frac{x.v}{v.v} v \tag{3.65}$$

where λ is a primitive nth root of unity.]

It is straightforward to verify that Ω is invariant under reflection in $(0, 0, 1, 1, 1, 1)$. [By symmetry it is only necessary to look at the images of $(0, 0, 2, 0, 0, 0)$, $(0, 0, -1, 1, 1, 1)$, $(1, 1, 1, 1, 0, 0)$ and $(0, 1, 0, -1, \omega, \overline{\omega})$, which we leave as an exercise for the reader.] Hence the set Ω is invariant under reflection in any element of Ω. Moreover, the group of symmetries is transitive on the 126 vectors (counting modulo scalars). Indeed, it is transitive on pairs of mutually orthogonal vectors (up to scalars), and even on triples of mutually orthogonal vectors (again up to scalars). Finally, each such triple extends to a unique coordinate frame of six mutually orthogonal vectors, and therefore the group of symmetries is transitive on these coordinate frames.

Now the number of these (ordered) triples is $126.45.12$ so the number of coordinate frames is $126.45.12/6.5.4 = 567$ and hence the order of the symmetry group is $567.2^6.3.360 = 2^9.3^7.5.7$. Taking the subgroup of index 2 consisting of the elements of determinant 1, and factoring out by the scalar group of order 6, we obtain a group of order $2^7.3^6.5.7 = 3265920$ which we shall show is isomorphic to $\text{PSU}_4(3)$ (or rather $\text{P}\Omega_6^-(3)$, which by the Klein correspondence (Section 3.11.2) is the same thing).

For if we take Λ to be the $\mathbb{Z}[\omega]$-span of the set Ω of vectors, and let $\theta = \omega - \overline{\omega} = \sqrt{-3}$, then as an additive group the quotient $\Lambda/\theta\Lambda$ is elementary abelian of order 3^6. Therefore it is naturally a 6-dimensional vector space over $\mathbb{F}_3$, and we may as well take as basis vectors the images modulo θ of the vectors of shape $(-2, 0^5)$ in Λ. With respect to this basis the complex inner product on Λ reduces to the symmetric bilinear form $\sum x_i y_i$, and the norm on Λ reduces to the quadratic form $\sum x_i^2$, which is of minus type (see Section 3.4.6). The set Ω of vectors maps to the set of all vectors of norm 1, that is vectors of shape $(\pm 1, 0^5)$ or $(\pm 1^4, 0^2)$, and the complex reflections induce $\mathbb{F}_3$-reflections in all these vectors. Moreover, the only symmetries of Λ which act trivially on $\Lambda/\theta\Lambda$ are the scalars $1, \omega, \overline{\omega}$, so the image of our reflection group in $\text{GO}_6^-(3)$ is a group of order $2^8.3^6.5.7$, and therefore of index 2. The claim follows.

A similar analysis of $\Lambda/2\Lambda$ reveals that $\Lambda/2\Lambda$ is naturally a 6-dimensional vector space over $\mathbb{F}_4$, where the complex scalars $\omega, \overline{\omega}$ map to $\omega, \overline{\omega} \in \mathbb{F}_4$ (hence our use of the same symbols for both concepts). The complex inner product then reduces to a non-singular conjugate-symmetric sesquilinear form over $\mathbb{F}_4$, so that the image group is a subgroup of $\text{SU}_6(2)$. Since only the scalars ± 1 act trivially on $\Lambda/2\Lambda$, we obtain an embedding of $3\cdot\text{P}\Omega_6^-(3){:}2$ into $\text{SU}_6(2)$, and hence an embedding of $\text{P}\Omega_6^-(3){:}2$ in $\text{PSU}_6(2)$, as a subgroup of index 1408. In this case the complex reflections induce (some of the) unitary transvections.

The Atlas [28] describes several different bases for the 6-dimensional representation of $6\cdot\text{PSU}_4(3).2$, exhibiting various different subgroups. In order to simplify the construction of $6\cdot\text{PSL}_3(4)$ in the next section, we briefly describe one more of these. We leave it as Exercise 3.40 to show that these constructions are equivalent.

In this case the monomial subgroup has the shape $6.2^4.S_6$, generated by scalars, sign-changes on an even number of coordinates, and all coordinate permutations. The 126 distinguished 1-spaces are spanned by $(2, 2, 0, 0, 0, 0)$ and $(\theta, 1, 1, 1, 1, 1)$ and their images under the monomial group. As before, the group is generated by the reflections in these vectors. This time when we reduce modulo θ, taking as basis the vectors of shape $(4, 0^5)$, we find that the distinguished vectors map to all the vectors of norm 2 over $\mathbb{F}_3$, that is the vectors of shape $(\pm 1^2, 0^4)$ and $(\pm 1^5, 0)$. More about this complex reflection group can be found in Section 5.7.2, where it is used in the construction of the Fischer group Fi_{22}.

3.12.3 Covering groups of $\text{PSL}_3(4)$

To see the subgroup $6\cdot\text{PSL}_3(4)$ of $6\cdot\text{PSU}_4(3)$, we fix a suitable partition of the 126 1-spaces given above into 21 'coordinate frames', each frame consisting of 6 mutually orthogonal 1-spaces. This is easiest to do in the second basis, given at the end of Section 3.12.2. First pick a subgroup $6.2^4.\text{PSL}_2(5)$ of the monomial group $6.2^4.S_6$, so that we can label the 6 coordinates with the points of the projective line $\{\infty, 0, 1, 2, 3, 4\}$ (in any order we like—so let us take this order). Now any pair of points determines a syntheme (a partition of the points into three pairs), so any vector of shape $(2, 2, 0, 0, 0, 0)$ determines six such vectors, up to scalar multiplications, forming an orthogonal basis. Since there are 30 such 1-spaces altogether, they form five coordinate frames.

Similarly, the other 96 1-spaces form 16 coordinate frames, each fixed by a complementary $PSL_2(5)$. However, we have to be careful here because there are four conjugacy classes of $PSL_2(5)$ in the monomial group $6.2^4.PSL_2(5)$, and we have to get the right one! In particular, we cannot use the $PSL_2(5)$ consisting of pure permutations. Instead, take the $PSL_2(5)$ generated by $(0, 1, 2, 3, 4)$ and the product of $(\infty, 0)(1, 4)$ with the sign-change on coordinates 2 and 3. This fixes the coordinate frame consisting of $(\theta, 1, 1, 1, 1, 1)$ and the images of $(1, \theta, 1, -1, -1, 1)$ under rotation of the last five coordinates. The other 15 coordinate frames are obtained by even sign-changes from this one.

Looking at the coordinate frames of vectors of shape $(2, 2, 0, 0, 0, 0)$ we see that the product of the reflections in two vectors in a frame is an element of our monomial group $6.2^4.PSL_2(5)$. So if we do the same for the other coordinate frames, we will obtain generators for $6.PSL_3(4)$. To prove this, we need to verify that such elements do indeed preserve the given coordinate frames. By symmetry, it is enough to consider the product of reflections in the two vectors $(\theta, 1, 1, 1, 1, 1)$ and $(1, \theta, 1, -1, -1, 1)$. This calculation is left to the reader.

Finally, we need to show that the 21 coordinate frames have the structure of a projective plane of order 4. The line through two points in the projective plane contains three more points, and the corresponding coordinate frames are obtained by applying the above operation defined by one coordinate frame, to the other. That is, pick two vectors in one frame, and apply the product of the reflections in these vectors to the other frame. Again, there is quite a lot of calculation involved in checking these assertions.

In particular we have an embedding of $2 \cdot PSL_3(4)$ into $2 \cdot PSU_4(3) \cong \Omega_6^-(3)$. This gives rise to an embedding of $4 \cdot PSL_3(4)$ in the corresponding spin group, which is isomorphic to $SU_4(3)$. (Note that the central involution lifts to an element of order 4 in the spin group.) Now we observe that this covering group $4 \cdot PSL_3(4)$ is not preserved by a diagonal automorphism of $PSL_3(4)$ (of order 3), and hence we get another (isomorphic) 4-fold cover. Putting these together we obtain $4^2 \cdot PSL_3(4)$.

The sextuple cover $6 \cdot PSL_3(4)$ is also visible inside the exceptional double cover $2 \cdot G_2(4)$, which is described in Section 5.6.8, using a quaternionic extension of the above construction. Note that this allows us to see both bases simultaneously: the basis described in this section is obtained by taking the vectors in (5.66) whose coordinates lie in $\mathbb{Z}[\omega]$, while the basis described in Section 3.12.2 is obtained by taking the vectors whose coordinates lie in $(1 + i)\mathbb{Z}[\omega]$.

3.12.4 The exceptional Weyl groups

In Chapter 4 we shall need a lot of detailed information about the exceptional Weyl groups, namely those of types G_2, F_4, E_6, E_7 and E_8, which were introduced briefly in Section 2.8.4. We collect some of this information here for convenience. Since Weyl groups are reflection groups, and reflections are orthogonal transformations of real Euclidean space, it is hardly surprising that orthogonal groups turn up again in this context.

The Weyl group of type G_2 is the group of symmetries of a regular hexagon, and is isomorphic to D_{12}. It is generated by reflections in two vectors at an angle of $5\pi/6$ to each other. One of these is a short root, and the other is a long root, whose length is $\sqrt{3}$ times that of the short roots. The six short roots are the vertices of the hexagon; the six long roots are twice the midpoints of the edges (i.e. the sums of two short roots adjacent on the hexagon).

The Weyl group of type F_4 is generated by the four reflections in the vectors $(-1,1,0,0)$, $(0,-1,1,0)$, $(0,0,-1,0)$ and $(\frac{1}{2},\frac{1}{2},\frac{1}{2},\frac{1}{2})$, corresponding to the nodes of the Dynkin diagram below.

$$(-1,1,0,0) \quad (0,-1,1,0) \quad (0,0,-1,0) \quad (\tfrac{1}{2},\tfrac{1}{2},\tfrac{1}{2},\tfrac{1}{2})$$

$$\bullet\!\!-\!\!-\!\!-\!\!-\!\!\bullet\!\!\Longrightarrow\!\!\bullet\!\!-\!\!-\!\!-\!\!-\!\!\bullet \tag{3.66}$$

The 24 long roots are all the vectors of shape $(\pm 1, \pm 1, 0, 0)$, and the 24 short roots are 8 of shape $(\pm 1, 0, 0, 0)$ and 16 of shape $(\pm\frac{1}{2}, \pm\frac{1}{2}, \pm\frac{1}{2}, \pm\frac{1}{2})$. If we reduce the coordinates modulo 3 we see that the generating reflections of $W(F_4)$ map to all the generating reflections of $GO_4^+(3)$, and therefore $W(F_4) \cong GO_4^+(3)$. This interpretation shows that $W(F_4)$ has the structure $(2 \cdot A_4 \circ 2 \cdot A_4):2^2$. This orthogonal group has an outer automorphism which multiplies the quadratic form by -1, and this can be lifted to an 'automorphism' of F_4 which swaps the long roots with the short roots, and which is crucial in the construction of the exceptional groups of type $^2F_4$ (see Section 4.9). Of course, this automorphism cannot be realised by an isometry of the Euclidean space, but it may be realised by applying a matrix such as

$$\begin{pmatrix} 1 & 1 & 0 & 0 \\ -1 & 1 & 0 & 0 \\ 0 & 0 & 1 & 1 \\ 0 & 0 & -1 & 1 \end{pmatrix}$$

and then dividing the long roots by 2 so they become short roots. Indeed, it is not uncommon to coordinatise F_4 so that the short roots have shape $(\pm 1, \pm 1, 0, 0)$ and the long roots have shape $(\pm 2, 0, 0, 0)$ and $(\pm 1, \pm 1, \pm 1, \pm 1)$.

Note that the subgroup of $W(F_4)$ generated by reflections in the short roots only (or the long roots only) is isomorphic to $W(D_4)$, and the quotient $W(F_4)/W(D_4) \cong S_3$. Indeed, these two normal subgroups isomorphic to $W(D_4)$

intersect in an extraspecial group 2_+^{1+4}, and $W(\mathrm{F}_4)$ has the shape $2_+^{1+4}{:}(S_3 \times S_3)$.

We treat the Weyl groups of types E_6, E_7, E_8 together. There are various different coordinate systems in common use, each of which has its own advantages and disadvantages. The easiest to describe is the one in which the reflecting vectors are $\binom{8}{2}.2^2 = 112$ of shape $(\pm 1, \pm 1, 0, 0, 0, 0, 0, 0)$ and $2^7 = 128$ of shape $(\pm\tfrac{1}{2}, \pm\tfrac{1}{2}, \pm\tfrac{1}{2}, \pm\tfrac{1}{2}, \pm\tfrac{1}{2}, \pm\tfrac{1}{2}, \pm\tfrac{1}{2}, \pm\tfrac{1}{2})$ with an odd number of $+$ signs. (Alternatively, take the ones with an even number of $+$ signs.) We can then label the 8 nodes of the Dynkin diagram with the roots as follows:

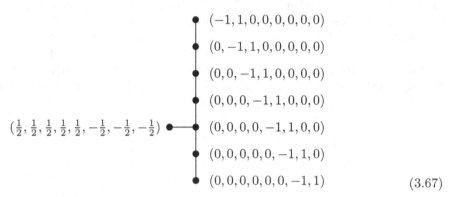

$$\begin{aligned}
&\bullet \quad (-1, 1, 0, 0, 0, 0, 0, 0)\\
&\bullet \quad (0, -1, 1, 0, 0, 0, 0, 0)\\
&\bullet \quad (0, 0, -1, 1, 0, 0, 0, 0)\\
&\bullet \quad (0, 0, 0, -1, 1, 0, 0, 0)\\
(\tfrac{1}{2}, \tfrac{1}{2}, \tfrac{1}{2}, \tfrac{1}{2}, \tfrac{1}{2}, -\tfrac{1}{2}, -\tfrac{1}{2}, -\tfrac{1}{2}) \;\bullet\!\!-\!\!\bullet \quad &(0, 0, 0, 0, -1, 1, 0, 0)\\
&\bullet \quad (0, 0, 0, 0, 0, -1, 1, 0)\\
&\bullet \quad (0, 0, 0, 0, 0, 0, -1, 1)
\end{aligned}$$
(3.67)

Another basis is described in Section 4.4.2, in which there are 16 roots of shape $(\pm 1, 0, 0, 0, 0, 0, 0, 0)$ and 224 of shape $(\pm\tfrac{1}{2}, \pm\tfrac{1}{2}, \pm\tfrac{1}{2}, \pm\tfrac{1}{2}, 0, 0, 0, 0)$.

Now consider the root lattice modulo 2: that is, we take all $\mathbb{Z}$-linear combinations of the roots, and define two such vectors to be equivalent if their difference is twice a lattice vector. The Euclidean norm (divided by 2) gives a non-degenerate quadratic form on the resulting 8-dimensional $\mathbb{F}_2$-vector space. The subspace consisting of the images of $(2, 0, 0, 0, 0, 0, 0, 0)$ and $(1, 1, 1, 0, \pm 1, 0, 0, 0)$, together with rotations of the last seven coordinates, is a totally isotropic 4-space, so this quadratic form is of plus type. Moreover, the 240 reflecting vectors map to the 120 vectors of norm 1, and the reflections induce orthogonal transvections in these 120 vectors. Thus the image of $W(\mathrm{E}_8)$ under reduction modulo 2 is $\mathrm{GO}_8^+(2)$. The kernel of this action is just $\{\pm 1\}$, and we have that $W(\mathrm{E}_8) \cong 2{\cdot}\mathrm{GO}_8^+(2) \cong 2{\cdot}\Omega_8^+(2){:}2$.

This gives us a double cover $2{\cdot}\Omega_8^+(2)$, acting as an orthogonal group in characteristic 0. Either by lifting to the characteristic 0 spin group, using a Clifford algebra construction analogous to that given in Section 3.9, or by first reducing modulo an odd prime p and then using the p-modular Clifford algebra, we obtain the full cover of shape $2^2{\cdot}\Omega_8^+(2)$. (Another approach is to use triality, as in Section 4.7.2.)

Similarly we find that $W(\mathrm{E}_7) \cong 2 \times \Omega_7(2) \cong 2 \times \mathrm{Sp}_6(2)$, represented as a 7-dimensional orthogonal group in characteristic 0 or in characteristic p for any odd prime p, and lifting to the spin group again gives us a construction of

$2 \cdot \mathrm{Sp}_6(2)$. Also $W(\mathrm{E}_6) \cong \mathrm{GO}_6^-(2) \cong \mathrm{PSU}_4(2){:}2$, and lifting to the spin group gives us $2 \cdot \mathrm{PSU}_4(2)$.

Finally we take $W(\mathrm{E}_6)$ and reduce modulo 3. If v and w are adjacent nodes in the diagram, i.e. $f(v, w) = -1$, then reflection in v maps w to $w + v$. If the vectors corresponding to the nodes of the diagram are a, b, c, d, e, f, as follows,

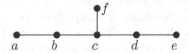

then it is easy to check that the vector $a - b + d - e$ is fixed modulo 3 by all the fundamental reflections. Since this vector is non-isotropic, its perpendicular space is non-singular, so there is a natural map $W(\mathrm{E}_6) \to \mathrm{GO}_5(3)$. We may readily check that this map is onto a subgroup isomorphic to $\mathrm{SO}_5(3)$ and its kernel is trivial, so that $W(\mathrm{E}_6) \cong \mathrm{SO}_5(3)$ and therefore

$$\mathrm{PSU}_4(2) \cong \Omega_6^-(2) \cong W(\mathrm{E}_6)' \cong \mathrm{P}\Omega_5(3) \cong \mathrm{PSp}_4(3). \qquad (3.68)$$

Further reading

The classic text on the classical groups is L. E. Dickson's 1901 book 'Linear groups with an exposition of the Galois field theory' [48], but of course it is rather dated. An often cited reference is Dieudonné's book 'La géométrie des groupes classiques' [52], which gives a concise and elegant treatment. It does not claim to give detailed proofs for all the results, but the 3rd edition does include a full discussion of the automorphism groups of the classical groups. More recent books which I have consulted, and where the reader may find some of the details which I have perhaps glossed over, include Don Taylor's book 'The geometry of the classical groups' [162], which is a thorough treatment of all the classical groups, and concludes with a construction of the Suzuki groups, as well as Peter Cameron's unpublished 'Notes on classical groups' [20] which emphasise the projective geometries and projective groups, and Larry Grove's book 'Classical groups and geometric algebra'. This last is written at a fairly elementary level and does not cover as much material as is in this Chapter, though it does discuss Clifford algebras in characteristic 2, and uses them to give a nice definition of the quasideterminant. Another useful book is Alperin and Bell's 'Groups and representations' [2]. The 'Atlas of finite groups' [28] is a good reference for the basic facts, uncluttered by any proofs.

For background in representation theory, which is needed in some of the later sections in this chapter, the reader wishing for an elementary but reasonably comprehensive introductory text will not be disappointed by James and Liebeck 'Representations and characters of groups' [98]. At a higher level of sophistication, I recommend Serre's 'Représentations linéaires des groupes finis' [152], or the English translation of the second edition. A more comprehensive treatment is given by Curtis and Reiner [47], but perhaps less daunting for the novice is Dornhoff [54]. The representation theory of the general linear groups is treated in the seminal work of Green 'Polynomial representations of GL_n' [68]. Feit's 'The representation theory of finite groups' is a comprehensive treatment of the state of the art of modular representation theory in the early 1980s.

The book 'The subgroup structure of the finite classical groups', by Kleidman and Liebeck [108], contains a comprehensive and detailed account of the Aschbacher–Dynkin theorem and its consequences for the subgroup structure of the classical groups over finite fields. It also contains a concise general introduction to the classical groups. For an approach to classical groups via Lie algebras, where the notion of BN-pair has its natural home, see Carter [21]. More abstractly, this leads on to the theory of buildings, see for example Brown [16] or Ronan [147]. Closely related to this is the theory of generalised polygons (roughly speaking, these are buildings of rank 2), for which see van Maldeghem's 'Generalized polygons' [168] and 'Moufang polygons' [167] by Tits and Weiss. In another direction, the recent book by Geck [65] approaches the finite classical and exceptional groups from the algebraic group point of view.

Exercises

3.1. Compute the addition and multiplication tables for the fields

(i) $\mathbb{F}_4 = \mathbb{F}_2[x]/(x^2 + x + 1)$;
(ii) $\mathbb{F}_8 = \mathbb{F}_2[x]/(x^3 + x + 1)$;
(iii) $\mathbb{F}_9 = \mathbb{F}_3[x]/(x^2 + 1)$.

In each case compute the Frobenius automorphism as a permutation of the elements.

3.2. Let $G = GL_n(q)$. Prove that $Z(G) = \{\lambda I_n \mid 0 \neq \lambda \in \mathbb{F}_q\}$, where I_n is the $n \times n$ identity matrix.

3.3. Prove the following variant of Iwasawa's Lemma: Suppose that G is a finite perfect group acting faithfully and primitively on a set Ω, and suppose that the stabiliser of a point has a normal soluble subgroup S, whose conjugates generate G. Then G is simple.

3.4. Use Iwasawa's Lemma to prove simplicity of the alternating groups A_n for $n \geqslant 5$. Where does your proof break down for $n \leqslant 4$?

[Hint: use the action of A_n on unordered triples from $\{1, 2, \ldots, n\}$, unless $n = 6$.]

3.5. Let A be a $k \times k$ matrix, B a $k \times m$ matrix and C an $m \times m$ matrix over a field F. Show that $\det \begin{pmatrix} A & 0 \\ B & C \end{pmatrix} = \det A . \det C$.

Deduce that the group G of invertible matrices of the shape $\begin{pmatrix} A & 0 \\ B & C \end{pmatrix}$ has a normal subgroup $Q = \left\{ \begin{pmatrix} I_k & 0 \\ B & I_m \end{pmatrix} \right\} \cong F^{km}$ and a subgroup $L = \left\{ \begin{pmatrix} A & 0 \\ 0 & C \end{pmatrix} \right\} \cong \mathrm{GL}_k(F) \times \mathrm{GL}_m(F)$.

Show also that G is a semidirect product of Q and L.

3.6. How many k-dimensional subspaces are there in a vector space of dimension n over the field of q elements?

3.7. Prove that the maximal parabolic subgroups of $\mathrm{GL}_n(q)$, as defined in Section 3.3.3, are maximal subgroups of $\mathrm{GL}_n(q)$.

3.8. Show that every parabolic subgroup in $\mathrm{GL}_n(q)$ is a semidirect product of a p-group by $\mathrm{GL}_{n_1}(q) \times \cdots \times \mathrm{GL}_{n_r}(q)$, where $n_1 + \cdots + n_r = n$.

3.9. Prove that $\Sigma\mathrm{L}_n(q)$ as defined in Section 3.3.4 is well-defined up to isomorphism.

3.10. Verify that conjugation by $\begin{pmatrix} 0 & 1 \\ -1 & 0 \end{pmatrix}$ maps every element of $\mathrm{SL}_2(q)$ to its inverse-transpose.

3.11. Show that $\mathrm{PSL}_2(q)$ is generated by the maps $z \mapsto z + 1$, $z \mapsto \lambda^2 z$ and $z \mapsto -1/z$ on $\mathrm{PL}(q)$, where λ is a generator for the multiplicative group of the field $\mathbb{F}_q$.

3.12. (i) Perform the calculations required in the proofs that $\mathrm{PSL}_2(5) \cong A_5$ and $\mathrm{PSL}_2(9) \cong A_6$ in Section 3.3.5.

(ii) Perform the calculations required in the construction of the actions of $\mathrm{PSL}_2(11)$ on 11 points in Section 3.3.5.

3.13. Show that in $V = \mathbb{F}_4{}^3$ there are 21 1-spaces ('points') and 21 2-spaces ('lines'), that every line contains 5 points and that every point lies in 5 lines.

3.14. Show that both A_8 and $\mathrm{PSL}_3(4)$ have order 20160, but that they are not isomorphic.

3.15. Show that if f is any bilinear or sesquilinear form on a vector space V, and $S^\perp = \{v \in V \mid f(u, v) = 0 \text{ for all } u \in S\}$, then $S^\perp$ is a subspace of V.

3.16. Let H be a Hermitian form on a vector space V over the field $\mathbb{F}_{q^2}$, and let $\overline{}$ denote the field automorphism $\lambda \mapsto \lambda^q$. Use the expansion in (3.18) of $H(u + v)$ and $H(u + \lambda v)$, where $\lambda \neq \overline{\lambda}$, to derive a formula for f in terms of H.

3.17. Show that if f is a non-singular bilinear or sesquilinear form, and U is a subspace of V, then $(U^\perp)^\perp = U$ and $\dim(U) + \dim(U^\perp) = \dim(V)$. Deduce that if $U \cap U^\perp = 0$ then $V = U \oplus U^\perp$.

3.18. Let f be a non-singular alternating bilinear form on a vector space V of dimension $2m$ over $\mathbb{F}_q$. If $k \leqslant m$, how many non-singular subspaces of dimension $2k$ are there in V? How many totally isotropic subspaces of dimension k are there?

3.19. Verify that S_6 and $\mathrm{Sp}_4(2)$ have the same order.

3.20. Show that the symplectic transvections $T_v(\lambda) : x \mapsto x + \lambda f(x, v)v$ preserve the alternating bilinear form f.

3.21. Verify that the symplectic transvections are commutators in $\mathrm{Sp}_4(3)$ and $\mathrm{Sp}_6(2)$.

3.22. Show that the transvections $T_v(\lambda) : x \mapsto x + \lambda f(x, v)v$ preserve the non-singular conjugate-symmetric sesquilinear form f if and only if $\lambda^{q-1} = -1$.

3.23. Let V be a 3-dimensional space over $\mathbb{F}_4 = \{0, 1, \omega, \omega^2\}$, and let f be a non-singular conjugate-symmetric sesquilinear form on V. Show that there are 21 one-dimensional subspaces of V, of which 9 contain isotropic vectors and 12 contain non-isotropic vectors.

3.24. From the previous question we get an action of $\mathrm{GU}_3(2)$ (and $\mathrm{PGU}_3(2)$) on the set of 9 isotropic 1-spaces in V. Show that this action is 2-transitive, and that $|\mathrm{PGU}_3(2)| = 216$.
   Deduce (from the O'Nan–Scott theorem, or otherwise) that the resulting subgroup of A_9 is the 'affine' subgroup $(C_3 \times C_3){:}\mathrm{SL}_2(3)$.

3.25. Prove the equivalence of the following alternative definitions of the Frattini subgroup $\Phi(G)$.

 (i) $\Phi(G)$ is the intersection of all the maximal subgroups of G.
(ii) $\Phi(G)$ is the set of non-generators of G, where $x \in G$ is a *non-generator* if for all subsets $X \subseteq G$, $\langle X \rangle = G \Rightarrow \langle X \setminus \{x\} \rangle = G$.

3.26. Prove that $\Phi(G)$ is nilpotent.

3.27. Prove that if G is nilpotent then $\Phi(G) \geqslant G'$.

3.28. Show that the 2-dimensional orthogonal groups are dihedral: $\mathrm{GO}_2^+(q) \cong D_{2(q-1)}$ and $\mathrm{GO}_2^-(q) \cong D_{2(q+1)}$, both for q odd and q even.

3.29. Show that the long root subgroup $\{T_{u,v}(\lambda) \mid \lambda \in F\}$ in an orthogonal group in odd characteristic depends only on the totally isotropic subspace $\langle u, v \rangle$ and not on the vectors u, v.

3.30. Prove the isomorphism $\Omega_{2m+1}(q) \cong \mathrm{Sp}_{2m}(q)$ for q even.

3.31. Prove Witt's Lemma for quadratic forms in characteristic 2 (see Section 3.4.7).

3.32. Prove that the group generated by the orthogonal transvections in $\mathrm{GO}^{\varepsilon}_{2n}(q)$, q even, is transitive on the vectors of norm 1, except in the case $n = 2$, $q = 2$. Deduce, by induction on n, that these transvections generate $\mathrm{GO}^{\varepsilon}_{2n}(q)$.

3.33. Prove that if q is even and $n \geqslant 3$ then $\Omega^{\varepsilon}_{2n}(q)$ is simple.

3.34. Show that the centre of the Clifford algebra $C(V)$ has dimension 1 if $\dim V$ is even, and dimension 2 if $\dim V$ is odd.

3.35. Prove that $Q(vw) = Q(v)Q(w)$ inside the Clifford algebra $C(V)$, for all $v, w \in V$. Generalise to $Q(xy) = Q(x)Q(y)$ for all $x, y \in C(V)$.

3.36. Compute the Clifford algebra $C(V, Q)$ where $\dim V = 4$ and Q is a quadratic form of plus type, as an algebra of 4×4 matrices.

Find non-singular subspaces W of dimension 3 and X of dimension 2 and minus type, inside V, and compute $C(W)$ and $C(X)$ as subalgebras of $C(V)$.

3.37. Write down (complex) 2×2 matrices generating (i) D_8 and (ii) Q_8. Hence obtain 4×4 matrices generating (iii) 2^{1+4}_+ and (iv) 2^{1+4}_-.

Now do the same over $\mathbb{F}_q$, for $q \equiv 1 \bmod 4$, and then for $q \equiv 3 \bmod 4$.

3.38. Find 3×3 matrices (i) over $\mathbb{F}_7$ generating 3^{1+2}_+, and (ii) over $\mathbb{F}_{19}$ generating 3^{1+2}_-.

3.39. By interpreting 0, 1, ω and $\overline{\omega}$ as the elements of the field of order 4 (i.e. by reducing $\mathbb{Z}[\omega]$ modulo 2), show that $3 \cdot A_6$ is a subgroup of $\mathrm{SU}_6(2)$. Show also that the set of 45 vectors given in Exercise 2.39 span a 3-dimensional subspace modulo 2, and deduce that $3 \cdot A_6$ embeds in $\mathrm{SL}_3(4)$.

3.40. By finding an explicit base-change matrix, or otherwise, show that the two sets of 6×126 vectors constructed in Section 3.12.2 are isometric to each other in $\mathbb{C}^6$ with its usual inner product (up to an overall scale factor).

3.41. Complete the construction of $6 \cdot \mathrm{PSL}_3(4)$ by computing the 21 coordinate frames mentioned in Section 3.12.3 and defining the projective plane structure.

3.42. Show that the short roots of the G_2 root system form a root system of type A_2. Do the same for the long roots.

3.43. Show that the short roots of the F_4 root system form a root system of type D_4. Do the same for the long roots.

Find a labelling of the D_4 diagram with short roots of F_4, and find three reflections in $W(\mathsf{F}_4)$ which generate the S_3 of diagram symmetries.

4

The exceptional groups

4.1 Introduction

It is the aim of this chapter to describe the ten families of so-called 'exceptional groups of Lie type'. There are three main ways to approach these groups. The first is via Lie algebras, as is wonderfully developed in Carter's book [21]. The second, more modern, approach is via algebraic groups (see for example Geck's book 'Introduction to algebraic geometry and algebraic groups' [65]). The third is via 'alternative' algebras, as in 'Octonions, Jordan algebras and exceptional groups' by Springer and Veldkamp [158]. I shall adopt the 'alternative' approach, for a number of reasons: although it lacks the elegance and uniformity of the other approaches, it gains markedly when it comes to performing concrete calculations. We obtain not only the smallest representations in this way, but also construct the (generic) covering groups, whereas the Lie algebra approach only constructs the simple groups.

Our approach is analogous to studying the special linear group $\mathrm{SL}_n(q)$ via its natural action on the vector space of dimension n, while the Lie algebra approach means studying $\mathrm{PSL}_n(q)$ via its action by conjugation on the $n \times n$ matrices of trace 0 (the Lie algebra, of dimension $n^2 - 1$), and the algebraic group approach studies $\mathrm{SL}_n(q)$ via its defining equations ($\det x = 1$, a polynomial equation of degree n in n^2 variables). The price we pay for this concrete approach is having to study six families of classical groups separately, as well as ten families of exceptional groups.

From the Lie algebra point of view, the exceptional groups are of three different types. Most straightforward are the five families of Chevalley (or untwisted) groups $G_2(q)$, $F_4(q)$, and $E_n(q)$ for $n = 6, 7, 8$. Next in difficulty are the Steinberg–Tits–Hertzig twisted groups $^3D_4(q)$ and $^2E_6(q)$ which also exist for any finite field $\mathbb{F}_q$, and whose construction is analogous to the construction of the unitary groups from the special linear groups. Finally there are the three families of Suzuki and Ree groups $^2B_2(2^{2n+1}) = \mathrm{Sz}(2^{2n+1})$, $^2G_2(3^{2n+1}) = R(3^{2n+1})$ and $^2F_4(2^{2n+1}) = R(2^{2n+1})$, which only exist in one characteristic. In each case the group is not simple if $n = 1$: the derived group $^2F_4(2)'$

R.A. Wilson, *The Finite Simple Groups*,
Graduate Texts in Mathematics 251,
© Springer-Verlag London Limited 2009

has index 2 in $^2F_4(2)$, and is a simple group known as the Tits group, not appearing elsewhere in the classification.

We shall treat these ten families in order of (untwisted) Lie rank, that is, the integer subscript in the name, as the smaller rank cases are generally easier to understand. We begin by giving a more-or-less complete and quite elementary description of the Suzuki groups as groups of 4×4 matrices over fields of characteristic 2, including the exceptional isomorphism $\mathrm{Sz}(2) \cong 5{:}4$. Then in Section 4.3 we describe the octonion algebra (also known as the Cayley algebra) over a field $\mathbb{F}_q$ of odd characteristic, and the automorphism group of the algebra, which is (for our purposes, by definition) $G_2(q)$, and calculate the order of $G_2(q)$. By a change of basis we obtain a definition which also works in characteristic 2, and use this basis to help us describe some of the subgroups. Then in Section 4.4 we prove that $G_2(2) \cong \mathrm{PSU}_3(3){:}2$ by constructing an octonion algebra on the E_8 lattice, and reducing modulo 2. This also provides an alternative definition of octonion algebras in characteristic 2.

The small Ree groups $^2G_2(3^{2n+1})$ are constructed from the octonion algebras in characteristic 3, as groups of 7×7 matrices, in an analogous way to the construction of the Suzuki groups given earlier. This construction includes a proof that $^2G_2(3) \cong \mathrm{PSL}_2(8){:}3$. The twisted groups $^3D_4(q)$ are constructed as the automorphism group of 'twisted' octonion algebras, following Springer and Veldkamp [158].

The fundamental concept of triality is introduced next, showing the relationship of $G_2(q)$ and $^3D_4(q)$ to the orthogonal groups $\mathrm{GO}_7(q)$ and $\mathrm{GO}_8^+(q)$, and leading naturally to the discussion of the exceptional Jordan algebras (also known as Albert algebras), and their automorphism groups $F_4(q)$, in Section 4.8. Here I sketch a proof of the orders of the groups, and of simplicity, and describe some of the maximal subgroups. Not all of the arguments work in the exceptional characteristics 2 and 3, but most can be modified to use the characteristic-free definition in terms of a quadratic form and a cubic form, which then leads to a proof in arbitrary characteristic. The large Ree groups $^2F_4(2^{2n+1})$ are constructed from $F_4(q)$ in an analogous way to the construction of the small Ree groups from $G_2(q)$, though the proofs in this case are even more sketchy.

This leads on to a description in Section 4.10 of the groups of type E_6 as the automorphism groups of a cubic form (the 'determinant'), which is not very different from Dickson's original construction of these groups in 1901. The other twisted groups $^2E_6(q)$ are briefly mentioned.

The chapter concludes with Section 4.12, which contains a brief introduction to Lie algebras and a sketch of a construction of $E_7(q)$ and $E_8(q)$, but with no attempt to prove anything.

4.2 The Suzuki groups

4.2.1 Motivation and definition

To motivate the construction of the Suzuki groups $\mathrm{Sz}(2^{2n+1})$ we revisit the outer automorphism of S_6 (see Section 2.4.2) in the light of the isomorphism $S_6 \cong \mathrm{Sp}_4(2)$ (see Section 3.5.2), and the Klein correspondence (see Section 3.11.2). In the proof of this isomorphism we saw that the permutation representation of S_6 over the field $\mathbb{F}_2$ has a natural symplectic form on it (just the ordinary inner product $\sum_{i=1}^{6} x_i y_i$). The fixed vector $v_0 = (1,1,1,1,1,1)$ gives rise to the 4-space $v_0^{\perp}/\langle v_0 \rangle$ which inherits the symplectic form. Indeed, we can take as symplectic basis the vectors

$$\{e_1 = (1,0,0,0,0,1), \quad f_1 = (1,1,0,0,0,0),$$
$$e_2 = (0,0,1,1,0,0), \quad f_2 = (0,0,0,1,1,0)\}. \tag{4.1}$$

As generators for the group we may take the following elements (where basis vectors not mentioned are fixed):

Transformation	Permutation
$e_1 \mapsto e_1 + f_1$	$(1,2)$
$e_1 \leftrightarrow f_1$	$(2,6)$
$e_2 \mapsto e_2 + f_2$	$(3,4)$
$e_2 \leftrightarrow f_2$	$(3,5)$
$f_1 \mapsto f_1 + e_2, f_2 \mapsto f_2 + e_1$	$(1,6)(2,5)(3,4)$

If we now take the exterior square of this 4-dimensional representation, spanned by the six vectors $e_1 \wedge e_2$, ..., $f_1 \wedge f_2$, there is again a natural symplectic form f (see the Klein correspondence in Section 3.11.2), under which $f(a \wedge b, c \wedge d) = 1$ if and only if $\{a,b,c,d\}$ is a basis. The vector $v_1 = e_1 \wedge f_1 + e_2 \wedge f_2$ is fixed by $\mathrm{Sp}_4(2)$, and the 4-space $v_1^{\perp}/\langle v_1 \rangle$ has a symplectic basis

$$\{e_1^* = e_1 \wedge e_2, \quad f_1^* = f_1 \wedge f_2,$$
$$e_2^* = e_1 \wedge f_2, \quad f_2^* = e_2 \wedge f_1\}. \tag{4.2}$$

(Strictly speaking, instead of saying $u^* = v \wedge w$ we should say that u^* is congruent to $v \wedge w$ modulo $e_1 \wedge f_1 + e_2 \wedge f_2$.)

It follows that the linear map $*$ defined by $e_i \mapsto e_i^*, f_i \mapsto f_i^*$ induces an automorphism of $\mathrm{Sp}_4(2)$, which we shall also denote by $*$. Moreover, the symplectic transvection $f_1 \mapsto f_1 + e_1$ induces $f_1^* \mapsto f_1^* + e_2^*, f_2^* \mapsto f_2^* + e_1^*$, which corresponds under this automorphism to $f_1 \mapsto f_1 + e_2, f_2 \mapsto f_2 + e_1$. Since the latter is not a transvection, the automorphism $*$ of $\mathrm{Sp}_4(2)$ is not inner. It is easy to see that $*$ commutes with the symplectic transformation $e_i \leftrightarrow f_i$ (corresponding to the coordinate permutation $(2,6)(3,5)$), and also with its square root $(2,5,6,3)$ given by

$$e_1 \mapsto e_1 + f_1 + f_2, \qquad f_1 \mapsto f_1 + e_1 + e_2,$$
$$e_2 \mapsto f_2 + e_1 + e_2, \qquad f_2 \mapsto e_2 + f_1 + f_2. \qquad (4.3)$$

We can also check that it commutes with the coordinate permutation $(2,3,4,5,6)$, which corresponds to the transformation

$$e_1 \mapsto f_1, \qquad f_1 \mapsto e_1 + f_1 + f_2,$$
$$e_2 \mapsto f_2, \qquad f_2 \mapsto e_2 + f_1. \qquad (4.4)$$

These two maps generate a group of order 20, which is in fact the full centraliser of the automorphism $*$. This is the group $\mathrm{Sz}(2) \cong 5{:}4$, which we now generalise to $\mathrm{Sz}(2^{2n+1})$ for all non-negative integers n.

We first show that the linear map $**$ commutes with the action of $\mathrm{Sp}_4(2)$ and so acts as the identity automorphism. Since $S_6 \cong \mathrm{Sp}_4(2)$ is generated by the permutations $(1,2)$, $(3,4)$, and $(2,5,6,3)$, it is enough to check that $**$ commutes with the transvections $e_1 \mapsto e_1 + f_1$ and $e_2 \mapsto e_2 + f_2$, in addition to the subgroup $\mathrm{Sz}(2)$ already checked. If $e_1 \mapsto e_1 + f_1$ then

$$
\begin{aligned}
e_1^{**} &= (e_1 \wedge e_2) \wedge (e_1 \wedge f_2) \\
&\mapsto ((e_1 + f_1) \wedge e_2) \wedge ((e_1 + f_1) \wedge f_2) \\
&= (e_1 \wedge e_2 + f_1 \wedge e_2) \wedge (e_1 \wedge f_2 + f_1 \wedge f_2) \\
&= e_1^{**} + f_1^{**} + e_1^* \wedge f_1^* + e_2^* \wedge f_2^* \\
&= e_1^{**} + f_1^{**}, \qquad (4.5)
\end{aligned}
$$

while f_1^{**}, e_2^{**} and f_2^{**} are fixed. Similarly, if $e_2 \mapsto e_2 + f_2$ then $e_2^{**} \mapsto e_2^{**} + f_2^{**}$.

Now if we extend the field from $\mathbb{F}_2$ to $\mathbb{F}_q$, and extend the map $*$ linearly, then the map $**$ commutes with the action of $\mathrm{Sp}_4(2)$ but not with the action of $\mathrm{Sp}_4(q)$. This is because the element $e_1 \mapsto \lambda e_1, f_1 \mapsto \lambda^{-1} f_1$ of $\mathrm{Sp}_4(q)$ maps $e_1^{**} = (e_1 \wedge e_2) \wedge (e_1 \wedge f_2)$ to $(\lambda e_1 \wedge e_2) \wedge (\lambda e_1 \wedge f_2) = \lambda^2 e_1^{**}$. Indeed, it is easy to see that the map $**$ takes the generic vector

$$\lambda_1 e_1 + \lambda_2 e_2 + \mu_1 f_1 + \mu_2 f_2 \mapsto \lambda_1^2 e_1^{**} + \lambda_2^2 e_2^{**} + \mu_1^2 f_1^{**} + \mu_2^2 f_2^{**}, \quad (4.6)$$

and therefore it *is* the field automorphism.

In the case when this field automorphism σ has odd order $k = 2n + 1$, it has a square root $\sigma^{n+1} = \sigma^{-n}$, and therefore the automorphism $* \circ \sigma^n$ squares to the identity. This automorphism of order 2 can be written as

$$\lambda_1 e_1 + \lambda_2 e_2 + \mu_1 f_1 + \mu_2 f_2 \mapsto \lambda_1^* e_1^* + \lambda_2^* e_2^* + \mu_1^* f_1^* + \mu_2^* f_2^*, \quad (4.7)$$

where $\lambda^* = \lambda^{2^{n+1}}$, and its centraliser in $\mathrm{Sp}_4(2^{2n+1})$ is by definition $\mathrm{Sz}(2^{2n+1})$.

Our next task is to determine some information about these groups, which have been defined in a rather abstract way. It is not even straightforward to determine the order of the group.

4.2.2 Generators for Sz(q)

Throughout this section we write $q = 2^{2n+1}$. Already in Sz(2) $\cong$ 5:4 we see that the permutation $(2,3,5,4)$ acts with respect to the ordered basis $\{e_1, e_2, f_2, f_1\}$ as

$$\begin{pmatrix} 1 & 0 & 0 & 0 \\ 1 & 1 & 0 & 0 \\ 1 & 1 & 1 & 0 \\ 1 & 0 & 1 & 1 \end{pmatrix} \tag{4.8}$$

and generates a Sylow 2-subgroup of Sz(2). Now look in Sz(q) to see which diagonal matrices diag($\alpha, \beta, \gamma, \delta$) belong to the group. By (4.7) and (4.2) we have that $\alpha\beta = \alpha^{2^{n+1}}$, $\gamma\delta = \delta^{2^{n+1}}$, $\alpha\gamma = \beta^{2^{n+1}}$ and $\beta\delta = \gamma^{2^{n+1}}$. Hence, once we know α we know also

$$\begin{aligned} \beta &= \alpha^{2^{n+1}-1}, \\ \gamma &= \beta^{2^{n+1}}\alpha^{-1} = \alpha^{-2^{n+1}+1} = \beta^{-1}, \\ \delta &= \gamma^{2^{n+1}}\beta^{-1} = \alpha^{-1}. \end{aligned} \tag{4.9}$$

Conversely, if α, β, γ, δ satisfy these equations, then $* \circ \sigma^n$ does indeed commute with diag($\alpha, \beta, \gamma, \delta$). Therefore the group of diagonal matrices is isomorphic to the multiplicative group of the field, and is generated by

$$\text{diag}(\alpha, \alpha^{2^{n+1}-1}, \alpha^{-2^{n+1}+1}, \alpha^{-1}). \tag{4.10}$$

Together with the lower unitriangular matrix given in (4.8) above, they generate a group of order $q^2(q-1) = 2^{4n+2}(2^{2n+1}-1)$.

To prove this, conjugate the above matrix (4.8) by diag($\alpha, \beta, \beta^{-1}, \alpha^{-1}$), where $\beta = \alpha^{2^{n+1}-1}$, to obtain the matrix

$$\begin{pmatrix} 1 & 0 & 0 & 0 \\ \alpha\beta^{-1} & 1 & 0 & 0 \\ \alpha\beta & \beta^2 & 1 & 0 \\ \alpha^2 & 0 & \alpha\beta^{-1} & 1 \end{pmatrix}, \text{ squaring to } \begin{pmatrix} 1 & 0 & 0 & 0 \\ 0 & 1 & 0 & 0 \\ \alpha\beta & 0 & 1 & 0 \\ \alpha^2 & \alpha\beta & 0 & 1 \end{pmatrix}.$$

It is now easy to check that these squares form a group isomorphic to the additive group of the field, and that modulo this group the elements themselves form another such group. In other words, the lower triangular group which we have constructed has the shape $E_q.E_q.C_{q-1}$, where E_q denotes an elementary abelian group of order q, and C_{q-1} denotes a cyclic group of order $q-1$. This group turns out to be a maximal subgroup of Sz(q), of index q^2+1. We proceed now to prove this.

We first need to show that the stabiliser in Sz(q) of the 1-space spanned by e_1 is the group of order $q^2(q-1)$ just constructed (and no bigger). If $\langle e_1 \rangle$ is fixed, then so is $\langle e_1^* \rangle = \langle e_1 \wedge e_2 \rangle$, and therefore the 2-space $\langle e_1, e_2 \rangle$ is fixed. It follows that $\langle e_1^*, e_2^* \rangle = \langle e_1 \wedge e_2, e_1 \wedge f_2 \rangle$ is fixed, and therefore so is the 3-space $\langle e_1, e_2, f_2 \rangle$. Thus the stabiliser consists of lower triangular matrices. We have already determined the diagonal elements of the stabiliser, and the same argument is easily seen to determine the diagonal entries of any lower triangular element. Therefore it only remains to determine the lower unitriangular elements. Even in $\mathrm{Sp}_4(q)$ there is just a group of q^4 such matrices (see Section 3.5.3), of the form

$$\begin{pmatrix} 1 & 0 & 0 & 0 \\ \lambda & 1 & 0 & 0 \\ \mu + \alpha\lambda & \alpha & 1 & 0 \\ \beta + \lambda\mu & \mu & \lambda & 1 \end{pmatrix}$$

for arbitrary $\alpha, \beta, \lambda, \mu \in \mathbb{F}_q$. Now by computing the image of $f_1^* = f_1 \wedge f_2$ in two different ways, we see that $\alpha = \lambda^{2^{n+1}}$ and $\alpha\lambda^2 + \beta = \mu^{2^{n+1}}$, so that α and β are determined by λ and μ. Therefore the stabiliser of $\langle e_1 \rangle$ in $\mathrm{Sz}(q)$ has order at most $q^2 \cdot (q-1)$, as required.

Next observe that the element $e_i \leftrightarrow f_i$ of $\mathrm{Sz}(q)$ maps e_1 to f_1, and that $\langle f_1 \rangle$ has q^2 images under the stabiliser of the point $\langle e_1 \rangle$, namely the 1-spaces spanned by $f_1 + \lambda f_2 + \mu e_2 + (\beta + \lambda\mu)e_1$, subject to the above equations. In other words, we take the 1-spaces spanned by $f_1 + xf_2 + ye_2 + ze_1$ where

$$z = xy + x^{2^{n+1}+2} + y^{2^{n+1}}. \tag{4.11}$$

We wish to show that the set of $q^2 + 1$ such 1-spaces is fixed by the map $e_i \leftrightarrow f_i$, and for this purpose it is sufficient to verify that

$$Z = XY + X^{2^{n+1}+2} + Y^{2^{n+1}},$$

where $Z = z^{-1}$, $X = yz^{-1}$, and $Y = xz^{-1}$. But

$$z^{2^{n+1}+2}(Z + XY + X^{2^{n+1}+2} + Y^{2^{n+1}})$$

evaluates to

$$z^{2^{n+1}+1} + xyz^{2^{n+1}} + y^{2^{n+1}+2} + x^{2^{n+1}}z^2,$$

and substituting $z = xy + x^{2^{n+1}+2} + y^{2^{n+1}}$ this expression simplifies to 0. (Exercise 4.2.)

It follows immediately that the given set of $q^2 + 1$ subspaces of dimension 1 is invariant under the whole group $\mathrm{Sz}(q)$, and so the order of this group is $(q^2 + 1)q^2(q-1)$.

An alternative argument, which we now give, and which proves rather more, is to show that there are just $q^2 + 1$ spaces $\langle v \rangle$ such that $v^* = v \wedge w$ for some w perpendicular to v. Suppose that

$$\begin{aligned} v &= \alpha f_1 + \beta f_2 + \gamma e_2 + \delta e_1, \\ w &= \varepsilon f_1 + \zeta f_2 + \eta e_2 + \theta e_1. \end{aligned} \tag{4.12}$$

Then by orthogonality we have $\alpha\theta + \beta\eta = \gamma\zeta + \delta\varepsilon$, and from the condition $v^* = v \wedge w$ we calculate

$$\alpha\zeta + \beta\varepsilon = \alpha^{2^{n+1}},$$
$$\gamma\varepsilon + \alpha\eta = \beta^{2^{n+1}},$$
$$\delta\zeta + \beta\theta = \gamma^{2^{n+1}},$$
$$\delta\eta + \gamma\theta = \delta^{2^{n+1}}. \tag{4.13}$$

We show first that unless v is a scalar multiple of e_1 then $\alpha \neq 0$. For if $\alpha = 0$ then $\beta\varepsilon = 0$, and if $\alpha = \varepsilon = 0$ then $\beta = 0$. Thus $\alpha = \beta = 0$ and $\gamma\varepsilon = 0$. Now if $\gamma = 0$ then $v = \delta e_1$, so we may assume $\gamma \neq 0$, so $\varepsilon = 0$ and $\zeta \neq 0$, contradicting the assumption that w is perpendicular to v.

Therefore $\alpha \neq 0$, and by applying elements of the lower triangular subgroup to v we may assume $\alpha = 1$ and $\beta = \gamma = 0$, so $v = f_1 + \delta e_1$ and $v^* = f_1 \wedge f_2 + \delta^{2^{n+1}} e_1 \wedge e_2$. Then the above equations (4.13) give $\zeta = 1$, $\eta = 0$ and $\delta\zeta = 0$ so $\delta = 0$ as required.

We have shown that $\mathrm{Sz}(q)$ has order $(q^2+1)q^2(q-1)$ and acts 2-transitively on q^2+1 points. The point stabiliser has shape $q^{1+1}.C_{q-1}$, and provided $q > 2$, the involutions are commutators and generate $\mathrm{Sz}(q)$. Therefore by Iwasawa's Lemma (Theorem 3.1) $\mathrm{Sz}(q)$ is simple for $q > 2$.

4.2.3 Subgroups

The subgroup structure of the Suzuki groups is relatively straightforward. Besides the (soluble) lower triangular subgroup of order $q^2(q - 1)$ already mentioned (also known as the Borel subgroup), and smaller Suzuki groups if q happens to be a proper power, the only other notable subgroups are a dihedral group of order $2(q - 1)$ (generated by the diagonal matrices and the map $e_i \leftrightarrow f_i$), and two classes of Frobenius groups of orders $4(2^{2n+1}\pm2^{n+1}+1)$. [Note that $q^2 + 1 = (2^{2n+1} + 2^{n+1} + 1)(2^{2n+1} - 2^{n+1} + 1)$.] Indeed, Suzuki proved the following.

Theorem 4.1. *If $q = 2^{2n+1}$, $n > 1$, then the maximal subgroups of $\mathrm{Sz}(q)$ are (up to conjugacy)*

(i) $E_q.E_q.C_{q-1}$,
(ii) $D_{2(q-1)}$,
(iii) $C_{q+\sqrt{2q}+1}{:}4$,
(iv) $C_{q-\sqrt{2q}+1}{:}4$,
(v) $\mathrm{Sz}(q_0)$, *where $q = q_0{}^r$, r is prime, and $q_0 > 2$.*

The condition $q_0 > 2$ in the last case arises because $\mathrm{Sz}(2) \cong 5{:}4$ and $q^2 + 1$ is divisible by 5 so one or other of $C_{q\pm\sqrt{2q}+1}{:}4$ contains $\mathrm{Sz}(2)$. In the language of Lie theory, the subgroup $D_{2(q-1)}$ is the normaliser N of the torus $T \cong C_{q-1}$, which consists of the diagonal matrices.

4.2.4 Covers and automorphisms

The simple Suzuki groups have no outer automorphisms except the field automorphisms. The only proper covers are the exceptional covers of Sz(8), whose full (i.e. universal) cover is a group $2^2 \cdot \mathrm{Sz}(8)$. The three (isomorphic) double covers have 8-dimensional orthogonal representations over $\mathbb{F}_5$, and indeed $2 \cdot \mathrm{Sz}(8) < \Omega_8^+(5)$.

We choose an orthonormal basis for the underlying orthogonal space, and then we may generate the double cover $2^{1+3+3} {:} 7$ of the Borel subgroup $2^{3+3} {:} 7$ with the monomial matrices

$$(x_\infty, x_0, \ldots, x_6) \mapsto (x_4, -x_5, x_2, x_1, -x_6, x_\infty, x_0, x_3),$$
$$(x_\infty, x_0, \ldots, x_6) \mapsto (x_\infty, x_1, x_2, x_3, x_4, x_5, x_6, x_0). \tag{4.14}$$

(The first of these two elements squares to the sign-change on the coordinates $0, 3, 5, 6$.) Next we adjoin an involution inverting the latter element (which clearly has order 7): it turns out that this involution may be taken to be

$$\begin{pmatrix} 2 & 1 & 1 & 1 & 1 & 1 & 1 & 1 \\ 1 & 4 & 3 & 3 & 4 & 1 & 0 & 3 \\ 1 & 3 & 3 & 4 & 1 & 0 & 3 & 4 \\ 1 & 3 & 4 & 1 & 0 & 3 & 4 & 3 \\ 1 & 4 & 1 & 0 & 3 & 4 & 3 & 3 \\ 1 & 1 & 0 & 3 & 4 & 3 & 3 & 4 \\ 1 & 0 & 3 & 4 & 3 & 3 & 4 & 1 \\ 1 & 3 & 4 & 3 & 3 & 4 & 1 & 0 \end{pmatrix}. \tag{4.15}$$

Since the Borel subgroup is monomial, it determines a coordinate frame, namely the standard coordinate frame. Therefore there are 65 such coordinate frames permuted by the group in the same way that it acts on the 65 points of the oval. One of the 64 non-standard frames is given by the rows of the matrix (4.15), and the others are the images of this under the action of the Borel subgroup. [We omit the proof that these matrices generate a double cover of Sz(8).]

4.3 Octonions and groups of type G_2

4.3.1 Quaternions

The quaternion group Q_8 consists of the 8 elements ± 1, $\pm i$, $\pm j$, $\pm k$ and is defined by the presentation

$$\langle i, j, k \mid ij = k, jk = i, ki = j \rangle, \tag{4.16}$$

from which it follows that $i^2 = j^2 = k^2 = -1$ and $ji = -k$, $kj = -i$, $ik = -j$. The (real) quaternion algebra $\mathbb{H}$ (named thus after Hamilton) consists of all real linear combinations of these elements (where -1 in the group is identified with -1 in $\mathbb{R}$). Thus

$$\mathbb{H} = \{a + bi + cj + dk \mid a, b, c, d \in \mathbb{R}\} \tag{4.17}$$

with the obvious addition, and multiplication defined by the above rules and the distributive law. This is a non-commutative algebra which has many applications in physics and elsewhere. Given a quaternion $q = a + bi + cj + dk$ we write $\bar{q} = a - bi - cj - dk$ for the (quaternion) *conjugate* of q, and $\mathrm{Re}\,(q) = a = \frac{1}{2}(q + \bar{q})$ is the *real part* of q. There is a natural norm N under which $\{1, i, j, k\}$ is an orthonormal basis, and this norm satisfies $N(q) = q\bar{q}$.

More generally, we may replace the real numbers in this definition by any field F of characteristic not 2. (Fields of characteristic 2 do not work: one difficulty is that $1 = -1$ in the field but $1 \neq -1$ in the group. See Sections 4.3.4 and 4.4 for two different ways to overcome this problem.) We obtain in this way a 4-dimensional non-commutative algebra over the field F, and we extend the definitions of $\bar{q}$, $\mathrm{Re}\,(q)$ and $N(q)$ in the obvious way to this algebra.

The group of automorphisms of this algebra must fix the identity element 1, and therefore fixes its orthogonal complement (the space of *purely imaginary* quaternions, spanned by i, j, k). Therefore it is a subgroup of the orthogonal group $\mathrm{GO}_3(F)$, and is in fact isomorphic to the group $\mathrm{SO}_3(F) \cong \mathrm{PGL}_2(F)$. To prove this we simply need to check that if x, y, z are any three mutually orthogonal purely imaginary quaternions of norm 1, then $xy = \pm z$. (See Exercise 4.3.)

In fact, the automorphism group of the quaternions consists entirely of *inner automorphisms* $\alpha_q : x \mapsto q^{-1}xq$ for invertible $q \in \mathbb{H}$ (see Exercise 4.4). Since $\alpha_{-q} = \alpha_q$ this gives a 2-to-1 map from the group of quaternions of norm 1 to $\mathrm{SO}_3(F)$. Indeed, this group of quaternions is a double cover of $\mathrm{SO}_3(F)$ and is isomorphic to the corresponding spin group (see Section 3.9). If F is finite this spin group has a subgroup of index 2 isomorphic to $\mathrm{SL}_2(F)$.

4.3.2 Octonions

The (real) octonion algebra $\mathbb{O}$ (also known as the Cayley numbers, even though, as Cayley himself admitted [22], they were first discovered by Graves [67]) can be built from the quaternions by taking 7 mutually orthogonal square roots of -1, labelled $i_0, \ldots, i_6$ (with subscripts understood modulo 7), subject to the condition that for each t, the elements i_t, i_{t+1}, i_{t+3} satisfy the same multiplication rules as i, j, k (respectively) in the quaternion algebra. It is easy to see that this defines all of the multiplication, and that this multiplication is non-associative. For example, $(i_0 i_1)i_2 = i_3 i_2 = -i_5$ but $i_0(i_1 i_2) = i_0 i_4 = i_5$. [Recall from Section 3.9 that, strictly speaking, an algebra is associative. We emphasise the generalisation by describing $\mathbb{O}$ as a *non-associative algebra*.]

For reference, here is the multiplication table:

1	i_0	i_1	i_2	i_3	i_4	i_5	i_6
i_0	-1	i_3	i_6	$-i_1$	i_5	$-i_4$	$-i_2$
i_1	$-i_3$	-1	i_4	i_0	$-i_2$	i_6	$-i_5$
i_2	$-i_6$	$-i_4$	-1	i_5	i_1	$-i_3$	i_0
i_3	i_1	$-i_0$	$-i_5$	-1	i_6	i_2	$-i_4$
i_4	$-i_5$	i_2	$-i_1$	$-i_6$	-1	i_0	i_3
i_5	i_4	$-i_6$	i_3	$-i_2$	$-i_0$	-1	i_1
i_6	i_2	i_5	$-i_0$	i_4	$-i_3$	$-i_1$	-1

$$(4.18)$$

It is worth pausing for a moment to consider the symmetries of this table. By definition it is invariant under the map $i_t \mapsto i_{t+1}$, and it is easy to check that it is invariant under $i_t \mapsto i_{2t}$, extending the automorphism $i \mapsto j \mapsto k \mapsto i$ of the quaternions (where $i = i_1$, $j = i_2$ and $k = i_4$). It is also invariant under the map

$$(i_0, \ldots, i_6) \mapsto (i_0, i_2, i_1, i_6, -i_4, -i_5, i_3) \tag{4.19}$$

which extends the automorphism $i \leftrightarrow j, k \mapsto -k$ of the quaternions. Ignoring the signs for the moment, we recognise the permutations $(0, 1, 2, 3, 4, 5, 6)$, $(1, 2, 4)(3, 6, 5)$ and $(1, 2)(3, 6)$ generating $\mathrm{GL}_3(2)$ as described in Section 3.3.5. Thus there is a homomorphism from the group of symmetries onto $\mathrm{GL}_3(2)$. The kernel is a group of order 2^3, since we may change sign independently on i_0, i_1 and i_2, and then the other signs are determined. In fact, the resulting group $2^3\mathrm{GL}_3(2)$ is a non-split extension (see Exercise 4.8).

Now the set $\{\pm 1, \pm i_0, \ldots, \pm i_6\}$ is closed under multiplication, but does not form a group, since the associative law fails. In fact it is a *Moufang loop*, which means that it is a *loop* (a set with a multiplication, such that left and right multiplication by a are permutations of the set, and with an identity element) which satisfies the *Moufang laws*

$$\begin{aligned} (xy)(zx) &= (x(yz))x, \\ x(y(xz)) &= ((xy)x)z, \text{ and} \\ ((yx)z)x &= y(x(zx)). \end{aligned} \tag{4.20}$$

In the loop, these laws may be verified directly: the symmetries given above reduce the work to checking the single case $x = i_0$, $y = i_1$, $z = i_2$. Since this loop has an identity element 1, the Moufang laws imply the *alternative* laws

$$\begin{aligned} (xy)x &= x(yx), \\ x(xy) &= (xx)y, \text{ and} \\ (yx)x &= y(xx). \end{aligned} \tag{4.21}$$

Indeed, the Moufang laws hold not just in the loop, but also in the algebra: this is not obvious, however, since the laws are not linear in x. It is sufficient to take $y = i_0$, $z = i_1$, and check that the cross terms cancel out in the cases $x = i_2 + i_t$, $t = 3, 4, 5, 6$ (see Exercise 4.6). It follows from the alternative laws that any 2-generator subalgebra is associative.

Just as with the quaternions, an octonion algebra may be defined by the same rules over any field F of characteristic not 2. There is again a natural norm N, under which $\{1, i_0, \ldots, i_6\}$ is an orthonormal basis, and $N(x) = x\overline{x}$, where $^-$ (called *octonion conjugation*) is the F-linear map fixing 1 and negating i_0, $\ldots$, i_6. Also we define the *real part* $\mathrm{Re}\,(x) = \frac{1}{2}(x + \overline{x})$, so that $\overline{x} = 2\mathrm{Re}\,(x) - x$. Since $\overline{xy} = \overline{y}.\overline{x}$, and $\overline{x}$ is expressible as a linear combination of 1 and x, the alternative laws imply immediately that $N(xy) = N(x)N(y)$. The norm N is a quadratic form as defined in (3.15), and the associated bilinear form is

$$f(x, y) = N(x + y) - N(x) - N(y) = 2\mathrm{Re}\,(x\overline{y}), \qquad (4.22)$$

which is twice the usual inner product. For some purposes, notably the discussion of triality in Section 4.7, it is useful to combine the algebra product with the inner product into a cyclically-symmetric trilinear form

$$t(x, y, z) = f(xy, \overline{z}) = 2\mathrm{Re}\,(xyz). \qquad (4.23)$$

See Section 4.7 for a proof that this form is cyclically-symmetric.

The automorphisms of the algebra again preserve the identity element 1, so live inside the orthogonal group $\mathrm{GO}_7(F)$ acting on the purely imaginary octonions. This time, however, it is clearly a proper subgroup of $\mathrm{SO}_7(F)$, since once we know the images of i_0 and i_1, we know the image of $i_3 = i_0 i_1$. Indeed, if we also know the image of i_2, then we know the images of all the basis vectors. This automorphism group is known as $G_2(F)$, or, if F is the field $\mathbb{F}_q$ of q elements, $G_2(q)$.

4.3.3 The order of $G_2(q)$

To calculate the order of $G_2(q)$, in the case q odd, we calculate the number of images under $G_2(q)$ of the list i_0, i_1, i_2 of generators for the algebra. The crux of the matter is to prove that the group is transitive on triples of elements satisfying the obvious properties of the triple (i_0, i_1, i_2), namely that they are mutually orthogonal purely imaginary octonions of norm 1, and that i_2 is orthogonal to $i_0 i_1$. We do this by showing that the multiplication is completely determined by these properties.

First, if i is pure imaginary of norm 1 then $i^2 = -i.\overline{i} = -N(i) = -1$. Second, if $i = \sum_{t=0}^{6} \lambda_t i_t$ and $j = \sum_{t=0}^{6} \mu_t i_t$ are perpendicular pure imaginary octonions of norm 1 then all terms in the expansion of ij anticommute, except the terms in $i_n.i_n$, which commute and sum to 0 (since i is orthogonal to j), so that $ij = -ji$. Third, if i, j, $k = ij$ and l are four mutually orthogonal norm 1 pure imaginary octonions, then, in the expansion of $(ij)l$, the terms which are associative correspond to the real parts of ij, jl, il or kl, and each of these sets of terms individually adds up to 0 (we have already seen the ij case: the others are similar), so that $(ij)l = -i(jl)$. Since $N(xy) = N(x)N(y)$, multiplication by an octonion of norm 1 preserves norms, and therefore inner products.

Thus we see that $\{1, i, j, ij, l, il, jl, (ij)l\}$ is an orthonormal basis. The entire multiplication table is now determined by these relations (see Exercise 4.7), and is visibly the same as given in (4.18).

Now we count the number of such triples (i, j, l): first i can be any vector of norm 1, and the number of such vectors is

$$|SO_7(q)|/|SO_6^\varepsilon(q)| = q^6 + \varepsilon q^3 = q^3(q^3 + \varepsilon) \tag{4.24}$$

(where $\varepsilon = \pm 1$ satisfies $\varepsilon \equiv q$ mod 4). Then j can be any of the $q^5 - \varepsilon q^2 = q^2(q^3 - \varepsilon)$ vectors of norm 1 in the orthogonal 6-space of type ε. Finally, l can be any vector of norm 1 orthogonal to i, j and $k = ij$, i.e. lying in the orthogonal 4-space of plus type spanned by l, il, jl, kl. There are $q^3 - q$ such vectors, so the order of $G_2(q)$ is $q^3(q^3 + \varepsilon).q^2(q^3 - \varepsilon).q(q^2 - 1)$, that is

$$|G_2(q)| = q^6(q^6 - 1)(q^2 - 1). \tag{4.25}$$

As an immediate corollary we obtain (for q odd) the transitivity of $G_2(q)$ on various other objects in the orthogonal 7-space. For example, $G_2(q)$ is transitive on 2-spaces spanned by two orthogonal vectors of norm 1, in other words on non-singular 2-spaces of type $\varepsilon = \pm$, where $\varepsilon \equiv q$ mod 4. If α is any non-square in the field, then any vector of norm α lies in such a 2-space. Moreover, this vector may be written as $\lambda i + \mu j$, where $\lambda^2 + \mu^2 = \alpha$ and i and j are two orthogonal vectors of norm 1. Therefore $G_2(q)$ is transitive on vectors of norm α. In Section 4.3.5 we show that $G_2(q)$ is also transitive on isotropic vectors. Thus $G_2(q)$ is transitive on non-zero vectors of any given norm. For the record there are $q^6 + q^3$ vectors of norm -1, spanning $\frac{1}{2}(q^6 + q^3)$ subspaces of dimension 1, and $q^6 - q^3$ vectors of norm $-\alpha$, spanning $\frac{1}{2}(q^6 - q^3)$ subspaces, as well as $q^6 - 1$ isotropic vectors, spanning $(q^6 - 1)/(q - 1)$ subspaces.

4.3.4 Another basis for the octonions

As we saw in Chapter 3 when looking at the unitary and orthogonal groups, an orthonormal basis may be convenient for defining the groups, but a symplectic basis is more useful for looking at subgroup structure, especially the so-called 'BN-pair' (see Section 4.3.5) and the parabolic subgroups. The same is true for $G_2(q)$, and as an added bonus we get a definition which works also in characteristic 2.

We change basis as follows. First pick elements $a, b \in \mathbb{F}_q$ such that $b \neq 0$ and $a^2 + b^2 = -1$ (this can be done for all odd q). Then our new basis is $\{x_1, \ldots, x_8\}$ defined by

$$
\begin{array}{ll}
2x_1 = i_4 + ai_6 + bi_0, & 2x_8 = i_4 - ai_6 - bi_0, \\
2x_2 = i_2 + bi_3 + ai_5, & 2x_7 = i_2 - bi_3 - ai_5, \\
2x_3 = i_1 - bi_6 + ai_0, & 2x_6 = i_1 + bi_6 - ai_0, \\
2x_4 = 1 + ai_3 - bi_5, & 2x_5 = 1 - ai_3 + bi_5.
\end{array} \tag{4.26}
$$

With respect to the norm N and the associated bilinear form f, the basis vectors are isotropic and mutually perpendicular, except that $f(x_i, x_{9-i}) = 1$. Rewriting the multiplication table with respect to this new basis we find that all entries are integers, as follows (where blank entries are 0).

	x_1	x_2	x_3	x_4	x_5	x_6	x_7	x_8
x_1					x_1	x_2	$-x_3$	$-x_4$
x_2			$-x_1$	x_2			$-x_5$	x_6
x_3		x_1		x_3	$-x_5$		$-x_7$	
x_4	x_1			x_4	x_6	x_7		
x_5		x_2	x_3	x_5			x_8	
x_6	$-x_2$			$-x_4$	x_6	x_8		
x_7	x_3	$-x_4$			x_7	$-x_8$		
x_8	$-x_5$	$-x_6$	x_7	x_8				

$$(4.27)$$

This multiplication can now be interpreted over any field whatsoever, for example the real numbers, in which case the resulting algebra is called the *split real form* of the octonion algebra: this is not the same as the *compact real form* defined in Section 4.3.2, since there are no solutions to $a^2 + b^2 = -1$ in $\mathbb{R}$.

In characteristic 2, of course, the signs disappear, but the algebra is still non-commutative (since $x_1 x_4 = 0$ but $x_4 x_1 = x_1$) and non-associative (since $(x_1 x_4) x_7 = 0$ but $x_1 (x_4 x_7) = x_1 x_7 = x_1$). The norm reduced modulo 2 is again a quadratic form with $N(x_i) = 0$. Clearly this quadratic form is of plus type (see Section 3.4.7). The 'real part' is $\mathrm{Re}\,(\sum_i \lambda_i x_i) = \lambda_4 + \lambda_5$. The automorphism group of the octonion algebra defined over $\mathbb{F}_{2^n}$ is denoted $G_2(2^n)$. In Section 4.4.3, I shall give another definition of this group, and prove that the two definitions are equivalent. Since the identity element $x_4 + x_5$ is fixed by every automorphism, and is non-isotropic, we have $G_2(q) < \mathrm{Sp}_6(q) < \Omega_8^+(q)$. It is still necessary to compute the order of $G_2(2^n)$ (for example by the method employed in Section 4.6 to compute the order of $^3D_4(q)$—see Exercise 4.11), but then the rest of Section 4.3 applies to the characteristic 2 case with very little alteration. It turns out that the formula

$$|G_2(q)| = q^6(q^6 - 1)(q^2 - 1) \tag{4.28}$$

applies also in the case $q = 2^n$.

4.3.5 The parabolic subgroups of $G_2(q)$

With respect to the basis $\{x_1, \ldots, x_8\}$ defined in (4.26) more symmetries become apparent. For example it is easy to check that the diagonal matrices which preserve the multiplication are precisely those of shape

$$\mathrm{diag}(\lambda, \mu, \lambda\mu^{-1}, 1, 1, \lambda^{-1}\mu, \mu^{-1}, \lambda^{-1}) \tag{4.29}$$

for any non-zero $\lambda, \mu \in \mathbb{F}_q$. These matrices form the (maximally split) *torus* T, which is normalised by a group $W \cong D_{12}$ generated by the maps

$$r : (x_1, \ldots, x_8) \mapsto (-x_1, -x_3, -x_2, x_4, x_5, -x_7, -x_6, -x_8),$$
$$s : (x_1, \ldots, x_8) \mapsto (-x_2, -x_1, -x_6, x_5, x_4, -x_3, -x_8, -x_7). \qquad (4.30)$$

It is left as Exercise 4.10 to verify that these maps preserve the multiplication table in (4.27). The product rs of these two maps is the coordinate permutation $(1, 2, 6, 8, 7, 3)(4, 5)$. Traditionally, the normaliser of a (maximally split) torus is denoted N, and is the 'N' part of the 'BN-pair' (see Section 3.3.3).

The relationship of this group to the Weyl group of type G_2 (see Section 2.8) may be seen by taking generators

$$h_1 = \mathrm{diag}(\alpha, \alpha^{-1}, \alpha^2, 1, 1, \alpha^{-2}, \alpha, \alpha^{-1}),$$
$$h_2 = \mathrm{diag}(1, \alpha, \alpha^{-1}, 1, 1, \alpha, \alpha^{-1}, 1), \qquad (4.31)$$

for the torus T, where α is a generator for the multiplicative group of the field $\mathbb{F}_q$. Then the two generators r, s for the Weyl group act by conjugation respectively as

$$r : h_1 \mapsto h_1 h_2^3,$$
$$h_2 \mapsto h_2^{-1}, \qquad (4.32)$$

(which may be thought of as a multiplicative version of 'reflection' in the 'short root' h_2), and

$$s : h_1 \mapsto h_1^{-1},$$
$$h_2 \mapsto h_1 h_2, \qquad (4.33)$$

(which may be thought of as 'reflection' in the 'long root' h_1).

The unipotent subgroup U has order q^6 and is generated by the following elements, where $\lambda \in \mathbb{F}_q$:

$$A(\lambda) : x_7 \mapsto x_7 - \lambda x_1, x_8 \mapsto x_8 + \lambda x_2,$$
$$B(\lambda) : x_6 \mapsto x_6 - \lambda x_1, x_8 \mapsto x_8 + \lambda x_3,$$
$$C(\lambda) : x_4 \mapsto x_4 - \lambda x_1, x_5 \mapsto x_5 + \lambda x_1, x_6 \mapsto x_6 - \lambda x_2,$$
$$\qquad x_7 \mapsto x_7 + \lambda x_3, x_8 \mapsto x_8 + \lambda x_5 - \lambda x_4 + \lambda^2 x_1,$$
$$D(\lambda) : x_3 \mapsto x_3 - \lambda x_1, x_4 \mapsto x_4 - \lambda x_2, x_5 \mapsto x_5 + \lambda x_2,$$
$$\qquad x_7 \mapsto x_7 - \lambda x_4 + \lambda x_5 + \lambda^2 x_2, x_8 \mapsto x_8 + \lambda x_6,$$
$$E(\lambda) : x_3 \mapsto x_3 - \lambda x_2, x_7 \mapsto x_7 + \lambda x_6,$$
$$F(\lambda) : x_2 \mapsto x_2 - \lambda x_1, x_4 \mapsto x_4 + \lambda x_3, x_5 \mapsto x_5 - \lambda x_3,$$
$$\qquad x_6 \mapsto x_6 + \lambda x_4 - \lambda x_5 + \lambda^2 x_3, x_8 \mapsto x_8 + \lambda x_7. \qquad (4.34)$$

Conjugates of $A(\lambda)$ are called *long root elements* and conjugates of $F(\lambda)$ are called *short root elements*. The long root elements are special cases of the (long) root elements (also called Eichler or Siegel transformations: see Section 3.7.3) in the orthogonal group $GO_7(q)$. The *Borel subgroup* is the subgroup $B = UT$ of order $q^6(q-1)^2$, and is the 'B' of the 'BN-pair'.

The subgroup of order q generated by the $A(\lambda)$ is a *long root subgroup*, and its normaliser is a group P_1 of shape q^{1+4}:$GL_2(q)$, in which the normal subgroup q^{1+4} is generated by the elements $B(\lambda)$, $C(\lambda)$, $D(\lambda)$, and $E(\lambda)$. In fact, P_1 is the stabiliser of the 2-space $\langle x_1, x_2 \rangle$, which is characterised up to automorphisms by the property of being an isotropic 2-space on which the algebra product is identically 0. To obtain generators for P_1 (one of the *maximal parabolic subgroups*) we take the Borel subgroup $B = UT$ and adjoin the element s defined in (4.30).

In the general theory, a parabolic subgroup is defined as any subgroup containing a Borel subgroup, which is in turn defined as the normaliser of a Sylow p-subgroup, where p is the characteristic of the underlying field. In the case of $G_2(q)$, the stabiliser of the 1-space $\langle x_1 \rangle$ is the other maximal parabolic subgroup P_2 containing B, and has shape q^{2+1+2}:$GL_2(q)$. It is the normaliser of the group $\langle A(\lambda), B(\lambda) \rangle$ of order q^2. Modulo this, the normal p-subgroup is generated by $C(\lambda)$, $D(\lambda)$ and $F(\lambda)$. This quotient is a special group of shape q^{1+2} if q is odd (it is elementary abelian if q is even). To obtain generators for P_2 we take the Borel subgroup $B = UT$ and adjoin the element r defined in (4.30). Notice that P_2 has index $(q^6 - 1)/(q - 1)$ in $G_2(q)$, which is equal to the total number of isotropic 1-spaces in $1^\perp$, and therefore $G_2(q)$ is transitive on these 1-spaces. It follows that $G_2(q)$ is transitive on the isotropic vectors in $1^\perp$.

It can be shown that if the characteristic of the field is 3, then these two maximal parabolic subgroups P_1 and P_2 are isomorphic, and have shape $(q^2 \times q^{1+2})$:$GL_2(q)$. [Actually the structure of these groups is quite subtle: each has normal subgroups of orders q and q^2, but has no normal subgroup of shape q^{1+2}.] Furthermore, there is an exceptional outer automorphism of $G_2(q)$ interchanging the two classes of maximal parabolic subgroups. This is described in more detail in Section 4.5.

4.3.6 Other subgroups of $G_2(q)$

The maximal parabolic subgroups, described above, may be generated by the Borel subgroup B and a suitable subgroup of N, the normaliser of a torus. Some other maximal subgroups may be obtained similarly, but using a proper subgroup of B. For example, the group generated by $A(\lambda)$, $F(\lambda)$, T, s and $(rs)^3 : x_i \mapsto x_{9-i}$ is a group isomorphic to $SO_4^+(q)$. The structure of this group is $SL_2(q) \times SL_2(q)$ if q is even and $2 \cdot (PSL_2(q) \times PSL_2(q)).2$ if q is odd. In the latter case it is the centraliser of the involution $\mathrm{diag}(-1, -1, 1, 1, 1, 1, -1, -1)$. In any case it fixes the 4-space $\langle x_3, x_4, x_5, x_6 \rangle$, which is a quaternion subalgebra, and it acts on this 4-space as $SO_3(q) \cong PGL_2(q)$, centralising the identity element $1 = x_4 + x_5$. The kernel of this action is $SL_2(q)$, and the full group acts naturally and faithfully as $SO_4^+(q)$ on the orthogonal complement $\langle x_1, x_2, x_7, x_8 \rangle$ of the quaternion subalgebra.

Another subgroup of interest is $SL_3(q)$:2, which may be generated by $A(\lambda)$, $B(\lambda)$, $E(\lambda)$ and N. If q is odd, this group fixes the vector $x_4 - x_5$ of norm -1, up to sign. (If q is even, it fixes x_4 modulo $1 = x_4 + x_5$.) It also preserves the pair of 3-spaces $\langle x_1, x_6, x_7 \rangle$ and $\langle x_2, x_3, x_8 \rangle$, with elements of $\mathrm{SL}_3(q)$ acting naturally on one of these 3-spaces and dually on the other. The outer half of the group $\mathrm{SL}_3(q){:}2$ acts by swapping the two 3-spaces. Since there are exactly $\frac{1}{2}(q^6 + q^3)$ such 1-spaces, and this is the index of $\mathrm{SL}_3(q){:}2$ in $G_2(q)$, it follows that this subgroup is the full stabiliser of this 1-space. Notice that if $q \equiv 1 \bmod 3$ then $\mathbb{F}_q$ has an element ω of order 3, and $\mathrm{SL}_3(q)$ has a central element $\mathrm{diag}(\omega, \omega^2, \omega^2, 1, 1, \omega, \omega, \omega^2)$ of order 3. Indeed, $\mathrm{SL}_3(q)$ is exactly the centraliser of this element of order 3.

Similarly, if α is a non-square in $\mathbb{F}_q$ and q is odd, then the stabiliser of a vector v of norm $-\alpha$ perpendicular to $1 = x_4 + x_5$ is a subgroup $\mathrm{SU}_3(q)$, contained in a maximal subgroup $\mathrm{SU}_3(q){:}2$. (If q is even take a suitable vector modulo 1 instead.) These groups are not so easy to see from our generators, as we need to extend the field to $\mathbb{F}_{q^2}$ to make the natural representation of $\mathrm{SU}_3(q)$. Nevertheless, the same counting argument as before shows that $\mathrm{SU}_3(q){:}2$ is the stabiliser of $\langle v \rangle$.

Complete determination of the maximal subgroups of $G_2(q)$ was obtained by Cooperstein [38] for q even and Kleidman [106] for q odd. For q even the only other generic maximal subgroups are the subfield groups $G_2(q_0)$, where $q = q_0{}^r$ and r is prime. Besides this there are the exceptional subgroups J_2 and $\mathrm{PSL}_2(13)$ in $G_2(4)$: see Sections 5.6.8 and 5.6.9 for much more on $G_2(4)$.

For q odd and not a power of 3 there is one more generic subgroup, namely $\mathrm{PGL}_2(q)$ acting irreducibly on the 7-space of pure imaginary octonions, provided $q \geqslant 11$ and the characteristic p is at least 7. There are also a number of exceptional subgroups when q is a small power of p. Of these, we have already seen $2^3{\cdot}\mathrm{GL}_3(2)$ and we shall see $\mathrm{PSU}_3(3){:}2$ in Section 4.4 (both these subgroups are maximal if and only if $q = p$). The containment of the sporadic Janko group J_1 in $G_2(11)$ is treated in Section 5.9.1. Other subgroups which can occur are subgroups $\mathrm{PSL}_2(13)$, which occur whenever 13 is a square in the field (that is, whenever $q \equiv \pm1, \pm3, \pm4 \bmod 13$), and $\mathrm{PSL}_2(8)$ which occurs whenever the polynomial $x^3 - 3x + 1$ has a root in the field.

In characteristic 3 things are rather different, as there is an exceptional outer automorphism of $G_2(q)$ in this case (see Sections 4.3.9 and 4.5.1). A number of other subgroups occur, including another class each of $\mathrm{SL}_3(q){:}2$ and $\mathrm{SU}_3(q){:}2$, acting irreducibly on the 7-space, and (if $q = 3^{2k+1}$) the Ree group ${}^2G_2(q)$ (see Section 4.5 below). The complete list of maximal subgroups of $G_2(q)$ is given in Table 4.1.

4.3.7 Simplicity of $G_2(q)$

In fact $G_2(q)$ is simple for all q except for $q = 2$, when we have $G_2(2) \cong \mathrm{PSU}_3(3){:}2$ (see Section 4.4.4). As usual, we can use Iwasawa's Lemma to prove simplicity. There are a number of primitive actions of the group which could be used. Probably the easiest is the action on isotropic 1-spaces perpendicular

Table 4.1. Maximal subgroups of $G_2(q)$, $q > 2$

$p = 2$	$p = 3$	$p > 3$
$q^{1+4}{:}\mathrm{GL}_2(q)$	$(q^2 \times q^{1+2}){:}\mathrm{GL}_2(q)$	$q^{1+4}{:}\mathrm{GL}_2(q)$
$q^{2+3}{:}\mathrm{GL}_2(q)$	$(q^2 \times q^{1+2}){:}\mathrm{GL}_2(q)$	$q^{2+1+2}{:}\mathrm{GL}_2(q)$
$\mathrm{SL}_3(q){:}2$	$\mathrm{SL}_3(q){:}2$	$\mathrm{SL}_3(q){:}2$
	$\mathrm{SL}_3(q){:}2$	
$\mathrm{SU}_3(q){:}2$	$\mathrm{SU}_3(q){:}2$	$\mathrm{SU}_3(q){:}2$
	$\mathrm{SU}_3(q){:}2$	
$\mathrm{L}_2(q) \times \mathrm{L}_2(q)$	$2^{\cdot}(\mathrm{L}_2(q) \times \mathrm{L}_2(q)){:}2$	$2^{\cdot}(\mathrm{L}_2(q) \times \mathrm{L}_2(q)){:}2$
$G_2(q_0)$	$G_2(q_0)$	$G_2(q_0)$
	${}^2G_2(q)$, m odd	
		$\mathrm{PGL}_2(q), q \geqslant 11, p \geqslant 7$
	$2^3{\cdot}\mathrm{L}_3(2), q = 3$	$2^3{\cdot}\mathrm{L}_3(2), q = p$
		$G_2(2), q = p$
$\mathrm{L}_2(13), q = 4$	$\mathrm{L}_2(13), q = 3$	$\mathrm{L}_2(13), p \neq 13, \mathbb{F}_q = \mathbb{F}_p[\sqrt{13}]$
		$\mathrm{L}_2(8), \mathbb{F}_q = \mathbb{F}_p[\gamma], \gamma^3 - 3\gamma + 1 = 0$
$\mathrm{J}_2, q = 4$		
		$\mathrm{J}_1, q = 11$

Note: $q = p^m = q_0{}^r$, with p and r prime.

to 1. In this case, there are $(q^6 - 1)/(q - 1)$ points, and the point stabiliser has the shape $q^2.q.q^2.\mathrm{GL}_2(q)$.

If we stabilise the point $\langle x_1 \rangle$ then the normal subgroup of order q^2 is generated by $A(\lambda)$ and $B(\lambda)$ (see Section 4.3.5). Notice that $E(\lambda) = A(\lambda)^{rs}$ is a long root element, and its conjugates generate a subgroup $q^2.q.q^2.\mathrm{SL}_2(q)$ (namely the stabiliser of an isotropic vector) of the point stabiliser (except when $q = 2$). Moreover, this subgroup contains an element of order $q - 1$ from the torus T, and powers of this element multiply certain isotropic vectors by all non-zero scalars. It follows that the subgroup generated by long root elements is transitive on the isotropic vectors, and is therefore the whole of $G_2(q)$.

Next we calculate the orbits of the point stabiliser $q^2.q.q^2.\mathrm{GL}_2(q)$ on the $q^5 + q^4 + q^3 + q^2 + q + 1$ points. We find that there are four orbits, of lengths 1, $q(q+1)$, $q^3(q+1)$ and q^5, represented by $\langle x_1 \rangle$, $\langle x_2 \rangle$, $\langle x_7 \rangle$ and $\langle x_8 \rangle$ respectively. Therefore the only possibility for a block system would be $q^3 + 1$ blocks of size $q^2 + q + 1$. Moreover, each block would correspond to an isotropic 3-space, for example $\langle x_1, x_2, x_3 \rangle$. However, $\langle x_1 \rangle$ is distinguished in this subspace as the image of the restriction of the product map. This contradiction implies that the action is primitive. Clearly long root elements are in the derived subgroup, so $G_2(q)$ is perfect. Now apply Iwasawa's Lemma (Theorem 3.1).

4.3.8 The generalised hexagon

There are $(q^6 - 1)/(q - 1)$ isotropic 1-spaces (points), and we can construct a graph with these points as vertices, joining two points whenever their product in the algebra is 0. From the information above we see that this graph is regular of degree $q(q + 1)$. [The *degree*, or *valency*, of a vertex is the number of edges incident with it. A graph is *regular* if all its vertices have the same degree, called the degree of the graph.] Moreover, if two points $\langle u \rangle$ and $\langle v \rangle$ are joined, then obviously all $q + 1$ points in the 2-space $\langle u, v \rangle$ are joined to each other. These cliques (a *clique* in a graph is a set of vertices all of which are joined to each other) are usually called *lines*, and it is easy to calculate that the number of lines is equal to the number of points. If we now draw a new graph, the *point–line incidence graph*, with vertices corresponding to both points and lines, and edges joining incident pairs, we obtain a bipartite graph with the following shape, where figures below the nodes denote the number of vertices, and figures above the edges denote the number of edges incident with each vertex.

$$
\begin{array}{ccccccccccccc}
& q+1 & & 1 & q & & 1 & q & & 1 & q & & 1 & q & & 1 & q & q+1 \\
\bullet & & \circ & & & \bullet & & & \circ & & & \bullet & & & \circ & & & \bullet \\
1 & & & q+1 & & & q(q+1) & & q^2(q+1) & & q^3(q+1) & & q^4(q+1) & & & & q^5
\end{array}
$$

There are two ways to interpret this picture: either the black vertices denote points and the white vertices denote lines, or vice versa. Note however that there is no automorphism of the graph which swaps the black vertices with the white vertices, except when q is a power of 3. (Compare the generalised quadrangles for $\mathrm{PSp}_4(q)$ described in Section 3.5.6 and for $\mathrm{PSU}_4(q)$ and $\mathrm{PSU}_5(q)$ described in Section 3.6.4, the generalised hexagon for $^3D_4(q)$ described in Section 4.6.4, and the generalised octagon for $^2F_4(q)$ described in Section 4.9.4.)

4.3.9 Automorphisms and covers

Except in characteristic 3, the only outer automorphisms of $G_2(q)$ are field automorphisms. In characteristic 3 there is also a 'twisting' automorphism (known as a *graph automorphism* because it derives from the automorphism of the Dynkin diagram) which squares to a generating field automorphism. This automorphism gives rise to the small Ree groups (see Section 4.5, especially 4.5.1, below).

The only proper covers of any group $G_2(q)$ are the exceptional covers $2 \cdot G_2(4)$ and $3 \cdot G_2(3)$. The former is visible inside the automorphism group of the Leech lattice, which can be used to prove the interesting inclusions $J_2 < G_2(4) < \mathrm{Suz}$ (see Chapter 5, especially Sections 5.6.8 and 5.6.3). The latter is seen in the chain of groups

$$3 \cdot G_2(3) < 3 \cdot \Omega_7(3) < 3 \cdot \mathrm{Fi}_{22} < 3 \cdot {}^2E_6(2)$$

(see Section 5.7).

4.4 Integral octonions

This section gives an alternative approach to constructing $G_2(q)$ in characteristic 2, and can safely be omitted at a first reading. Its main purpose here is to provide a mechanism for proving the exceptional isomorphism $G_2(2) \cong \mathrm{PSU}_3(3){:}2$, and showing in particular that $G_2(2)$ is not simple.

4.4.1 Quaternions in characteristic 2

Over fields of characteristic 2, the quaternions i, j, k generate a commutative ring which is not very interesting. A more interesting quaternion algebra may be obtained by first constructing the ring generated (over $\mathbb{Z}$) by i and $\omega = \frac{1}{2}(-1 + i + j + k)$. This ring is described in more detail in Section 5.6.8, and consists of the integral linear combinations of 1, i, j and $\overline{\omega} = \frac{1}{2}(-1 - i - j - k)$. [These four vectors, incidentally, suitably scaled, form a set of fundamental roots for the root system of type D_4.] Reducing modulo 2 we now have the 16 elements $1, i, j, k, \omega, i\omega, j\omega, k\omega, \overline{\omega}, i\overline{\omega}, j\overline{\omega}, k\overline{\omega}, 0, 1+i, 1+j, 1+k$, with addition and multiplication induced from those on the real quaternion algebra. Just the first 12 elements listed are invertible, and form a group isomorphic to A_4. The integral norm, reduced modulo 2, gives rise to a (singular) quadratic form. The last four elements listed comprise the radical of the quadratic form, modulo which is an orthogonal 2-space of minus type.

This set of 16 elements we take as 'the' quaternion algebra over $\mathbb{F}_2$. Its automorphism group is rather larger than you might expect by comparing with the odd characteristic case. It is $2^2{:}\mathrm{SL}_2(2) \cong S_4$, generated by the A_4 of inner automorphisms $\alpha_q : x \mapsto q^{-1}xq$ together with the outer automorphism $i \leftrightarrow j$, $\omega \leftrightarrow \overline{\omega}$. Now to get a quaternion algebra over an arbitrary field F of characteristic 2, we simply take linear combinations of 1, i, j, and $\overline{\omega}$ with coefficients in F. The automorphism group of the algebra is then an affine group $F^2{:}\mathrm{SL}_2(F)$.

4.4.2 Integral octonions

The corresponding process for the octonion algebras in characteristic 2 is considerably more complicated. In essence we have to put an octonion multiplication onto the E_8 lattice, in order to obtain an orthogonal space of plus type when the lattice is reduced modulo 2 (see Section 3.12.4). This is tricky as there is not as much symmetry of the coordinates as we would like. One might expect, for example, that it is sufficient to adjoin the appropriate elements $\overline{\omega_t} = \frac{1}{2}(-1 - i_t - i_{t+1} - i_{t+3})$ to all the quaternion subalgebras. However, this does not work, as one can check by calculating for example $(\overline{\omega_0}.\overline{\omega_1})i_1$ (see Exercise 4.12).

One solution is to take the ordinary integral octonions and adjoin instead $\omega = \frac{1}{2}(-1 + i_0 + i_1 + i_3)$, $\psi = \frac{1}{2}(i_0 + i_1 + i_2 + i_4)$, and close under multiplication and addition. Now it is easy to see that this algebra contains elements corresponding to the fundamental roots of E_8, thus:

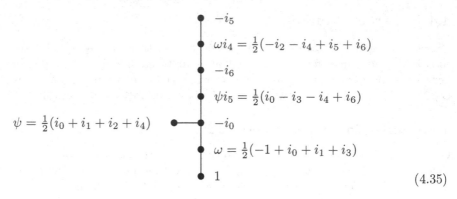

$$-i_5$$
$$\omega i_4 = \tfrac{1}{2}(-i_2 - i_4 + i_5 + i_6)$$
$$-i_6$$
$$\psi i_5 = \tfrac{1}{2}(i_0 - i_3 - i_4 + i_6)$$
$$\psi = \tfrac{1}{2}(i_0 + i_1 + i_2 + i_4) \qquad -i_0$$
$$\omega = \tfrac{1}{2}(-1 + i_0 + i_1 + i_3)$$
$$1 \tag{4.35}$$

and therefore contains all the 240 roots of E_8. Writing $1 = i_\infty$ for convenience, a straightforward calculation shows that these roots are the 16 vectors $\pm i_t$, together with $16.14 = 224$ vectors $\tfrac{1}{2}(\pm i_r \pm i_s \pm i_t \pm i_i)$ where $\{r, s, t, u\}$ is one of the 4-sets obtained from $\{\infty, t, t+1, t+3\}$ or its complement by swapping ∞ with 0. [This swap is vital but easy to forget: failure to make it has been labelled Kirmse's mistake. Of course, before making this swap there is a symmetry $i_t \mapsto i_{t+1}$, and therefore we could equally well make the swap of ∞ with t, for any fixed $t \in \mathbb{F}_7$. Thus there are seven different non-associative rings of integral octonions we could construct in this way. Kirmse claimed there were eight, and described the only one which does not exist. Dickson constructed three of the seven, but Coxeter was the first to construct them all.]

Conversely we can check that the product of any two fundamental roots is again a root: products with 1, $-i_0$, $-i_5$ and $-i_6$ are easy to check, and for the rest it is sufficient to check the four cases $\omega\psi$, $\omega(\psi i_5)$, $\psi(\omega i_4)$ and $(\psi i_5)(\omega i_4)$. It follows that the E_8 lattice is closed under the multiplication. Indeed, the 240 roots form an interesting Moufang loop.

The notions of octonion conjugate, norm and inner product still make sense, but the usual inner product is not necessarily an integer, so it is important to use the associated bilinear form

$$f(x, y) = N(x + y) - N(x) - N(y) \tag{4.36}$$

as the inner product instead. Similarly the real part of x can no longer be defined as $\tfrac{1}{2}(x + \overline{x})$, so let us define $\operatorname{Re}(x) = f(1, x)$ and then define $\overline{x} = f(1, x) - x$. This agrees with the usual definition of $\overline{x}$, and $\operatorname{Re}(x) = x + \overline{x}$ is twice the usual real part. In particular, it is still true that $N(x) = x\overline{x}$.

The swap of ∞ with 0 (i.e. of 1 with i_0) implies that i_0 is somehow special, and therefore the visible (monomial) symmetry group is now a subgroup $2^3 \cdot S_4$ of the original $2^3 \cdot \mathrm{GL}_3(2)$. It is generated by the maps

$$a : i_t \mapsto i_{-t} \text{ if } t = 0, 3, 5, 6,$$
$$b : i_t \mapsto i_{2t},$$
$$c : (i_0, \ldots, i_6) \mapsto (-i_0, -i_1, i_6, i_3, i_5, i_4, i_2),$$
$$d : (i_0, \ldots, i_6) \mapsto (i_0, -i_1, -i_4, -i_3, -i_2, -i_6, -i_5). \tag{4.37}$$

In addition, one can check that the map e defined by

$$e : i_0 \mapsto i_6 \mapsto i_2 \mapsto i_0,$$
$$i_1 \mapsto \tfrac{1}{2}(i_1 + i_3 - i_4 + i_5),$$
$$i_3 \mapsto \tfrac{1}{2}(-i_1 + i_3 + i_4 + i_5),$$
$$i_4 \mapsto \tfrac{1}{2}(i_1 - i_3 + i_4 + i_5),$$
$$i_5 \mapsto \tfrac{1}{2}(-i_1 - i_3 - i_4 + i_5) \tag{4.38}$$

is an automorphism of this 'non-associative ring' of integral octonions. (We shall loosely describe the integral octonions as a ring, although strictly speaking a ring is associative.)

The full symmetry group of this ring is a group of order 12096. To see this, first note (see Exercise 4.12) that e fuses the four orbits of the monomial group on the 126 pure imaginary roots (i.e. octonions of norm 1), so we may choose any of these to call i_0. There are two orbits of the monomial symmetry group on roots orthogonal to i_0, with representatives i_1 and $\tfrac{1}{2}(i_2 + i_4 + i_5 + i_6)$. However, if i_1 is mapped to the latter, then $i_3 = i_0 i_1$ is mapped to

$$\tfrac{1}{2} i_0 (i_2 + i_4 + i_5 + i_6) = \tfrac{1}{2}(-i_2 - i_4 + i_5 + i_6), \tag{4.39}$$

so by linearity, $\tfrac{1}{2}(1 + i_0 + i_1 + i_3)$ is mapped to $\tfrac{1}{2}(1 + i_0 + i_5 + i_6)$, which is not in the lattice. This contradiction implies that i_1 must be mapped to $\pm i_t$ (12 choices). Finally, fixing i_1, and therefore i_3, we see that i_2 is mapped to one of the remaining 8 vectors $\pm i_t$, since these are the only norm 1 vectors perpendicular to $1, i_0, i_1$, and i_3. Hence the automorphism group, which I shall denote $G_2(\mathbb{Z})$ (although there may be other groups which have an equal or greater claim to this name), has order $126 \times 12 \times 8 = 12096 = 2^6.3^3.7$. We shall show in Section 4.4.4 that this group is isomorphic to both $G_2(2)$ and $\mathrm{PSU}_3(3){:}2$.

We have verified that with respect to a basis of fundamental roots of E_8, the multiplication table consists of integers. Therefore it can be interpreted over any field whatsoever. If the field has odd characteristic, however, $\tfrac{1}{2}$ is in the field, so the vectors $\tfrac{1}{2}(i_r + i_s + i_t + i_u)$ are already in the $\mathbb{F}_q$-span of the i_t, so we get nothing new. Indeed, the resulting homomorphism from $G_2(\mathbb{Z})$ to $G_2(q)$ is injective, so $G_2(\mathbb{Z})$ is a subgroup of $G_2(q)$ for every odd q. It turns out to be maximal whenever q is prime.

4.4.3 Octonions in characteristic 2

Interpreting the multiplication table over the field $\mathbb{F}_2$, we are in effect reducing the ring of integral octonions modulo 2. This gives a set of $2^8 = 256$ elements which form what I shall call 'the' octonions over $\mathbb{F}_2$. They may be thought of as all $\mathbb{F}_2$-linear combinations of the 8 fundamental roots of the E_8 lattice given above. Of these, just 120 are invertible, and arise from the 240 roots of the E_8-lattice. The multiplicative structure they form is again a Moufang loop, and its automorphism group is (by definition) $G_2(2)$. It turns out that this group has order 12096 and is isomorphic to $G_2(\mathbb{Z})$ and $\mathrm{SU}_3(3){:}2$. (See Section 4.4.4 below for a proof.)

As in the case of the quaternions (see Section 4.4.1), we extend to arbitrary fields F of characteristic 2 by taking F-linear combinations of the fundamental roots illustrated in (4.35) above. The automorphism group of this algebra is called $G_2(F)$, or $G_2(q)$ if F has order q.

To see that this definition of $G_2(2^n)$ is equivalent to our earlier one, given in Section 4.3.4, we choose a basis for the 8-dimensional algebra (over $\mathbb{F}_{2^n}$) as follows. Let

$$
\begin{aligned}
x_1 &= 1 + i_0, \\
x_2 &= 1 + i_1, \\
x_3 &= 1 + i_2, \\
x_4 &= \tfrac{1}{2}(1 + i_0 + i_1 + i_2 + i_3 - i_4 + i_5 + i_6), \\
x_5 &= 1 + x_4, \\
x_6 &= 1 + \tfrac{1}{2}(i_2 + i_4 + i_5 + i_6), \\
x_7 &= 1 + \tfrac{1}{2}(i_1 - i_3 + i_4 - i_5), \\
x_8 &= 1 + \tfrac{1}{2}(-i_0 + i_3 + i_4 - i_6).
\end{aligned}
\tag{4.40}
$$

You can now check for yourself that these are isotropic vectors forming a symplectic basis, and that the multiplication table with respect to this basis is exactly the same as that given in (4.27) for the algebra in odd characteristic (ignoring signs). We also define conjugates, norms, inner products and real parts by reducing modulo 2 the corresponding concepts in the integral octonions. In particular $1 = x_4 + x_5$ has norm 1 and (counter-intuitively!) we have $\mathrm{Re}\,(1) = 0$. Notice also that $\langle x_4 + x_5 \rangle^\perp = \langle x_1, x_2, x_3, x_4 + x_5, x_6, x_7, x_8 \rangle$. Therefore $G_2(2^n)$ acts on the symplectic 6-space $\langle x_4 + x_5 \rangle^\perp / \langle x_4 + x_5 \rangle$. Indeed, the octonion multiplication gives rise to a symmetric trilinear form t defined on this 6-space by

$$
t(x_1, x_6, x_7) = t(x_2, x_3, x_8) = 1
\tag{4.41}
$$

and otherwise t is zero on the basis vectors. Moreover, $G_2(2^a)$ is the subgroup of $\mathrm{Sp}_6(2^a)$ consisting of all elements which preserve this symmetric trilinear form.

4.4.4 The isomorphism between $G_2(2)$ and $\mathrm{PSU}_3(3){:}2$

To show that $G_2(2) \cong \mathrm{SU}_3(3){:}2$ (which is isomorphic to $\mathrm{PSU}_3(3){:}2$) we show that both are isomorphic to the group $G_2(\mathbb{Z})$ of automorphisms of the integral octonions constructed in Section 4.4.2.

First let this group $G_2(\mathbb{Z})$ act on the 7-space of pure imaginary integral octonions reduced modulo 3. We shall explicitly construct an isomorphism with the faithful irreducible part of the so-called 'adjoint' representation of $\mathrm{SU}_3(3){:}2$. Recall that if F is a field with an automorphism $x \mapsto \overline{x}$ of order 2, then a matrix M with entries in F is called *Hermitian* if $M^\top = \overline{M}$. The set of 3×3 Hermitian matrices over $\mathbb{F}_9 = \{0, \pm 1, \pm i, \pm 1 \pm i\}$ forms a 9-dimensional space over $\mathbb{F}_3$, and if $SU_3(3)$ is written with respect to an orthonormal basis, then $g \in SU_3(3)$ acts on this space by $M \mapsto \overline{g}^\top M g$, and the field automorphism $i \mapsto -i$ of $\mathbb{F}_9$ acts as $M \mapsto -\overline{M}$. The subspace of trace 0 matrices is invariant under this action, as is the subspace of scalar matrices. Thus there is an induced action of $SU_3(3){:}2$ on the 7-dimensional space of trace 0 matrices modulo scalars.

We map each i_t to the matrix J_t, modulo $\langle I_3 \rangle$, where $J_0 = \begin{pmatrix} 1 & 0 & 0 \\ 0 & 0 & 0 \\ 0 & 0 & -1 \end{pmatrix}$,

$$J_1 = \begin{pmatrix} 0 & 0 & i \\ 0 & 0 & 0 \\ -i & 0 & 0 \end{pmatrix}, \; J_2 = \begin{pmatrix} 0 & -i & 0 \\ i & 0 & 0 \\ 0 & 0 & 0 \end{pmatrix}, \; J_4 = \begin{pmatrix} 0 & 0 & 0 \\ 0 & 0 & -i \\ 0 & i & 0 \end{pmatrix},$$

$$J_3 = \begin{pmatrix} 0 & 0 & 1 \\ 0 & 0 & 0 \\ 1 & 0 & 0 \end{pmatrix}, \; J_6 = \begin{pmatrix} 0 & 1 & 0 \\ 1 & 0 & 0 \\ 0 & 0 & 0 \end{pmatrix}, \; J_5 = \begin{pmatrix} 0 & 0 & 0 \\ 0 & 0 & 1 \\ 0 & 1 & 0 \end{pmatrix},$$

so that the map $i_t \mapsto J_t \bmod \langle I_3 \rangle$ is an isomorphism of the two $\mathbb{F}_3$-vector spaces of dimension 7.

It remains to check that this map induces an isomorphism of the groups $G_2(\mathbb{Z})$ and $SU_3(3){:}2$. First check that the monomial subgroup $4^2 D_{12} \cong 2^3 S_4$ of $SU_3(3){:}2$, generated by $\mathrm{diag}(-i, i, 1)$, the coordinate permutation $(1, 2, 3)$ and $-(2, 3)$ and the field automorphism, acts on these matrices in the same way that the monomial automorphisms act on the integral octonions, and then that the automorphism e given in (4.38) above corresponds to the matrix $\frac{1}{2}\begin{pmatrix} 1+i & 1+i & 0 \\ -1+i & 1-i & 0 \\ 0 & 0 & 2 \end{pmatrix}$ in $SU_3(3)$. These are straightforward calculations: for example $\mathrm{diag}(-i, i, 1)$ conjugates $J_0, \ldots, J_6$ to $J_0, -J_3, -J_2, J_1, -J_5, J_4, -J_6$ respectively.

Since the two groups $G_2(\mathbb{Z})$ and $SU_3(3){:}2$ have the same order, it follows that they are isomorphic.

Next we need to show that $G_2(\mathbb{Z})$ and $G_2(2)$ are isomorphic. The definition of $G_2(2)$ in Section 4.4.3 shows that there is a natural map $G_2(\mathbb{Z}) \to G_2(2)$. The kernel of this map is trivial, since any element of the kernel fixes the 63 images of the roots, so acts as ± 1 (and therefore as $+1$, since -1 does not preserve the multiplication) on the 7-space. In particular, the order of $G_2(2)$ is at least 12096. We have already stated without proof, in (4.28), the fact that $G_2(2)$ has order $2^6(2^6 - 1)(2^2 - 1) = 12096$, but for the sake of completeness we give a proof here.

Any element of $G_2(2)$ is determined by the images of i_0, $\frac{1}{2}(1 + i_0 + i_1 + i_3)$ and $\frac{1}{2}(i_0 + i_1 + i_2 + i_4)$, since these generate the whole algebra. But i_0 has at most 63 images, and having fixed i_0 there are at most 32 possibilities for the image of $\frac{1}{2}(i_0 + i_1 + i_2 + i_4)$, namely the 32 non isotropic vectors which are orthogonal to 1 but not to i_0. Having fixed these two elements, we have fixed the algebra A they generate, which is spanned as a vector space by 1, $\frac{1}{2}(1 + i_3 + i_5 + i_6)$, i_0, $\frac{1}{2}(i_0 + i_1 + i_2 + i_4)$, and is therefore an orthogonal 4-space of plus type. The image of $\frac{1}{2}(1 + i_0 + i_1 + i_3)$ can then be written as $a + b$, with $a \in A$, $b \in A^{\perp}$, where $a = \frac{1}{2}(-1 - i_0 + i_1 + i_2 + i_3 + i_4 + i_5 + i_6)$ is determined by the inner products with the basis vectors of A. Hence b is one of the six non-isotropic vectors of $A^{\perp}$. Therefore $G_2(2)$ has order at most $63.32.6 = 12096$ and hence the two groups $G_2(2)$ and $G_2(\mathbb{Z})$ are isomorphic as claimed.

4.5 The small Ree groups

The small Ree groups ${}^2G_2(3^{2n+1})$ bear the same relation to the groups $G_2(3^{2n+1})$ as the Suzuki groups ${}^2B_2(2^{2n+1})$ bear to the symplectic groups $\mathrm{Sp}_4(2^{2n+1})$. There are many parallels between the two cases, as we shall see as we go along. Our treatment of the Ree groups is however a little less elementary than our treatment of the Suzuki groups.

4.5.1 The outer automorphism of $G_2(3)$

First we construct the exceptional outer automorphism of $G_2(3)$. To do this we take the exterior square of the natural 7-dimensional module, defined in Section 4.3.2 with respect to an orthonormal basis $\{i_t \mid t \in \mathbb{F}_7\}$. This gives us a 21-dimensional space spanned by $i_s \wedge i_t = -i_t \wedge i_s$ for $s \neq t$. There is an invariant 7-space W_0 spanned by the vector

$$i_1 \wedge i_3 + i_2 \wedge i_6 + i_4 \wedge i_5 \qquad (4.42)$$

and its images under $i_t \mapsto i_{t+1}$. Modulo W_0 there is another 7-space V^* spanned by the vectors

$$i_t^* = i_{t+1} \wedge i_{t+3} - i_{t+2} \wedge i_{t+6} + W_0, \qquad (4.43)$$

which is the space we want. I claim that the linear map defined by $* : i_t \mapsto i_t^*$ induces an automorphism of $G_2(3)$.

To see this, note first that by construction the map $*$ commutes with the subscript permutations $(0, 1, 2, 3, 4, 5, 6)$ and $(1, 2, 4)(3, 6, 5)$, and it is easy to see that it commutes with the sign-change on coordinates 0, 3, 5, 6. Next we calculate that the map

$$(i_0, \ldots, i_6) \mapsto (i_0, -i_1, i_4, -i_3, i_2, i_6, i_5) \qquad (4.44)$$

takes $(i_0^*, \ldots, i_6^*)$ to $(-i_0^*, -i_1^*, i_4^*, i_3^*, i_2^*, -i_6^*, -i_5^*)$, so that $*$ acts as an outer automorphism of the monomial subgroup $2^3 \cdot \mathrm{GL}_3(2)$. But the octonion multiplication is determined (up to an overall scalar multiple) by this subgroup, and $G_2(3)$ is defined as the automorphism group of the octonion product, so $*$ acts as an (outer) automorphism of $G_2(3)$. Indeed, the same argument shows that if we extend the field to $\mathbb{F}_{3^k}$, then $*$ acts as an outer automorphism of $G_2(3^k)$.

It is easy to show that $G_2(3)$ commutes with the map $**$: we need only check that the map (4.44) induces the map

$$(i_0^{**}, \ldots, i_6^{**}) \mapsto (i_0^{**}, -i_1^{**}, i_4^{**}, -i_3^{**}, i_2^{**}, i_6^{**}, i_5^{**}). \tag{4.45}$$

We shall see in Section 4.5.2 that over arbitrary fields of characteristic 3, the map $**$ can be identified with the field automorphism of $G_2(3^k)$, induced by the map

$$\sum_{t=0}^{6} \lambda_t i_t \mapsto \sum_{t=0}^{6} \lambda_t^3 i_t.$$

We now use this to motivate the definition of the Ree groups. We know that $*$ is an automorphism of $G_2(3^k)$, squaring to the field automorphism which maps all matrix entries to their cubes. In order to turn $*$ into an automorphism of order 2, therefore, we need k to be odd, say $k = 2n + 1$, so that the map

$$\sum_{t=0}^{6} \lambda_t i_t \mapsto \sum_{t=0}^{6} \lambda_t^{3^{n+1}} i_t^* \tag{4.46}$$

is such an automorphism. Its centraliser is defined to be the Ree group $^2G_2(3^{2n+1})$.

4.5.2 The Borel subgroup of $^2G_2(q)$

To investigate the structure of these groups, it is best to change basis so that a Sylow 3-subgroup becomes a group of lower triangular matrices. The new basis is essentially the same basis as described for $G_2(q)$ in Section 4.3.4. We take $a = b = 1$ and replace x_4 and x_5 by $x_4 - x_5$ to get a pure imaginary octonion. Thus our basis consists of the vectors

$$\begin{aligned}
v_1 &= i_3 + i_5 + i_6, \\
v_2 &= i_1 + i_2 + i_4, \\
v_3 &= -i_0 - i_3 + i_6, \\
v_4 &= i_2 - i_1, \\
v_5 &= -i_0 + i_3 - i_6, \\
v_6 &= -i_1 - i_2 + i_4, \\
v_7 &= -i_3 + i_5 - i_6.
\end{aligned} \tag{4.47}$$

(Note: the version in [182] differs from this by changing sign on v_1, v_2, v_4, v_6 and v_7.) The octonion multiplication gives rise to a multiplication on this 7-space by projecting modulo $1 = x_4 + x_5$, as follows.

	v_1	v_2	v_3	v_4	v_5	v_6	v_7
v_1				$-v_1$	v_2	$-v_3$	v_4
v_2			$-v_1$	v_2		$-v_4$	v_5
v_3		v_1		v_3	$-v_4$		$-v_6$
v_4	v_1	$-v_2$	$-v_3$		v_5	v_6	$-v_7$
v_5	$-v_2$		v_4	$-v_5$		v_7	
v_6	v_3	v_4			$-v_6$	$-v_7$	
v_7	$-v_4$	$-v_5$	v_6	v_7			

$$(4.48)$$

Now we calculate (modulo the 7-space W_0 defined in (4.42))

$$v_1^* = i_3^* + i_5^* + i_6^* \equiv i_4 \wedge i_6 - i_5 \wedge i_2 + i_2 \wedge i_3 - i_6 \wedge i_1 + i_1 \wedge i_5 - i_3 \wedge i_4$$

and also

$$v_2 \wedge v_1 = i_1 \wedge i_3 + i_2 \wedge i_6 + i_4 \wedge i_5 + i_1 \wedge i_5 + i_4 \wedge i_3 + i_2 \wedge i_3 + i_1 \wedge i_6 + i_4 \wedge i_6 + i_2 \wedge i_5$$

so that $v_1^* \equiv v_2 \wedge v_1 \bmod W_0$. Similarly (see Exercise 4.16) we find expressions for the other vectors v_j^*:

$$v_2^* \equiv v_3 \wedge v_1,$$
$$v_3^* \equiv v_5 \wedge v_2,$$
$$v_4^* \equiv v_7 \wedge v_1 + v_6 \wedge v_2,$$
$$v_5^* \equiv v_3 \wedge v_6,$$
$$v_6^* \equiv v_5 \wedge v_7,$$
$$v_7^* \equiv v_6 \wedge v_7. \tag{4.49}$$

It is now elementary but tedious to show by direct calculation that with respect to the ordered basis $(v_1, \ldots, v_7)$ the matrices

$$
\begin{pmatrix}
1 & & & & & & \\
 & 1 & & & & & \\
1 & & 1 & & & & \\
 & 1 & & 1 & & & \\
2 & & & & 1 & & \\
 & 1 & & 2 & & 1 & \\
1 & & 1 & & 2 & & 1
\end{pmatrix}
\quad \text{and} \quad
\begin{pmatrix}
1 & & & & & & \\
2 & 1 & & & & & \\
1 & 2 & 1 & & & & \\
 & & 1 & 1 & & & \\
2 & & & 1 & 2 & 1 & \\
1 & & & 1 & 2 & 1 & 1 \\
2 & 1 & 1 & & & 1 & 1
\end{pmatrix}, \tag{4.50}
$$

where blank entries are 0, belong to $^2G_2(3)$, and generate a subgroup of order 27. Other useful elements are the sign change on i_0, i_3, i_5 and i_6, which is the sign change on v_1, v_3, v_5, v_7; and the sign change on i_1, i_2, i_3, and i_6, which maps v_4 to $-v_4$ and v_j to v_{8-j} for all $j \neq 4$.

Now consider which diagonal matrices can be in the Ree group. As we saw in Section 4.3.5, the torus in $G_2(q)$ consists of diagonal elements $\mathrm{diag}(\lambda, \mu, \lambda\mu^{-1}, 1, \lambda^{-1}\mu, \mu^{-1}, \lambda^{-1})$. Using the equations (4.49) we see that this map induces on $\{v_1^*, \ldots, v_7^*\}$ the map

$$\mathrm{diag}(\lambda\mu, \lambda^2\mu^{-1}, \lambda^{-1}\mu^2, 1, \lambda\mu^{-2}, \lambda^{-2}\mu, \lambda^{-1}\mu^{-1}), \tag{4.51}$$

and therefore induces on $\{v_1^{**}, \ldots, v_7^{**}\}$ the map

$$\mathrm{diag}(\lambda^3, \mu^3, \lambda^3\mu^{-3}, 1, \lambda^{-3}\mu^3, \mu^{-3}, \lambda^{-3}). \tag{4.52}$$

Since $**$ commutes with the irreducible subgroup $2^3{\cdot}\mathrm{GL}_3(2)$, this tells us that $**$ induces on $G_2(q)$ the field automorphism which replaces every matrix entry by its cube.

Since the Ree group consists of those matrices whose action on V^* is obtained by applying the field automorphism $x \mapsto x^{3^{n+1}}$ to each matrix entry, we have $\lambda^{3^{n+1}} = \mu.\lambda$, so $\mu = \lambda^{-3^{n+1}-1}$. In other words the diagonal elements of the Ree group are

$$\mathrm{diag}(\lambda, \lambda^{3^{n+1}-1}, \lambda^{-3^{n+1}+2}, 1, \lambda^{3^{n+1}-2}, \lambda^{-3^{n+1}+1}, \lambda^{-1}). \tag{4.53}$$

Now we can use this diagonal element in the same way as for the Suzuki group to make the Sylow 3-normaliser (or Borel subgroup) of order $q^3(q-1)$, which is generated by the matrices in (4.50) and (4.53). It can be shown by arguments similar to those used in Section 4.2.2 that the fixed 1-space $\langle v_1 \rangle$ of the Borel subgroup has $q^3 + 1$ images under the group $^2G_2(q)$, which therefore has order $(q^3 + 1)q^3(q - 1)$. Details of the calculation can be found in [182]. (See also Exercise 4.17.)

Indeed, $^2G_2(q)$ acts 2-transitively (and therefore primitively) on these $q^3 + 1$ points, and the point stabiliser is just the Borel subgroup. The latter has a normal abelian subgroup of order q, whose conjugates generate $^2G_2(q)$ provided $q > 3$. Moreover, these elements are easily seen to be commutators already inside the Borel subgroup, so we conclude that $^2G_2(q)$ is simple if $q > 3$, by Iwasawa's Lemma (Theorem 3.1).

As a final remark, note that another (less explicit, and less illuminating) construction of the small Ree groups is given in Section 5.9.1.

4.5.3 Other subgroups

The torus, a cyclic group of order $q - 1$ consisting of the diagonal elements (4.53), is normalised by the involution $v_4 \mapsto -v_4$, $v_j \mapsto v_{8-j}$ ($j \neq 4$), so that the normaliser of the torus is $D_{2(q-1)} \cong 2 \times D_{q-1}$, since $(q - 1)/2$ is odd. Another subgroup which can be fairly easily seen is the involution centraliser $2 \times \mathrm{PSL}_2(q)$. The involution $\mathrm{diag}(-1, 1, -1, 1, -1, 1, -1)$, which is the case $\lambda = -1$ in the above torus (4.53), is centralised by $2 \times D_{q-1}$. We can extend this to the full centraliser by adjoining the element which negates i_0, i_2, i_3, i_4 and fixes i_1, i_5, i_6. This element acts on our new basis as $\begin{pmatrix} 1 & -1 & 1 \\ 1 & 0 & -1 \\ 1 & 1 & 1 \end{pmatrix}$ on

$\langle v_2, v_4, v_6 \rangle$ and as $\begin{pmatrix} -1 & -1 & 1 & -1 \\ -1 & 1 & 1 & 1 \\ 1 & 1 & 1 & -1 \\ -1 & 1 & -1 & -1 \end{pmatrix}$ on $\langle v_1, v_3, v_5, v_7 \rangle$.

The other maximal subgroups (for $q > 3$) are in fact $^2G_2(q_0)$ whenever $q = q_0{}^r$ with r prime, the four-group normaliser $(2^2 \times D_{(q+1)/2}){:}3$, and two classes of Frobenius groups $(3^{2n+1} \pm 3^{n+1} + 1){:}6$, where $q = 3^{2n+1}$. These Frobenius groups are the stabilisers of certain 1-dimensional subspaces.

The four-group normaliser is most easily seen in the original basis. For example we may take the four-group generated by negation on two of the three 2-spaces $\langle i_1, i_3 \rangle$, $\langle i_2, i_6 \rangle$, $\langle i_4, i_5 \rangle$. As this four-group fixes a unique 1-space (spanned by i_0), the four-group normaliser is equal to the 1-space stabiliser. It may be generated by adjoining to the four-group itself the subscript permutation $(1, 2, 4)(3, 6, 5)$ and a suitable reflection inside the orthogonal space $\langle i_1, i_3 \rangle$ (which can be extended in a unique way to the whole space). Indeed, it is not hard to verify that the $\frac{1}{2}q^3(q^3 - 1)$ 1-spaces spanned by vectors of norm 1 fall into exactly three orbits under $^2G_2(q)$, with stabilisers $(2^2 \times D_{(q+1)/2}){:}3$ and $C_{q \pm \sqrt{3q}+1}{:}6$. On the other hand, $^2G_2(q)$ is transitive on the $\frac{1}{2}q^3(q^3 + 1)$ 1-spaces spanned by vectors of norm -1, and the stabiliser is the non-maximal subgroup $D_{2(q-1)}$, namely the normaliser of a maximal split torus. There are three orbits on isotropic 1-spaces, of lengths $q^3 + 1$, $q(q^3 + 1)$ and $q^2(q^3 + 1)$, represented respectively by $\langle x_1 \rangle$, $\langle x_2 \rangle$ and $\langle x_3 \rangle$.

To summarise (see Kleidman [106] and/or Levchuk and Nuzhin [119]):

Theorem 4.2. *If $q = 3^{2n+1}$ with $n \geqslant 1$, then the maximal subgroups of $^2G_2(q)$ are (up to conjugacy)*

(i) $q^{1+1+1}{:}C_{q-1}$,
(ii) $2 \times \mathrm{PSL}_2(q)$,
(iii) $(2^2 \times D_{(q+1)/2}){:}3$,
(iv) $C_{q+\sqrt{3q}+1}{:}6$,
(v) $C_{q-\sqrt{3q}+1}{:}6$,
(vi) $^2G_2(q_0)$, where $q = q_0{}^r$ and r is prime.

There are no non-trivial covers of the small Ree groups $^2G_2(3^{2n+1})$, and the only outer automorphisms are the field automorphisms, giving a cyclic outer automorphism group of order $2n + 1$.

4.5.4 The isomorphism $^2G_2(3) \cong \mathrm{P\Gamma L}_2(8)$

The smallest Ree group $^2G_2(3)$ turns out to be isomorphic to $\mathrm{PGL}_2(8){:}3$, and the natural 7-dimensional representation of the Ree group corresponds to the doubly-deleted permutation representation of $\mathrm{PGL}_2(8) \cong \mathrm{PSL}_2(8)$ on the 9-point projective line. (This is called 'doubly-deleted' because there is a fixed vector $v_0 = (1, 1, 1, 1, 1, 1, 1, 1, 1)$ and a fixed hyperplane $v_0{}^\perp$ which contains v_0, so that $v_0{}^\perp / \langle v_0 \rangle$ has dimension 7.)

We construct this representation first. The field of order 8 is obtained by adjoining to $\mathbb{F}_2$ an element η satisfying $\eta^3 + \eta + 1 = 0$. Thus the elements of $\mathbb{F}_8$ are 0, 1, η, η^2, $\eta^3 = \eta + 1$, $\eta^4 = \eta^2 + \eta$, $\eta^5 = \eta^2 + \eta + 1$, $\eta^6 = \eta^2 + 1$. Adjoining ∞ to this set, we obtain the projective line $\mathrm{PL}(8) = \mathbb{F}_8 \cup \{\infty\}$, whose automorphism group is $P\Gamma L_2(8) \cong PGL_2(8){:}3 \cong PSL_2(8){:}3$. This group may be generated by the following elements

$$
\begin{aligned}
a &: z \mapsto z+1, \\
b &: z \mapsto \eta z, \\
c &: z \mapsto z^2, \\
d &: z \mapsto z^{-1}.
\end{aligned} \tag{4.54}
$$

The stabiliser of ∞ is a group $2^3{:}7{:}3$ generated by a, b, and the field automorphism c. A Sylow 3-subgroup (of order 27) may be generated by c and $z \mapsto (\eta z + 1)/(z + \eta + 1)$ (i.e. the permutation $(\infty, \eta, \eta^6, 0, \eta^4, \eta^5, 1, \eta^2, \eta^3)$) of the projective line.

Now the 7-dimensional representation over $\mathbb{F}_3$ can be obtained by taking a 9-space spanned by vectors e_z for $z \in \mathbb{F}_8 \cup \{\infty\}$, and taking the subspace of vectors with coordinate sum 0, modulo the 1-space spanned by $(1,1,1,1,1,1,1,1,1)$. The orthonormal basis $\{i_0, \ldots, i_6\}$ can be calculated as a basis of eigenvectors of the 2^3 subgroup $\langle a, a^b, a^{b^2} \rangle$, and we find that (writing the coordinates in the order $\infty, 0, 1, \eta, \eta^2, \eta^3, \eta^4, \eta^5, \eta^6$)

$$
\begin{aligned}
i_0 &= (0,1,-1,1,1,-1,1,-1,-1), \\
i_1 &= (0,1,1,1,-1,1,-1,-1,-1), \\
i_2 &= (0,1,1,-1,1,-1,-1,-1,1),
\end{aligned} \tag{4.55}
$$

and so on (we fix the first two coordinates, and rotate the last seven coordinates backwards each time the subscript increases by 1). In other words, $b : i_t \mapsto i_{t-1}$. It is easy to verify that c acts on this basis as $i_t \mapsto i_{2t}$.

To prove that $PSL_2(8){:}3$ is contained in $G_2(3)$, therefore, it suffices to check that the map

$$
i_t \mapsto -i_{-t} + i_{1-t} + i_{2-t} + i_{4-t} \tag{4.56}
$$

corresponding to $z \mapsto z^{-1}$ is an automorphism of the octonion algebra. By symmetry, we need only check the product

$$
\begin{aligned}
i_0.i_1 &\mapsto (-i_0 + i_1 + i_2 + i_4)(-i_6 + i_0 + i_1 + i_3) \\
&= (-i_4 + i_5 + i_6 + i_1),
\end{aligned} \tag{4.57}
$$

which is the image of i_3, as required.

Finally we calculate that the map (4.56) takes i_0^* to $-i_0^* + i_1^* + i_2^* + i_4^*$, and hence, by symmetry, induces the map

$$
i_t^* \mapsto -i_{-t}^* + i_{1-t}^* + i_{2-t}^* + i_{4-t}^*. \tag{4.58}
$$

Thus the group $PSL_2(8){:}3$ constructed above commutes with the automorphism $*$, so $PSL_2(8){:}3 \subseteq {}^2G_2(3)$. But the two groups both have order 1512, so they are equal.

4.6 Twisted groups of type $^3D_4$

4.6.1 Twisted octonion algebras

The group $^3D_4(q)$ is usually defined as the centraliser in $\mathrm{P\Omega}_8^+(q^3)$ of a certain automorphism of order 3, namely the product of a field automorphism of order 3 with a 'triality' automorphism (see the end of Section 3.9 and Section 4.7). I prefer a more concrete approach, obtained by 'twisting' the octonion algebra (see Section 4.3.2) over $\mathbb{F}_{q^3}$ by the field automorphism $x \mapsto x^q$ of order 3. Because the triality automorphism rotates x, y and $\overline{xy}$ (rather than xy) we also need to take the octonion conjugate. That is, replace the ordinary octonion product by a new bi-additive product $*$ which takes the value $\overline{xy} = \overline{y}.\overline{x}$ for x and y in our standard basis (either $\{1, i_0, \ldots, i_6\}$ or $\{x_1, \ldots, x_8\}$), and instead of the bilinearity condition $(\lambda a)(\mu b) = (\lambda\mu)(ab)$ we now have

$$(\lambda a) * (\mu b) = (\lambda^q \mu^{q^2})(a * b) \tag{4.59}$$

for all $\lambda, \mu \in \mathbb{F}_{q^3}$ and all a, b in the algebra. Notice that the twisted algebra has no identity element, since $1 * x = x$ would imply $1 * (\lambda x) = \lambda^{q^2} x \neq \lambda x$. Indeed, we shall see that the automorphism group of the twisted algebra acts irreducibly on it. It is immediate from this construction that $^3D_4(q)$ contains $G_2(q)$.

The algebra still has a norm and inner product defined over $\mathbb{F}_{q^3}$. The norm is still a quadratic form but now it satisifies

$$N(a * b) = N(a)^q N(b)^{q^2} \tag{4.60}$$

instead of $N(ab) = N(a)N(b)$.

Warning. Springer and Veldkamp [158] define $x * y = \overline{yx} = \overline{x}.\overline{y}$ on the standard basis instead of $\overline{xy}$. This different convention does not affect the general picture, but does change some of the technical details.

4.6.2 The order of $^3D_4(q)$

In order to calculate the order of $^3D_4(q)$ we study the isotropic vectors v such that $v * v = 0$. We shall show that the group acts transitively on them, and calculate the number of them, and the stabiliser of one of them. Recall that in the untwisted octonions an isotropic vector v satisfies $v\overline{v} = 0$, so satisfies $vv = 0$ if and only if v is purely imaginary. But in the absence of an identity element, this no longer makes sense. First we do some preliminary investigations into the structure of the group.

Using the basis $\{x_1, \ldots, x_8\}$ defined for $G_2(q)$ in Section 4.3.4, and remembering that $\overline{x_4} = x_5$, it is easy to calculate that the diagonal elements in $^3D_4(q)$ are just

$$\mathrm{diag}(\alpha, \beta\alpha^{-1}, \beta^{-1}\alpha^{q^2+q}, \alpha^{q^2-q}, \alpha^{q-q^2}, \beta\alpha^{-q^2-q}, \beta^{-1}\alpha, \alpha^{-1}) \tag{4.61}$$

where $\alpha \in \mathbb{F}_{q^3}$ and $\beta \in \mathbb{F}_q$. These elements constitute a maximal torus $T \cong C_{q^3-1} \times C_{q-1}$. Notice that if α is a generator (of order q^3-1) for the multiplicative group of $\mathbb{F}_{q^3}$ then α^{q^2-q} has order $(q^3-1)/(q-1) = q^2+q+1$. Therefore the action of T on the middle two coordinates is only C_{q^2+q+1}, while it is a full C_{q^3-1} on each of the other coordinates. This torus is normalised by a group D_{12} (this is the Weyl group of type G_2 again) generated by the elements given in (4.30), which map $(x_1, \ldots, x_8)$ to $(-x_1, -x_3, -x_2, x_4, x_5, -x_7, -x_6, -x_8)$ and $(-x_2, -x_1, -x_6, x_5, x_4, -x_3, -x_8, -x_7)$.

To make the Borel subgroup B we simply take the elements $A(\lambda), \ldots, F(\lambda)$ generating the unipotent subgroup of order q^6 in $G_2(q)$ (see (4.34)), and conjugate them by the torus T to make the unipotent subgroup U of order q^{12} in $^3D_4(q)$. We find that the 'long root elements' $A(\lambda)$, $B(\lambda)$ and $E(\lambda)$ are essentially fixed (as a set) by this conjugation, whereas the 'short root elements' $C(\lambda)$, $D(\lambda)$ and $F(\lambda)$ are conjugated to 'twisted' elements defined for all $\lambda \in \mathbb{F}_{q^3}$. So for example,

$$
\begin{aligned}
C(\lambda) : x_4 &\mapsto x_4 - \lambda^{1+q-q^2} x_1, \\
x_5 &\mapsto x_5 + \lambda^{1-q+q^2} x_1, \\
x_6 &\mapsto x_6 + \lambda^{q+q^2-1} x_2, \\
x_7 &\mapsto x_7 - \lambda^{q+q^2-1} x_3, \\
x_8 &\mapsto x_8 + \lambda^{1+q-q^2} x_5 - \lambda^{1-q+q^2} x_4 + \lambda^2 x_1.
\end{aligned}
\tag{4.62}
$$

To summarise, the Borel subgroup of $^3D_4(q)$ has shape $[q^{12}]{:}(C_{q^3-1} \times C_{q-1})$.

Next we determine the stabiliser of the vector x_1. This subgroup fixes various subspaces defined by x_1, including

$$
\begin{aligned}
\langle x_1, x_2, x_3, x_4 \rangle &= \{v \mid x_1 * v = 0\}, \\
\langle x_1, x_2, x_3, x_5 \rangle &= \{v \mid v * x_1 = 0\},
\end{aligned}
\tag{4.63}
$$

their sum and intersection, and the orthogonal complements of all of these. Since $x_2 * x_3 = x_1$, the action of the stabiliser of x_1 on $\langle x_2, x_3 \rangle$ is at most $\mathrm{SL}_2(q)$. On the other hand, we already see a group $\mathrm{SL}_2(q)$ acting faithfully on it inside $G_2(q)$, so the action is exactly $\mathrm{SL}_2(q)$. Moreover, if we also fix x_2 then the resulting stabiliser consists of lower triangular matrices.

We must now show that with respect to the basis $\{x_1, \ldots, x_8\}$ any automorphism which is represented by a lower triangular matrix is in the group $[q^{12}]{:}(C_{q^3-1} \times C_{q-1})$ just described. First, the same argument that produced the torus shows that the diagonal entries of the matrix are as in (4.61), so we may assume the matrix is lower unitriangular. In particular, multiplying by suitable elements $F(\lambda_1)$ and $D(\lambda_2)$ we may assume that $x_1 \mapsto x_1$, $x_2 \mapsto x_2$ and $x_3 \mapsto x_3 + \mu x_2$. Now

$$
0 = (x_3 + \mu x_2) * (x_3 + \mu x_2) = (\mu^q - \mu^{q^2})x_1
\tag{4.64}
$$

so $\mu \in \mathbb{F}_q$, and multiplying by $E(\mu)$ we may assume $x_3 \mapsto x_3$. Next the products $x_4 * x_1 = -x_1$, $x_2 * x_4 = -x_2$ and $x_3 * x_4 = -x_3$ imply that $x_4 \mapsto x_4$.

Similar arguments complete the proof. It follows that the stabiliser in $^3D_4(q)$ of x_1 has the shape $q^{2+3+6}{:}SL_2(q)$. Indeed, if we adjoin the full torus we obtain the stabiliser of the 1-space $\langle x_1 \rangle$, namely the maximal parabolic subgroup

$$q^{2+3+6}{:}SL_2(q).C_{q^3-1}. \tag{4.65}$$

To complete the calculation of the order of $^3D_4(q)$ it only remains to count the number of images of x_1. We shall do more than this: we shall show that the images of x_1 are precisely the isotropic vectors v satisfying $v*v = 0$, and that there are precisely $(q^8 + q^4 + 1)(q^6 - 1)$ of the latter. We already know that the Weyl group maps x_1 to x_2, x_3, x_6, x_7 and x_8. The unipotent subgroup U takes $x_2, \ldots, x_8$ respectively to $q^3, q^4, q^7, q^8, q^{11}$ vectors, making

$$1 + q^3 + q^4 + q^7 + q^8 + q^{11} = (q^8 + q^4 + 1)(q^3 + 1)$$

in all (or $q^3 - 1$ times this if we count all non-zero multiples). Conversely, it is not difficult to show that if $x = x_k + \sum_{i<k} \lambda_i x_i$ is an isotropic vector with $x * x = 0$, then the coefficients λ_i satisfy various equations, such that the numbers of such vectors are exactly as above. [For example, if the leading term is x_6 we may use suitable $F(\lambda)$ and $C(\lambda)$ to reduce to $x = x_6 + \lambda_4 x_4 + \lambda_3 x_3 + \lambda_1 x_1$, whence

$$\begin{aligned} x * x = -\lambda_4{}^q x_6 - \lambda_3{}^{q^2} x_5 - \lambda_3{}^q x_4 - \lambda_3{}^q \lambda_4{}^{q^2} x_3 + \\ (\lambda_1{}^q - \lambda_1{}^{q^2}) x_2 - \lambda_4{}^q \lambda_1{}^{q^2} x_1 \end{aligned} \tag{4.66}$$

and therefore $\lambda_4 = 0$, $\lambda_3 = 0$ and $\lambda_1{}^q = \lambda_1{}^{q^2}$, so $\lambda_1 \in \mathbb{F}_q$, so by using a suitable element $B(\lambda)$ we may assume $x = x_6$.]

In other words, modulo a certain amount of explicit calculation, we have shown that

  (i) there are exactly $(q^8 + q^4 + 1)(q^6 - 1)$ isotropic vectors v with $v*v = 0$,
 (ii) the group $^3D_4(q)$ acts transitively on such vectors, and
(iii) the stabiliser of such a vector has order $q^{12}(q^2 - 1)$.

It follows immediately that the order of $^3D_4(q)$ is

$$|{}^3D_4(q)| = q^{12}(q^8 + q^4 + 1)(q^6 - 1)(q^2 - 1). \tag{4.67}$$

4.6.3 Simplicity

To prove simplicity of $^3D_4(q)$ we use the variant of Iwasawa's Lemma given in Exercise 3.3, applied to the action of $^3D_4(q)$ on the isotropic 1-spaces $\langle v \rangle$ with $v * v = 0$. The point stabiliser $q^{2+3+6}{:}SL_2(q).C_{q^3-1}$ has orbits of lengths 1, $q^3(q+1)$, $q^7(q+1)$ and q^{11} on the points, and it is easy to check that none of the putative block sizes $1 + q^3 + q^4$, $1 + q^7 + q^8$ or $1 + q^3 + q^4 + q^7 + q^8$ divides the total number of points, so the action is primitive (see Exercise 4.20). The point stabiliser has a normal soluble subgroup q^{2+3+6} which is contained in the derived subgroup. Let H be the subgroup of $^3D_4(q)$ generated by the conjugates of q^{2+3+6}: it remains to show that $H = {}^3D_4(q)$.

Since $E(\lambda)$ is conjugate to $A(\lambda)$, which is in q^{2+3+6}, it follows that H contains $q^{2+3+6}{:}\mathrm{SL}_2(q)$. Similarly, H contains $q^{1+8}{:}\mathrm{SL}_2(q^3)$, which is generated by conjugates of $C(\lambda)$. Therefore H strictly contains the point stabiliser $q^{2+3+6}{:}\mathrm{SL}_2(q).C_{q^3-1}$. But the latter subgroup is maximal, since the action on its cosets is primitive, and therefore $H = {}^3D_4(q)$. We conclude that $^3D_4(q)$ is simple for all q.

There are no proper covers of any of the groups $^3D_4(q)$. The only automorphisms are field automorphisms, induced by automorphisms of the field of order q^3.

4.6.4 The generalised hexagon

We showed that there are exactly $(q^6 - 1)(q^8 + q^4 + 1)$ non-zero vectors v with $v * v = 0$, and therefore there are $(q^3 + 1)(q^8 + q^4 + 1)$ 1-spaces spanned by such vectors. These 1-spaces are called *points*. Two points are called *adjacent* if their product in the twisted octonion algebra is 0. If two points are adjacent, then the 2-space they span contains $q^3 + 1$ subspaces of dimension 1, all of which are points adjacent with the first two. These 2-spaces are called *lines*, and it is not hard to see that each point is in $q + 1$ lines, and therefore the total number of lines is $(q + 1)(q^8 + q^4 + 1)$.

If we now draw the *point–line incidence graph* (as in Section 4.3.8), we obtain a bipartite graph with the following shape, where figures below the nodes denote the number of vertices, and figures above the edges denote the number of edges incident with each vertex. Notice that, unlike the case of the generalised hexagon for $G_2(q)$, the numbers of points and lines are different. Therefore I have drawn the graph twice, once from the point of view of a point stabiliser, and then from the point of view of a line stabiliser.

(As well as the generalised hexagon for $G_2(q)$ described in Section 4.3.8, compare the generalised quadrangles for $\mathrm{PSp}_4(q)$ described in Section 3.5.6 and for $\mathrm{PSU}_4(q)$ and $\mathrm{PSU}_5(q)$ described in Section 3.6.4, and the generalised octagon for $^2F_4(q)$ described in Section 4.9.4.)

4.6.5 Maximal subgroups of $^3D_4(q)$

To make the maximal parabolic subgroups, we adjoin to B suitable subgroups of the Weyl group, just as in the case of $G_2(q)$. Thus we adjoin the element

$(x_1, \ldots, x_8) \mapsto (x_2, x_1, -x_6, x_5, x_4, -x_3, x_8, x_7)$ to obtain

$$q^{2+3+6}{:}\mathrm{SL}_2(q).C_{q^3-1}, \tag{4.68}$$

and adjoin $(x_1, \ldots, x_8) \mapsto (-x_1, -x_3, -x_2, x_4, x_5, -x_7, -x_6, -x_8)$ to obtain

$$q^{1+8}{:}\mathrm{SL}_2(q^3).C_{q-1}. \tag{4.69}$$

In characteristic 2 the maximal parabolic subgroups have the slightly simpler shapes

$$\begin{aligned} q^{1+8}{:}(C_{q-1} \times \mathrm{SL}_2(q^3)), \\ q^{2+3+6}{:}(C_{q^3-1} \times \mathrm{SL}_2(q)). \end{aligned} \tag{4.70}$$

The involution centraliser (for q odd) has the shape

$$2{\cdot}(\mathrm{PSL}_2(q^3) \times \mathrm{PSL}_2(q)).2. \tag{4.71}$$

If we take the involution which negates x_1, x_3, x_6, x_8, its centraliser is generated by T, $B(\lambda)$, $D(\lambda)$ and the elements of the Weyl group mapping

$$\begin{aligned} (x_1, \ldots, x_8) &\mapsto (-x_6, -x_2, -x_8, x_4, x_5, -x_1, -x_7, -x_3), \\ (x_1, \ldots, x_8) &\mapsto (x_3, -x_7, x_1, x_5, x_4, x_8, -x_2, x_6). \end{aligned} \tag{4.72}$$

For q even there is an analogous subgroup $\mathrm{SL}_2(q^3) \times \mathrm{SL}_2(q)$, generated in the same way.

Other subgroups of interest include $\mathrm{PGL}_3(q)$ when $q \equiv 1 \bmod 3$, and $\mathrm{PGU}_3(q)$ when $q \equiv 2 \bmod 3$. A complete list of maximal subgroups is given by Kleidman (see [106]).

Theorem 4.3. *The maximal subgroups of $^3D_4(q)$ are as follows:*

(i) $q^{1+8}{:}\mathrm{SL}_2(q^3).C_{q-1}$,
(ii) $q^{2+3+6}{:}\mathrm{SL}_2(q).C_{q^3-1}$,
(iii) $G_2(q)$,
(iv) $\mathrm{PGL}_3(q)$ *if* $q \equiv 1 \bmod 3$,
(v) $\mathrm{PGU}_3(q)$ *if* $2 < q \equiv 2 \bmod 3$,
(vi) $^3D_4(q_0)$ *if* $q = q_0{}^r$, *for* r *prime*,
(vii) $\mathrm{SL}_2(q^3) \times \mathrm{SL}_2(q)$ *if* q *is even*,
(viii) $2{\cdot}(\mathrm{PSL}_2(q^3) \times \mathrm{PSL}_2(q)){:}2$ *if* q *is odd*,
(ix) $\mathrm{SL}_3(q).C_{q^2+q+1}.C_2$,
(x) $\mathrm{SU}_3(q).C_{q^2-q+1}.C_2$,
(xi) $(C_{q^2+q+1} \times C_{q^2+q+1}){:}\mathrm{SL}_2(3)$,
(xii) $(C_{q^2-q+1} \times C_{q^2-q+1}){:}\mathrm{SL}_2(3)$,
(xiii) $C_{q^4-q^2+1}{:}4$.

4.7 Triality

The phenomenon known as triality plays an important role in many of the exceptional groups, especially $F_4(q)$, $E_6(q)$ and $^3D_4(q)$, but is 'really' a property of the orthogonal groups $\mathrm{P}\Omega_8^+(q)$, or rather of the spin groups $\mathrm{Spin}_8^+(q)$. We saw at the end of Section 3.9 that triality arises naturally from the Clifford algebra. In this section we take a different approach, and show how to derive triality from the octonions. We work with an octonion algebra $\mathbb{O}$, which may be the real octonion algebra, or an octonion algebra over a finite field. Most of our arguments will apply equally to all cases, but sometimes there are extra difficulties in characteristic 2 (and occasionally in characteristic 3).

To understand what triality is, it is useful first to explore what we mean by duality of vector spaces. This word is used to describe a number of related phenomena. Given a vector space V over a field F, an (external) dual space V^* may be defined as the space of linear maps from V to F (see Section 3.3.4). If the dimension of V is finite, then $\dim V^* = \dim V$, so V and V^* are isomorphic as vector spaces. Thus the bilinear 'evaluation' map $V^* \times V \to F$ defined by $(f, v) \mapsto f(v)$ gives rise to a bilinear 'inner product' $V \times V \to F$. This inner product can be regarded as an 'internal' version of duality. The natural action of $\mathrm{GL}(V)$ induces a 'dual' action on V^* which is different from its action on V.

Similarly, an internal version of triality is given by the product on the octonion algebra, or more precisely by the trilinear form $t : \mathbb{O} \times \mathbb{O} \times \mathbb{O} \to F$ defined by $t(u, v, w) = \mathrm{Re}\,(uvw)$, where $\mathrm{Re}\,(v)$ denotes the real part of v, i.e. $\mathrm{Re}\,(v) = \frac{1}{2}(v + \bar{v})$. This trilinear form is *cyclically symmetric*, that is, $t(u, v, w) = t(w, u, v)$. To prove this, note first that if u, v, w are three octonions, then $\mathrm{Re}\,(uv) = \mathrm{Re}\,(vu)$ and $\mathrm{Re}\,((uv)w) = \mathrm{Re}\,(u(vw))$. This is because $\mathrm{Re}\,((uv)w)$ and $\mathrm{Re}\,(u(vw))$ are trilinear in u, v and w, and both are zero on the basis $1, i_0, \ldots, i_6$ unless u, v, w lie in a quaternion subalgebra, which is associative. Thus we have

$$
\begin{aligned}
uv + \overline{uv} &= 2\mathrm{Re}\,(uv) \\
&= 2\mathrm{Re}\,(vu) \\
&= vu + \overline{vu}
\end{aligned}
\tag{4.73}
$$

and therefore

$$
\begin{aligned}
(uv)w + \overline{(uv)w} &= 2\mathrm{Re}\,((uv)w) \\
&= 2\mathrm{Re}\,(u(vw)) \\
&= u(vw) + \overline{u(vw)}
\end{aligned}
\tag{4.74}
$$

as required. It follows that $\mathrm{Re}\,(u(vw)) = \mathrm{Re}\,((vw)u)$ so

$$
\mathrm{Re}\,(uvw) = \mathrm{Re}\,(vwu) = \mathrm{Re}\,(wuv).
\tag{4.75}
$$

[On the other hand, $\mathrm{Re}\,(uvw)$ is not in general equal to $\mathrm{Re}\,(wvu)$, even in the quaternions: for example, $\mathrm{Re}\,(ijk) = -1$ but $\mathrm{Re}\,(kji) = 1$.]

An 'external' version of triality may be constructed on three related orthogonal 8-spaces, V, V', V'', say, with the orthogonal group (or rather its double cover the spin group) acting in three different ways on these three spaces. In the next section we describe in detail this external manifestation of triality. We identify V, V' and V'' with $\mathbb{O}$ for convenience in describing the action of the spin group, but we are regarding them only as vector spaces with quadratic forms, not as algebras.

4.7.1 Isotopies

An *isotopy* of $\mathbb{O}$ is a map $(\alpha,\beta,\gamma):(x,y,z)\mapsto(x^\alpha,y^\beta,z^\gamma)$ from $\mathbb{O}\times\mathbb{O}\times\mathbb{O}$ to itself, where α, β, γ are orthogonal transformations, and which preserves the set of triples (x,y,z) with $xyz=1$. (Here no brackets are needed as $(xy)z=1$ implies that z is in the quaternion subalgebra generated by x and y).

Let $L_u:x\mapsto ux$ denote left multiplication by u, let $R_u:x\mapsto xu$ denote right multiplication by u, and $B_u:x\mapsto uxu$ denote bimultiplication by u. If $u\in\mathbb{O}$ has norm 1 (i.e. $u\bar{u}=1$) then $(L_u,R_u,B_{\bar{u}})$ is an isotopy. For if $xyz=1$ then the subalgebra generated by u and $xy=z^{-1}$ is associative (all 2-generator subalgebras are associative), so by the Moufang identity

$$\begin{aligned}((ux)(yu))(\bar{u}z\bar{u})&=(u(xy)u)(\bar{u}z\bar{u})\\&=u(xy)u\bar{u}z\bar{u}\\&=u(xy)z\bar{u}\\&=u\bar{u}\\&=1.\end{aligned}\tag{4.76}$$

Indeed, it is easy to show that the maps $(L_u,R_u,B_{\bar{u}})$ for $u\bar{u}=1$ generate the full group of isotopies. First assume that the characteristic is not 2, and note that if u is purely imaginary ($\bar{u}=-u$), then $B_{\bar{u}}$ negates 1 and u, and fixes the orthogonal complement of $\langle 1,u\rangle$. Thus in $\mathrm{GO}_7(q)$ acting on the purely imaginary octonions these maps are reflections in vectors of norm 1, and therefore generate a group $2\times\Omega_7(q)$. Similarly, in characteristic 2, the map $B_{\bar{u}}$ acts on $1^\perp$ as an orthogonal transvection, and these maps generate $\Omega_7(q)\cong\mathrm{Sp}_6(q)$. Also, if $\omega=\frac{1}{2}(-1+i_0+i_1+i_3)$ then B_ω maps the identity element to $\bar{\omega}$ so extends the group to $\Omega_8^+(q)$.

Next we need to look at the group homomorphism $(\alpha,\beta,\gamma)\mapsto\gamma$ from the group of isotopies to the orthogonal group, and show that its kernel is the group of order 2 generated by $(-1,-1,1)=(L_{-1},R_{-1},B_{-1})$. In other words, given an isotopy $(\alpha,\beta,1)$ we must show that $\alpha=\beta=\pm1$.

Suppose that $1^\alpha=a$, necessarily of norm 1. Applying the definition of isotopy to the triple $(x,y,z)=(1,1,1)$ we have $1=1^\alpha1^\beta1=a1^\beta$ so $1^\beta=\bar{a}$. Next, taking $(x,1,x^{-1})$ we have

$$1=x^\alpha1^\beta x^{-1}=x^\alpha\bar{a}x^{-1}$$

so $x^\alpha=xa$ for all x, and therefore $\alpha=R_a$. Similarly, taking $(1,y,y^{-1})$ we have

$$1 = 1^\alpha y^\beta y^{-1} = a y^\beta y^{-1}$$

so $y^\beta = \bar{a}y$ for all y, and therefore $\beta = L_{\bar{a}}$.

We wish to show that a is real, and therefore $a = \pm 1$. First define the *nucleus* of $\mathbb{O}$ to be the set of elements $a \in \mathbb{O}$ such that $(xa)w = x(aw)$ for all $x, w \in \mathbb{O}$. Clearly the nucleus is a subspace and is invariant under the automorphism group. Equally clearly, it contains 1 but not i_0. Hence the nucleus is exactly the subspace $\langle 1 \rangle$.

Therefore if a is not real, we may find x and w such that $(xa)w \neq x(aw)$. There exists y such that $\bar{a}y = w$, so

$$\begin{aligned} xy &= x((a\bar{a})y) \\ &= x(a(\bar{a}y)) \\ &\neq (xa)(\bar{a}y). \end{aligned} \tag{4.77}$$

In other words we have found x, y, z with $xyz = 1$ but $(xa)(\bar{a}y)z \neq 1$, which contradicts the assumption that $(R_a, L_{\bar{a}}, 1)$ is an isotopy. It follows that in odd characteristic the group of all isotopies is a double cover of $\Omega_8^+(q)$, namely the spin group $2{\cdot}\Omega_8^+(q)$ (see Section 3.9.2).

A similar argument works in characteristic 2 (see Exercise 4.21).

4.7.2 The triality automorphism of $\mathrm{P}\Omega_8^+(q)$

The group of isotopies may be extended to the group $2{\cdot}\mathrm{SO}_8^+(q) \cong 2{\cdot}\Omega_8^+(q){:}2$ in odd characteristic (or $\mathrm{GO}_8^+(q) \cong \Omega_8^+(q){:}2$ in characteristic 2) by adjoining the duality automorphism $(x, y, z) \mapsto (\bar{y}, \bar{x}, \bar{z})$.

There is another obvious automorphism, called *triality*, which maps (x, y, z) to (z, x, y). [Clearly $xyz = 1$ implies $z = (xy)^{-1}$ and so $zxy = 1$, since inverses are 2-sided.] This extends the spin group to a group of shape $2^2{\cdot}\mathrm{P}\Omega_8^+(q){:}S_3$ (or $\Omega_8^+(q){:}S_3$ in characteristic 2). The centraliser of the triality automorphism consists of all isotopies of the form (α, α, α). This means that if $z^{-1} = xy$ then $(z^{-1})^\alpha = x^\alpha y^\alpha$: in other words, α is an automorphism of the octonion algebra. Thus this centraliser is exactly $G_2(q)$.

Moreover, the set of isotopies preserving the subset

$$\{(1, y, y^{-1}) \mid y \in \mathbb{O} \text{ invertible}\}$$

is the spin group $2{\cdot}\Omega_7(q)$ (or just $\Omega_7(q) \cong \mathrm{Sp}_6(q)$ in characteristic 2). The stabiliser in this group of the triple $(1, 1, 1)$ consists of isotopies (α, β, γ) which simultaneously map $(1, y, y^{-1})$ to $(1, y^\beta, (y^{-1})^\gamma)$ (so that $\beta = \gamma$) and map $(x, x^{-1}, 1)$ to $(x^\alpha, (x^{-1})^\beta, 1)$ (so that $\alpha = \beta$), so is again equal to $G_2(q)$. This leads to an alternative description of $G_2(q)$ as the stabiliser of a non-isotropic vector in the 8-dimensional spin representation of $2{\cdot}\Omega_7(q)$ (or $\Omega_7(q) \cong \mathrm{Sp}_6(q)$ in characteristic 2).

This idea of triality gives us another way of looking at the groups $^3D_4(q)$. Recall that the unitary groups may be defined by identifying the duality (or inverse-transpose) automorphism $x \mapsto (x^{-1})^\top$ of the general linear groups $GL_n(q^2)$ with the field automorphism $x \mapsto \overline{x} = x^q$ of order 2, so that the group consists of the matrices x satisfying $x^{-1} = \overline{x}^\top$. We may apply the same principle to the groups $P\Omega_8^+(q^3)$, identifying the triality automorphism with the field automorphism $x \mapsto x^q$ of order 3.

In other words, we consider those isotopies (α, β, γ) on the octonion algebra over $\mathbb{F}_{q^3}$ which commute with the map $(x, y, z) \mapsto (y^q, z^q, x^q)$. The group of such isotopies is denoted $^3D_4(q)$, because the triality automorphism can be thought of as an automorphism of order 3 of the Dynkin diagram D_4.

4.7.3 The Klein correspondence revisited

The geometry of triality gives us another way of deriving the Klein correspondence (see Section 3.7). Suppose for simplicity that $q \equiv 1 \bmod 4$. Then there is an element $i \in \mathbb{F}_q$ with $i^2 = -1$, so that the map $L_{i_0} : x \mapsto i_0 x$ is diagonalisable with eigenvalues $\pm i$, each of multiplicity 4. One of the eigenspaces is spanned by $1 - ii_0, i_1 - ii_3, i_2 - ii_6, i_4 - ii_5$, and the other is obtained by replacing $-i$ by $+i$. Both eigenspaces are totally isotropic and in the spin group have a full $GL_4(q)$ acting on them. On the other hand, the map $B_{i_0} : x \mapsto i_0 x i_0$ negates $\langle 1, i_0 \rangle$ and centralises the orthogonal 6-space $\langle 1, i_0 \rangle^\perp$. But triality maps L_{i_0} to B_{i_0} (modulo signs), and thereby induces an isomorphism between $PSL_4(q)$ and $P\Omega_6^+(q)$.

4.8 Albert algebras and groups of type F_4

4.8.1 Jordan algebras

The algebra of $n \times n$ matrices is defined by the well-known matrix product, which is associative but non-commutative, as long as $n > 1$. We can derive a commutative product from it, by defining $A \circ B = \frac{1}{2}(AB + BA)$. This is called the *Jordan product*, and is easily shown to be non-associative whenever $n > 1$. It does however satisfy the so-called *Jordan identity*

$$((A \circ A) \circ B) \circ A = (A \circ A) \circ (B \circ A). \tag{4.78}$$

(See Exercise 4.22.)

A *Jordan algebra* over a field of characteristic not 2 is defined abstractly to be a (non-associative) algebra with a (bilinear) commutative Jordan product $\circ$, which satisfies the Jordan identity.

Simple Jordan algebras (i.e. those with no non-trivial proper ideals) over finite or algebraically closed fields are completely classified (at least if the characteristic is not 2 or 3), and it turns out that apart from those which arise from associative algebras in the manner just described, there is just one other Jordan algebra. It is called the *exceptional Jordan algebra*, or *Albert algebra*, and has dimension 27. It may be constructed as the algebra of 3×3 Hermitian matrices (i.e. matrices x such that $x^\top = \bar{x}$) over the octonions. For brevity let us define

$$(d, e, f \mid D, E, F) = \begin{pmatrix} d & F & \overline{E} \\ \overline{F} & e & D \\ E & \overline{D} & f \end{pmatrix}, \tag{4.79}$$

where d, e, f lie in the ground field, and $\overline{}$ denotes octonion conjugation, that is, the linear map fixing 1 and negating i_n for all n.

Multiplication of such matrices makes sense, and the Jordan product is defined in the same way as before. It can be readily checked that the 27-dimensional space of octonion Hermitian matrices is closed under multiplication. The identity matrix acts as an identity element in this algebra, and its perpendicular space (with respect to the natural inner product) is the 26-dimensional subspace of matrices of trace 0.

4.8.2 A cubic form

The natural inner product $\mathrm{Tr}(x\bar{y}^\top)$ on the real vector space of octonion Hermitian matrices can be expressed as

$$\mathrm{Tr}(x\bar{y}^\top) = \mathrm{Tr}(xy) = \mathrm{Tr}(yx) = \mathrm{Tr}(x \circ y). \tag{4.80}$$

Indeed, there are at least three natural invariant forms on the Albert algebra: a linear form $L(x) = \mathrm{Tr}(x)$, a bilinear form $B(x, y) = \mathrm{Tr}(x \circ y)$, and a trilinear form $T(x, y, z) = \mathrm{Tr}((x \circ y) \circ z)$. It is clear that the bilinear form is symmetric, i.e. $B(x, y) = B(y, x)$. It is also clear that the trilinear form satisfies $T(x, y, z) = T(y, x, z)$. What is much less obvious, but crucial, is that $T(x, y, z) = T(y, z, x)$, so that T is a symmetric trilinear form.

To prove this, recall from (4.75) that if u, v, w are three octonions, then $\mathrm{Re}(uvw) = \mathrm{Re}(vwu) = \mathrm{Re}(wuv)$. Now we calculate the trilinear form at the three matrices $(d, e, f \mid D, E, F)$, $(g, h, j \mid G, H, J)$ and $(k, l, m \mid K, L, M)$ as follows

$$\begin{aligned}
& dgk + ehl + fjm \\
& + k\mathrm{Re}\,(F\overline{J} + \overline{E}H) + l\mathrm{Re}\,(D\overline{G} + \overline{F}J) + m\mathrm{Re}\,(E\overline{H} + \overline{D}G) \\
& + g\mathrm{Re}\,(M\overline{F} + \overline{L}E) + h\mathrm{Re}\,(K\overline{D} + \overline{M}F) + j\mathrm{Re}\,(L\overline{E} + \overline{K}D) \\
& + d\mathrm{Re}\,(J\overline{M} + \overline{H}L) + e\mathrm{Re}\,(G\overline{K} + \overline{J}M) + f\mathrm{Re}\,(H\overline{L} + \overline{G}K) \\
& + \mathrm{Re}\,(MGE + LJD + KHF + DHM + FGL + EJK)
\end{aligned} \tag{4.81}$$

and observe that this is unchanged under all permutations of the three matrices.

This enables us to replace the trilinear form by a cubic form $C(x) = \frac{1}{6}T(x, x, x)$ in the same way that we replace the bilinear form $B(x, y)$ by the quadratic form $Q(x) = \frac{1}{2}B(x, x)$. Thus

$$Q(d, e, f \mid D, E, F) = \tfrac{1}{2}(d^2 + e^2 + f^2) + (D\overline{D} + E\overline{E} + F\overline{F}) \qquad (4.82)$$

and

$$C(d, e, f \mid D, E, F) = \tfrac{1}{6}(d^3 + e^3 + f^3) + \tfrac{1}{2}d(E\overline{E} + F\overline{F}) + \tfrac{1}{2}e(D\overline{D} + E\overline{E})$$
$$+ \tfrac{1}{2}f(D\overline{D} + E\overline{E}) + \tfrac{1}{2}\mathrm{Re}\,(DEF + D\overline{F}E). \qquad (4.83)$$

We recover the original forms by

$$B(x, y) = Q(x + y) - Q(x) - Q(y),$$
$$T(x, y, z) = C(x + y + z)$$
$$- C(x + y) - C(y + z) - C(x + z)$$
$$+ C(x) + C(y) + C(z). \qquad (4.84)$$

Indeed, we can also recover the Jordan product from these forms. For $B(x, y)$ is a non-degenerate form, so that $x \circ y$ is uniquely determined by evaluating the identity

$$B(x \circ y, z) = \mathrm{Tr}(x \circ y \circ z) = T(x, y, z)$$

as z runs over a basis.

4.8.3 The automorphism groups of the Albert algebras

An analogous construction can be performed using the octonions over any finite field $\mathbb{F}_q$ of characteristic not 2 or 3. [In these two characteristics the arguments of Section 4.8.2 break down and different definitions are required in order to overcome this problem: see Section 4.8.4.] We obtain in this way a finite Albert algebra, whose automorphism group we call $F_4(q)$. This group in fact acts irreducibly on the 26-dimensional space of trace 0 matrices, except if the field has characteristic 3, when the identity matrix has trace 0, and $F_4(q)$ acts irreducibly on the 25-space of trace 0 matrices modulo the identity.

Now if α is any element of $G_2(q)$, i.e. an automorphism of the octonions, then it induces a map $(d, e, f \mid D, E, F) \mapsto (d, e, f \mid D^\alpha, E^\alpha, F^\alpha)$ on the Albert algebra over $\mathbb{F}_q$, and it is easy to see that this map is an automorphism of the Albert algebra. It follows that $G_2(q)$ is a subgroup of $F_4(q)$.

Indeed, the double cover of $\Omega_8^+(q)$, i.e. the spin group $2 \cdot \Omega_8^+(q) \cong 2^2 \cdot \mathrm{P}\Omega_8^+(q)$, is generated by the maps

$$(d, e, f \mid D, E, F) \mapsto (d, e, f \mid uD, Eu, \overline{u}F\overline{u}) \qquad (4.85)$$

where u is an octonion of norm 1. It is a straightforward calculation to show that this map preserves the quadratic and cubic forms (4.82) and (4.83), and therefore preserves the Jordan multiplication.

More generally, as we saw in Section 4.7.1, these maps generate the group of isotopies

$$(\alpha, \beta, \gamma) : (d, e, f \mid D, E, F) \mapsto (d, e, f \mid D^\alpha, E^\beta, F^\gamma), \qquad (4.86)$$

which have the property that $DEF = 1$ implies $D^\alpha E^\beta F^\gamma = 1$. It is easy to see that these maps preserve the quadratic form, so (unless the characteristic is 2 or 3) to verify that they preserve the Jordan algebra structure, it suffices to verify that they preseve the cubic form (4.83). But α, β and γ are orthogonal transformations, so $D\overline{D}$, $E\overline{E}$ and $F\overline{F}$ are fixed by any isotopy. Also we can write $F = F_1 + F_2$ in such a way that $(DE)F_1$ is real and $(DE)F_2$ is purely imaginary, and therefore by the definition of isotopy, $(D^\alpha E^\beta)F_1^\gamma = (DE)F_1 = \mathrm{Re}\,((DE)F)$. Moreover, left octonion multiplication by $D^\alpha E^\beta$ is an orthogonal similarity, so $(D^\alpha E^\beta)F_2^\gamma$ is purely imaginary, and $\mathrm{Re}\,((D^\alpha E^\beta)F^\gamma) = (D^\alpha E^\beta)F_1^\gamma = \mathrm{Re}\,((DE)F)$. Therefore, substituting into (4.83), we have $C(d, e, f \mid D^\alpha, E^\beta, F^\gamma) = C(d, e, f \mid D, E, F)$.

This group of isotopies is normalised by the triality automorphism

$$(d, e, f \mid D, E, F) \mapsto (e, f, d \mid E, F, D), \tag{4.87}$$

as well as the duality automorphism

$$(d, e, f \mid D, E, F) \mapsto (d, f, e \mid \overline{D}, \overline{F}, \overline{E}). \tag{4.88}$$

These elements extend $2^2{\cdot}\mathrm{P}\Omega_8^+(q)$ to a group of shape $2^2{\cdot}\mathrm{P}\Omega_8^+(q){:}S_3$, which is (in fact) the full stabiliser of the 2-dimensional space of diagonal matrices of trace 0.

4.8.4 Another basis for the Albert algebra

In order to exhibit the parabolic subgroups, as well as to define $F_4(q)$ in characteristics 2 and 3, we change basis in the same way as we did for $G_2(q)$. Thus we take the basis $\{x_1, \dots, x_8\}$ for the octonions given in (4.26), and define a basis $\{w_i, w_i', w_i'' \mid 0 \leqslant i \leqslant 8\}$ for the Albert algebra by

$$
\begin{aligned}
w_0 &= (1,0,0 \mid 0,0,0) \quad \text{and} \quad w_i = (0,0,0 \mid x_i, 0, 0) \text{ for } i > 0, \\
w_0' &= (0,1,0 \mid 0,0,0) \quad \text{and} \quad w_i' = (0,0,0 \mid 0, x_i, 0) \text{ for } i > 0, \\
w_0'' &= (0,0,1 \mid 0,0,0) \quad \text{and} \quad w_i'' = (0,0,0 \mid 0, 0, x_i) \text{ for } i > 0. \quad (4.89)
\end{aligned}
$$

The multiplication table is now easy to write down. We double the multiplication in order to get rid of the factors of $\frac{1}{2}$. Then multiplication by w_0 is defined by

$$
\begin{aligned}
w_0 \circ w_0 &= 2w_0, \\
w_0 \circ w_0' &= 0, \\
w_0 \circ w_i &= 0 \text{ for } i > 0, \\
w_0 \circ w_i' &= w_i' \text{ for } i > 0, \\
w_0 \circ w_i'' &= w_i'' \text{ for } i > 0, \quad (4.90)
\end{aligned}
$$

and images under the triality map $w_i \mapsto w_i' \mapsto w_i'' \mapsto w_i$. Multiplication by w_i for $i > 0$ is given by

$$w_i \circ w_{9-i} = w_0' + w_0'',$$
$$w_i \circ w_j = 0 \text{ for } j \neq 9 - i,$$
$$w_i \circ w_j' = \varepsilon w_k'' \text{ where } x_i x_j = \varepsilon \overline{x_k} \text{ (see (4.27))}, \tag{4.91}$$

and images under triality. For convenience we re-write (4.27) in this new notation.

	w_1''	w_2''	w_3''	w_4''	w_5''	w_6''	w_7''	w_8''
w_1'					$-w_1$	$-w_2$	w_3	$-w_5$
w_2'			w_1	$-w_2$			$-w_4$	$-w_6$
w_3'		$-w_1$		$-w_3$		$-w_4$		w_7
w_4'	$-w_1$			w_5		$-w_6$	$-w_7$	
w_5'		$-w_2$	$-w_3$		w_4			$-w_8$
w_6'	w_2		$-w_5$		$-w_6$		$-w_8$	
w_7'	$-w_3$	$-w_5$			$-w_7$	w_8		
w_8'	$-w_4$	w_6	$-w_7$	$-w_8$				

$$\tag{4.92}$$

Now the quadratic form at a typical vector $v = \sum_{t=0}^{8}(\lambda_t w_t + \lambda_t' w_t' + \lambda_t'' w_t'')$ is

$$Q(v) = \tfrac{1}{2}(\lambda_0^{\,2} + \lambda_0'^{\,2} + \lambda_0''^{\,2}) + \sum_{t=1}^{4}(\lambda_t \lambda_{9-t} + \lambda_t' \lambda_{9-t}' + \lambda_t'' \lambda_{9-t}''). \tag{4.93}$$

On the subspace of vectors of trace 0 we can re-write $\tfrac{1}{2}(d^2 + e^2 + (d+e)^2) = d^2 + de + e^2$, and therefore

$$Q(v) = \lambda_0^{\,2} + \lambda_0 \lambda_0' + \lambda_0'^{\,2} + \sum_{t=1}^{4}(\lambda_t \lambda_{9-t} + \lambda_t' \lambda_{9-t}' + \lambda_t'' \lambda_{9-t}'') \tag{4.94}$$

which can be interpreted in any characteristic.

We double the cubic form C, and, since $\tfrac{1}{3}(d^3 + e^3 - (d+e)^3) = -de(d+e)$, on the subspace of trace 0 vectors we re-write (4.83) as

$$C(d, e, f \mid D, E, F) = def - dD\overline{D} - eE\overline{E} - fF\overline{F} + \mathrm{Re}\,(DEF + DFE)$$

and therefore (see Exercise 4.26)

$$
\begin{aligned}
C(v) = &\sum_{t=0,4,5} \lambda_t \lambda_t' \lambda_t'' - \sum_{t=1}^{4}(\lambda_0 \lambda_t \lambda_{9-t} + \lambda_0' \lambda_t' \lambda_{9-t}' + \lambda_0'' \lambda_t'' \lambda_{9-t}'') \\
&- \sum(\lambda_1 \lambda_5' \lambda_8'' + \lambda_1 \lambda_8' \lambda_4'' + \lambda_2 \lambda_4' \lambda_7'' + \lambda_2 \lambda_7' \lambda_5'' + \lambda_3 \lambda_4' \lambda_6'' + \lambda_3 \lambda_6' \lambda_5'') \\
&+ \sum(\lambda_1 \lambda_7' \lambda_6'' - \lambda_1 \lambda_6' \lambda_7'' + \lambda_2 \lambda_3' \lambda_8'' - \lambda_2 \lambda_8' \lambda_3'')
\end{aligned}
\tag{4.95}
$$

where the last two sums are over all cyclic permutations of the subscripts. This description of the cubic form can be interpreted in any characteristic. Indeed, $F_4(q)$ may be defined for arbitrary q as the set of linear maps on $\mathbb{F}_q^{\,26}$ which preserve both the quadratic form Q defined in (4.94) and the cubic form C defined in (4.95).

4.8.5 The normaliser of a maximal torus

In this section we construct the maximal split torus T and find its normaliser N, and the Weyl group N/T. It is straightforward to show that a diagonal element is completely specified by its eigenvalues $\alpha, \beta, \gamma, \delta$ on the eigenvectors w_1, w_1', w_1'', w_2. For the quadratic form gives the eigenvalues $\alpha^{-1}, \beta^{-1}, \gamma^{-1}, \delta^{-1}$ on w_8, w_8', w_8'', w_7 and then the product $w_1 \circ w_8' = -w_5''$ gives eigenvalue $\alpha\beta^{-1}$ on w_5'', and $w_2 \circ w_3' = w_1''$ gives eigenvalue $\gamma\delta^{-1}$ on w_3', and so on. Conversely, we can check that for any choices of $\alpha, \beta, \gamma, \delta$ the resulting diagonal element does indeed preserve the quadratic form Q defined in (4.94) and the cubic form C defined in (4.95). The eigenvalues are as follows:

Vector	Eigenvalue	Vector	Eigenvalue	Vector	Eigenvalue
w_1	α	w_1'	β	w_1''	γ
w_2	δ	w_2'	$\alpha\beta^{-1}\delta$	w_2''	$\alpha\gamma^{-1}\delta$
w_3	$\alpha^{-1}\beta\gamma\delta^{-1}$	w_3'	$\gamma\delta^{-1}$	w_3''	$\beta\delta^{-1}$
w_4	$\beta^{-1}\gamma$	w_4'	$\alpha\gamma^{-1}$	w_4''	$\alpha^{-1}\beta$
w_5	$\beta\gamma^{-1}$	w_5'	$\alpha^{-1}\gamma$	w_5''	$\alpha\beta^{-1}$
w_6	$\alpha\beta^{-1}\gamma^{-1}\delta$	w_6'	$\gamma^{-1}\delta$	w_6''	$\beta^{-1}\delta$
w_7	δ^{-1}	w_7'	$\alpha^{-1}\beta\delta^{-1}$	w_7''	$\alpha^{-1}\gamma\delta^{-1}$
w_8	α^{-1}	w_8'	β^{-1}	w_8''	γ^{-1}

$$(4.96)$$

Note that the eigenvalue on w_{9-i} (respectively, w_{9-i}', w_{9-i}'') is the inverse of the eigenvalue on w_i (respectively, w_i', w_i'').

The 24 non-trivial eigenspaces of the torus can be identified with the 24 short roots in the root system of type F_4 as follows. We label the eigenspace with eigenvalues $\alpha^w \beta^x \gamma^y \delta^z$ by the vector (w, x, y, z) and determine the (unique up to scalar multiplication) quadratic form which makes these integral vectors all have norm 1. Transforming to the basis for the Weyl group of type F_4 described in Section 3.12.4, and writing $+$ and $-$ for $\frac{1}{2}$ and $-\frac{1}{2}$ for simplicity of notation, we find that the eigenvectors $w_1, w_1', w_1'', w_2, \ldots, w_8''$ may be taken to correspond to the short roots as follows:

Vector	Root	Vector	Root	Vector	Root
w_1	1000	w_1'	$+++-$	w_1''	$++++$
w_2	0100	w_2'	$++-+$	w_2''	$++--$
w_3	0010	w_3'	$+-++$	w_3''	$+-+-$
w_4	0001	w_4'	$+---$	w_4''	$-++-$
w_5	$-(0001)$	w_5'	$-+++$	w_5''	$+--+$
w_6	$-(0010)$	w_6'	$-+--$	w_6''	$-+-+$
w_7	$-(0100)$	w_7'	$--+-$	w_7''	$--++$
w_8	$-(1000)$	w_8'	$---+$	w_8''	$----$

$$(4.97)$$

Note that the root corresponding to w_{9-i} (respectively w_{9-i}', w_{9-i}'') is the negative of the root corresponding to w_i (respectively w_i', w_i'').

Now we can calculate the action of the Weyl group on the basis $w_1, \ldots, w_8''$. In fact, N is contained in the subgroup $2^2 \cdot P\Omega_8^+(q){:}S_3$. This is 'explained' by the fact that the short roots in F_4 form a root system of type D_4 (corresponding to groups of type $P\Omega_8^+(q)$) and so $W(D_4)$ is a normal subgroup of $W(F_4)$, with quotient $W(F_4)/W(D_4) \cong S_3$.

Assume for the moment that $q \equiv 1 \bmod 8$ so that $\mathbb{F}_q$ contains square roots of ± 2. Pick an element of the Weyl group of type D_4, such as the permutation $(x_3, x_5)(x_4, x_6)$ of the octonion coordinates. This can be expressed as the product of the reflections in the unit vectors $u = (x_3 + x_4 - x_5 - x_6)/\sqrt{-2}$ and $v = (x_3 - x_4 - x_5 + x_6)/\sqrt{2}$. Since reflection in u is the map $x \mapsto -u\bar{x}u$, the product of the two reflections is the map

$$x \mapsto v(\overline{u}x\overline{u})v. \tag{4.98}$$

The images of this map under triality are therefore

$$x \mapsto \overline{v}(ux)$$
$$\text{and } x \mapsto (xu)\overline{v} \tag{4.99}$$

which we calculate as $\sqrt{-1}$ times, respectively,

$$(x_2, -x_2)(x_8, -x_8)(x_3, x_4)(x_5, -x_6)$$
$$\text{and } (x_5, -x_5)(x_6, -x_6)(x_1, -x_2)(x_7, x_8). \tag{4.100}$$

By (4.85) there is an element of $F_4(q)$ which acts on the w_k'', w_k, w_k' respectively as these three monomial permutations, i.e. replacing x_k by w_k'', w_k, w_k' respectively.

Now to get rid of the $\sqrt{-1}$, pre-multiply by a suitable element of the torus, such as $\alpha = \beta = \delta = \sqrt{-1}$, $\gamma = 1$, to get the map

$$(w_1, -w_1)(w_8, -w_8)(w_3, w_4)(w_5, w_6)$$
$$(w_4', -w_4')(w_5', -w_5')(w_1', w_2')(w_7', w_8')$$
$$(w_2'', -w_2'')(w_7'', -w_7'')(w_3'', w_5'')(w_4'', w_6''). \tag{4.101}$$

This map fixes w_0, w_0' and w_0'', and now makes sense for arbitrary q.

To generate the rest of the Weyl group of type F_4, modulo the torus, we may take the triality element

$$w_k \mapsto w_k' \mapsto w_k'' \mapsto w_k \tag{4.102}$$

for all k, $0 \leqslant k \leqslant 8$, and the duality element

$$w_k \leftrightarrow \overline{w_k}, \qquad w_k' \leftrightarrow \overline{w_k''} \tag{4.103}$$

(also for all k), and the symmetry, obtained from the Weyl group of type G_2 (see (4.30)), which acts as $(w_1, w_6, w_7)(w_2, w_8, w_3)$ on the w_k and similarly on the w_k' and w_k''. Here $\overline{w_k} = (0, 0, 0 \mid \overline{x_k}, 0, 0)$, and so on.

Note that in characteristic 2 there is no sign problem, and the Weyl group simply permutes the given basis vectors.

4.8.6 Parabolic subgroups of $F_4(q)$

In order to generate the group $F_4(q)$, and to describe its parabolic subgroups, we must first describe the long and short root elements. The *long root elements* are the conjugates of the map $t_r(\lambda)$ defined by

$$t_r(\lambda) : w_7 \mapsto w_7 - \lambda w_1, w_7' \mapsto w_7' - \lambda w_1', w_7'' \mapsto w_7'' - \lambda w_1'',$$
$$w_8 \mapsto w_8 + \lambda w_2, w_8' \mapsto w_8' + \lambda w_2', w_8'' \mapsto w_8'' + \lambda w_2''. \quad (4.104)$$

(As usual, basis vectors not explicitly mentioned are fixed.) This element is labelled by the long root $r = (-1, -1, 0, 0)$, which is the difference of the roots corresponding to each of the pairs of vectors $(w_7, w_1), \ldots, (w_7'', w_2'')$. It is easy to check that this element preserves the structure of the Albert algebra. Since $t_r(\lambda)t_r(\mu) = t_r(\lambda + \mu)$, the elements $t_r(\lambda)$ as λ ranges over $\mathbb{F}_q$ form a group, called a *long root subgroup*, isomorphic to the additive group of $\mathbb{F}_q$.

We also need the *short root elements*. An example is the following element.

$$
\begin{aligned}
w_0 &\mapsto w_0 - \lambda w_5'', & w_0' &\mapsto w_0' + \lambda w_5'', \\
w_1' &\mapsto w_1' + \lambda w_1, & w_8 &\mapsto w_8 - \lambda w_8', \\
w_2 &\mapsto w_2 - \lambda w_2', & w_7' &\mapsto w_7' + \lambda w_7, \\
w_3 &\mapsto w_3 - \lambda w_3', & w_6' &\mapsto w_6' + \lambda w_6, \\
w_5 &\mapsto w_5 + \lambda w_4', & w_5' &\mapsto w_5' - \lambda w_4, \\
w_4'' &\mapsto w_4'' + \lambda w_0 - \lambda w_0' - \lambda^2 w_5''. &&
\end{aligned} \quad (4.105)
$$

It is a somewhat more tedious exercise to prove that this element preserves the structure of the Albert algebra (see Exercise 4.27). The corresponding short root in this case is $(-++-)$, corresponding to the basis vector w_4''.

It can be shown without too much difficulty that the stabiliser of the 1-space $\langle w_1 \rangle$ has the shape $q^{7+8}{:}2{\cdot}\Omega_7(q).C_{q-1}$ (for q odd). Similarly for q even it has shape $(q^6 \times q^{1+8})\mathrm{Sp}_6(q).C_{q-1}$. Now w_1 has the properties $w_1 \circ w_1 = 0$ and $\mathrm{Tr}(w_1) = 0$. In fact, for q odd, these properties characterise the orbit of w_1 under $F_4(q)$, as we now show.

Let v be any non-zero vector such that $v \circ v = 0$ and $\mathrm{Tr}(v) = 0$. We show that there are exactly $(q^{12} - 1)(q^4 + 1)$ such vectors v, provided q is odd. Write $v = (d, e, f \mid D, E, F)$ as usual, so that the conditions on v are expressed by the equations

$$
\begin{aligned}
d + e + f &= 0, \\
d^2 + E\overline{E} + F\overline{F} &= 0, \\
(d + e)\overline{F} + DE &= 0, \quad (4.106)
\end{aligned}
$$

and the equations derived from these by cycling d, e, f and D, E, F. Eliminating D and E from the three images of the middle equation, and substituting $f = -d - e$, gives $de = F\overline{F}$. The third equation simplifies to $f\overline{F} = DE$. Now if $f \neq 0$, we can multiply by $\overline{E}$ to get $f\overline{F}.\overline{E} = DE\overline{E} = fdD$, and cancelling the factor of f gives us $d\overline{D} = EF$. Similarly we get $e\overline{E} = FD$. Moreover,

$f^2 F\overline{F} = (\overline{E}.\overline{D})DE = ef.fd$ so $F\overline{F} = de$. In other words, in the case $f \neq 0$ the above equations are equivalent to

$$d + e + f = 0,$$
$$ef = D\overline{D},$$
$$fd = E\overline{E},$$
$$f\overline{F} = DE. \qquad (4.107)$$

Note that because the norm on the octonions induces a quadratic form of plus type, there are $q^7 + q^4 - q^3$ octonions of norm 0, and $q^7 - q^3$ of any other norm. If d and e are both non-zero, then there are $q-1$ choices for f, and $q-2$ choices for e, and then $q^7 - q^3$ choices for each of D and E independently (since they are of specified non-zero norm). This gives us $(q-1)(q-2)(q^7 - q^3)^2$ such vectors v. On the other hand, if $d = 0$ and $e = -f$ we have $q-1$ choices for f, and $q^7 - q^3$ choices for D and $q^7 + q^4 - q^3$ choices for E. Multiplying by 3 for the choice of which of d, e, f is zero, we get $3(q-1)(q^7 - q^3)(q^7 + q^4 - q^3)$ such vectors.

In all the remaining cases $d = e = f = 0$, and therefore the equations reduce to $D\overline{D} = E\overline{E} = F\overline{F} = 0$ and $DE = EF = FD = 0$. If two of these octonions are zero, there are $(q^3 + 1)(q^4 - 1)$ choices for the other one. If two of them are non-zero, say D and E, then there are $(q^3 + 1)(q^4 - 1)$ choices for D, and then the condition $DE = 0$ implies there are exactly $q^4 - 1$ choices for E. Finally there are q^3 choices for F, one of which is 0. By symmetry, this gives $(q^3 + 1)(q^4 - 1)^2(q^3 + 2)$ choices altogether.

Finally we compute the sum of these three expressions and obtain the total number of such vectors v as $(q^4 - 1)(q^{12} + 2q^8 + 2q^4 + 1) = (q^{12} - 1)(q^4 + 1)$, as claimed.

Now it is easy to see that the group $F_4(q)$ permutes these vectors transitively. Translating to our new basis, w_1 is such a vector, and hence so are the 24 coordinate vectors w_i, w_i' and w_i'' with $i \neq 0$. Transforming by the root elements corresponding to all the positive roots, these lie in 24 orbits of lengths various powers of q, corresponding to the monomials in the expansion of $(q^5 + q^4 + q^3 + q^2 + q + 1)(q^6 + 1)(q^4 + 1)$, which is another way of writing $(q^{12} - 1)(q^4 + 1)/(q - 1)$.

Hence, for q odd, the order of $F_4(q)$ is $(q^{12} - 1)(q^4 + 1)q^{15}|2{\cdot}\Omega_7(q)|$, that is

$$|F_4(q)| = q^{24}(q^{12} - 1)(q^8 - 1)(q^6 - 1)(q^2 - 1). \qquad (4.108)$$

In fact the same formula holds for q even, but we shall not prove this here.

Notice that a long root has inner product 1 with exactly 6 short roots. Let V_6 be the 6-space spanned by the corresponding eigenvectors of the torus acting on the Albert algebra. For example, if the long root is $r = (1, 1, 0, 0)$ then V_6 is the space spanned by $w_1, w_1', w_1'', w_2, w_2', w_2''$. The stabiliser of V_6 in $F_4(q)$ is one of the maximal parabolic subgroups, and has shape $q^{1+14}{:}\mathrm{Sp}_6(q).C_{q-1}$. The normal subgroup of order q is a long root subgroup. The action of the group $\mathrm{Sp}_6(q)$ on V_6 indicates that there is a natural symplectic form defined (uniquely up to scalar multiplication) on V_6. Abstractly, this symplectic form on V_6 may be described by noting that the perpendicular space with respect to this form of a vector v is the intersection of V_6 with the image of the map 'multiplication by v' in the Albert algebra. This orthogonality relation determines the form up to scalar multiplication.

The other maximal parabolic subgroups are the stabilisers in $F_4(q)$ of an i-dimensional subspace V_i of V_6 which is totally isotropic with respect to this symplectic form. The stabiliser of a 2-space V_2 and the stabiliser of a 3-space each have shape $[q^{20}]{:}(\mathrm{SL}_2(q) \times \mathrm{SL}_3(q)).C_{q-1}$, but they are not isomorphic unless q is even.

When q is odd the structure of the maximal parabolic subgroup of shape $q^{1+14}{:}\mathrm{Sp}_6(q).C_{q-1}$ is specified by the fact that the action of $\mathrm{Sp}_6(q)$ on the q^{14} factor is given by taking the exterior cube of the natural module, and factoring out a copy of the natural module. The action of this parabolic subgroup on the trace 0 part of the Albert algebra is as a uniserial module of shape 6.14.6. The 14-space $V_6^{\perp}/V_6$ is an irreducible module for $\mathrm{Sp}_6(q)$, obtained from the exterior square of the natural module by factoring out the trivial submodule. The long root elements corresponding to this 6-space can then be identified with the long root elements in $G_2(q)$, and thereby with the Siegel transformations in $\mathrm{SO}_7(q)$.

4.8.7 Simplicity of $F_4(q)$

To prove simplicity of $F_4(q)$ we can use Iwasawa's Lemma in the usual way. First, it is straightforward to show that $F_4(q)$ acts primitively on the set of $(q^{12} - 1)(q^4 + 1)/(q - 1)$ images of $\langle w_1 \rangle$. The stabiliser of $\langle w_1 \rangle$ is the subgroup $q^{7+8}{:}2{\cdot}\Omega_7(q).C_{q-1}$ (for q odd). The derived group $q^{7+8}{:}2{\cdot}\Omega_7(q)$ is generated by long and short root elements. It follows after a little more calculation that $F_4(q)$ is generated by long and short root elements. Therefore $F_4(q)$ is perfect, and all the hypotheses of Iwasawa's Lemma (Theorem 3.1) are satisfied. We deduce that $F_4(q)$ is simple (for q odd). In fact, essentially the same argument works for q even.

4.8.8 Primitive idempotents

Another important aspect of the Albert algebra is the concept of a *primitive idempotent*, that is, an element w with $w \circ w = w$ and $\mathrm{Tr}(w) = 1$. First we determine how many such elements there are, when q is odd. Straightforward calculation shows that the primitive idempotents are precisely the elements $(d, e, f \mid D, E, F)$ which satisfy

$$
\begin{aligned}
d + e + f &= 1, \\
d^2 + E\overline{E} + F\overline{F} &= d, \\
(d + e)\overline{F} + DE &= \overline{F},
\end{aligned}
\tag{4.109}
$$

and the equations derived from these by cycling d, e, f and D, E, F. Substituting $d+e = 1-f$, the last equation can be re-written as $DE = f\overline{F}$. Eliminating $E\overline{E}$ and $F\overline{F}$ from the three images of the middle equation gives

$$
\begin{aligned}
2D\overline{D} &= (f - f^2) + (e - e^2) - (d - d^2) \\
&= 2ef,
\end{aligned}
\tag{4.110}
$$

(by substituting $d = 1 - e - f$) so since the characteristic is not 2 we have $D\overline{D} = ef$, $E\overline{E} = df$ and $F\overline{F} = de$. We now divide into three cases according as one, two or three of d, e, f are non-zero. The number of possibilities for (d, e, f) in these three cases is 3, $3(q - 2)$ and $q^2 - 3q + 3$ respectively.

In each case, without loss of generality $f \neq 0$, so that F is determined by D, E and the equation $f\overline{F} = DE$. Therefore we only need to determine the number of possibilities for D and E. In the first case, $f = 1$ and $d = e = 0$, so D and E have norm 0, and there are $(q^7 + q^4 - q^3)^2$ possibilities. In the second case, without loss of generality $e \neq 0$ and $d = 0$, so that $D\overline{D} = ef \neq 0$ and E has norm 0, and there are $(q^7 - q^3)(q^7 + q^4 - q^3)$ possibilities. In the last case, both D and E have fixed non-zero norm, so there are $(q^7 - q^3)^2$ possibilities. Therefore the total number of primitive idempotents is

$$
3(q^7 + q^4 - q^3)^2 + 3(q - 2)(q^7 - q^3)(q^7 + q^4 - q^3) + (q^2 - 3q + 3)(q^7 - q^3)^2,
$$

which simplifies to

$$
q^8(q^8 + q^4 + 1).
$$

The stabiliser of a primitive idempotent is now seen to be the spin group $2{\cdot}\Omega_9(q)$. Taking for example the primitive idempotent $w = (1, 0, 0 \mid 0, 0, 0)$ it is easy to see that it determines the 9-space $\{(0, e, -e \mid D, 0, 0)\}$ of elements y of trace 0 with $w \circ y = 0$, and the 16-space $\{(0, 0, 0 \mid 0, E, F)\}$ of elements z of trace 0 with $w \circ z = \frac{1}{2}z$ (these are just the eigenspaces of the action of w on the algebra by Jordan multiplication). In fact the orthogonal group $\Omega_9(q)$ acts on this 9-space, and its double cover acts as the spin group on the 16-space. To see that the stabiliser of w is no bigger, it suffices to check that the pointwise stabiliser of the 9-space consists only of the group of order 2 generated by the element $(d, e, f \mid D, E, F) \mapsto (d, e, f \mid D, -E, -F)$ (see Exercise 4.28). Since we know the order of $F_4(q)$, it is easy to see that the group acts transitively on the primitive idempotents.

Finally notice that if w is a primitive idempotent, then the map

$$
t_w : x \mapsto x + 4(\mathrm{Tr}(x \circ w))w - 4x \circ w
\tag{4.111}
$$

is an automorphism of the algebra. For if $w = (1, 0, 0 \mid 0, 0, 0)$ then

$$
t_w : (d, e, f \mid D, E, F) \mapsto (d, e, f \mid D, -E, -F)
\tag{4.112}
$$

which is easily seen to be an automorphism of the algebra, so the result follows by transitivity. Similarly, if $w = \frac{1}{2}(0, 1, 1 \mid u, 0, 0)$, where $u\overline{u} = 1$, then

$$t_w : (d, e, f \mid D, E, F) \mapsto (d, f, e \mid u\overline{D}u, -\overline{u}\overline{F}, -\overline{E}\overline{u}). \qquad (4.113)$$

As we saw in Section 4.8.3, these maps generate a group of shape $2 \cdot \Omega_8^+(q).2$ as u runs over all the octonions of norm 1. Adjoining the triality automorphism gives a group $2^2 \cdot \mathrm{P}\Omega_8^+(q).S_3$ which is in fact maximal in $F_4(q)$.

4.8.9 Other subgroups of $F_4(q)$

We have already seen (when q is odd) a subgroup $2 \cdot \Omega_9(q)$ (the spin group) which can be described as the centraliser of the involution

$$(d, e, f \mid D, E, F) \mapsto (d, e, f \mid D, -E, -F), \qquad (4.114)$$

or as the stabiliser of the vector $(1, 0, 0 \mid 0, 0, 0)$. We have also seen the triality group $2^2 \cdot \mathrm{P}\Omega_8^+(q){:}S_3$. Another interesting maximal subgroup is $^3D_4(q){:}3$ (see Sections 4.6 and 4.7.2).

There is a second conjugacy class of involutions (when q is odd), and the centraliser of such an involution is

$$(\mathrm{Sp}_6(q) \circ \mathrm{SL}_2(q)){:}2 \cong 2 \cdot (\mathrm{PSp}_6(q) \times \mathrm{PSL}_2(q)){:}2. \qquad (4.115)$$

Since the subgroup $\mathrm{Sp}_6(q)$ is in the maximal parabolic subgroup described in Section 4.8.6 above, it acts on the trace 0 part of the Albert algebra as two copies of the natural representation and one copy of a 14-dimensional representation, which is obtained by factoring out a 1-dimensional module from the exterior square of the natural module. It follows easily that the representation of this involution centraliser has the structure $(6 \otimes 2) \oplus (14 \otimes 1)$. Thus the involution centraliser is also the stabiliser of a certain 14-dimensional subspace. An example is the 14-space

$$\{(d, e, f \mid D, E, F) \mid d + e + f = 0, D, E, F \in \langle 1, i_1, i_2, i_4 \rangle\} \qquad (4.116)$$

of matrices writable over a fixed quaternion subalgebra of the octonions. The subgroup is generated by the involution centraliser $2 \cdot (\mathrm{PSL}_2(q) \times \mathrm{PSL}_2(q)){:}2$ in $G_2(q)$ together with the transformations t_r for r a primitive idempotent lying in the given 14-space.

Another maximal subgroup (in all characteristics except 3) is the normaliser of a subgroup of order 3, and has shape

$$\begin{aligned} 3 \cdot (\mathrm{PSL}_3(q) \times \mathrm{PSL}_3(q)){:}S_3 \quad &\text{if } q \equiv 1 \bmod 3, \text{ and} \\ 3 \cdot (\mathrm{PSU}_3(q) \times \mathrm{PSU}_3(q)){:}S_3 \quad &\text{if } q \equiv 2 \bmod 3. \end{aligned} \qquad (4.117)$$

Similarly there are subgroups

$$\begin{aligned} (\mathrm{SU}_3(q) \times \mathrm{SU}_3(q)){:}2 \quad &\text{when } q \equiv 0, 1 \bmod 3, \text{ and} \\ (\mathrm{SL}_3(q) \times \mathrm{SL}_3(q)){:}2 \quad &\text{when } q \equiv 0, 2 \bmod 3. \end{aligned} \qquad (4.118)$$

Despite the apparent symmetry the two factors in these groups are quite different: only in characteristic 2 is there an (outer) automorphism of $F_4(q)$ interchanging them. The 26-dimensional representation restricted to any of these subgroups is a direct sum of two irreducibles of dimensions 8 and 18. The invariant 8-space consists of matrices of the form $(d, e, f \mid D, E, F)$ where D, E and F lie in a 1-generator subalgebra $\langle 1, v \rangle$ of the octonions, with v purely imaginary. The isomorphism type of the group depends on the norm of v and on q.

Along similar lines, there is a subgroup $SO_3(q) \times G_2(q)$. We already saw the subgroup $G_2(q)$ acting as the group of automorphisms of the octonions, simultaneously on the three coordinates D, E, F of $(d, e, f \mid D, E, F)$. This commutes with a group $SO_3(q)$ acting on 3×3 matrices by conjugation.

A subgroup of particular interest in $F_4(p)$, for $p \neq 3$, is of shape $3^3{:}SL_3(3)$. This is somewhat analogous to the subgroup $2^3{\cdot}GL_3(2)$ in $G_2(p)$ for odd p. The latter is monomial with respect to a suitable basis $\{1, i_0, \ldots, i_6\}$ of the octonions. Similarly, provided $p \equiv 1 \bmod 3$, there is a basis of the Albert algebra such that $3^3{:}SL_3(3)$ acts monomially. An explicit form of the Jordan product with respect to such a basis is given by Griess [71].

It is possible to find a copy of the normal 3^3 inside the group $2^2{\cdot}P\Omega_8^+(q){:}S_3$ of isotopies extended by duality and triality. For example, take the following three elements, where ω is an element of order 3 in the field $\mathbb{F}_p$:

(i) a 'twisted' triality automorphism, namely the product of the triality automorphism $w_i \mapsto w_i' \mapsto w_i'' \mapsto w_i$ with the subscript permutation $(1, 7, 6)(8, 2, 3)$;

(ii) the product of the isotopies $(B_{\overline{u}}, L_u, R_u)$ where u is successively -1, $x_1 + x_8$, $\omega x_1 + \overline{\omega} x_8$, $x_7 + x_2$, $\omega x_7 + \overline{\omega} x_2$, $x_6 + x_3$, $\omega x_6 + \overline{\omega} x_3$;

(iii) ab^{-1} where a is the product of the isotopies $(B_{\overline{u}}, L_u, R_u)$ for $u = x_1 + x_8$ and $\omega x_1 + \overline{\omega} x_8$, and b is the product of the isotopies $(R_u, B_{\overline{u}}, L_u)$ for $u = x_7 + x_2$ and $\omega x_7 + \overline{\omega} x_2$.

Diagonalising these three elements of order 3 simultaneously gives a basis of 27 eigenvectors, for all 27 possible combinations of eigenvalues. The product of two eigenvectors is clearly an eigenvector for the product of the eigenvalues. However, the precise scalar multiples take some effort to compute. It turns out that these multiples involve powers of 2, so this basis does not generalise easily to characteristic 2.

This construction needs some modification if $q \equiv 2 \bmod 3$, as then there is no element of order 3 in the field. Similarly, there is a suitable modification in characteristic 2. We omit the details, which can be found in [184], but note that this underlies the construction of the large Ree groups in Section 4.9 below.

A more or less complete determination of the maximal subgroups of $F_4(q)$ in characteristic not 2 or 3 is obtained by Kay Magaard in his Ph. D. thesis [127]. The following is a summary:

Theorem 4.4. *If q is a power of the prime p, $p \geqslant 5$, then the following are maximal subgroups of $F_4(q)$:*

(i) $q^{1+14}{:}\mathrm{Sp}_6(q).C_{q-1}$,
(ii) $q^{2+6+12}{:}(\mathrm{SL}_2(q) \times \mathrm{SL}_3(q)).C_{q-1}$,
(iii) $q^{3+2+9+6}{:}(\mathrm{SL}_3(q) \times \mathrm{SL}_2(q)).C_{q-1}$,
(iv) $q^{7+8}{:}2{\cdot}\Omega_7(q).C_{q-1}$,
(v) $2{\cdot}\Omega_9(q)$,
(vi) $2^2{\cdot}\mathrm{P}\Omega_8^+(q){:}S_3$,
(vii) ${}^3D_4(q){:}3$,
(viii) $(\mathrm{Sp}_6(q) \circ \mathrm{SL}_2(q)).2$,
(ix) $(\mathrm{SL}_3(q) \circ \mathrm{SL}_3(q)).C_{(q-1,3)}.2$,
(x) $(\mathrm{SU}_3(q) \circ \mathrm{SU}_3(q)).C_{(q+1,3)}.2$,
(xi) $\mathrm{SO}_3(q) \times G_2(q)$,
(xii) $F_4(q_0)$, if $q = q_0{}^r$, r prime,
(xiii) $3^3{:}\mathrm{SL}_3(3)$, if $q = p$,
(xiv) $G_2(q)$, if $p = 7$,
(xv) $\mathrm{PGL}_2(q)$, if $p \geqslant 13$ and $q \geqslant 17$.

Every other maximal subgroup of $F_4(q)$, for such q, is the normaliser of a simple subgroup S with trivial centraliser, with S isomorphic to one of the groups ${}^3D_4(2)$, $\mathrm{PSL}_3(3)$, $\mathrm{PSU}_3(3)$, or $\mathrm{PSL}_2(r)$ for $r = 7, 8, 9, 13, 17, 25$ or 27.

4.8.10 Automorphisms and covers of $F_4(q)$

In odd characteristic, the only outer automorphisms of $F_4(q)$ are the field automorphisms. In characteristic 2 there is an exceptional 'graph' automorphism which can be used to construct the large Ree groups ${}^2F_4(q^{2n+1})$ (see Section 4.9). The graph automorphism squares to the field automorphism $x \mapsto x^2$, so that the full outer automorphism group is cyclic of order $2e$, where $q = p^e$. This graph automorphism interchanges the long roots with the short roots, so maps the basis vectors $w_1, \ldots, w_8''$ to certain 6-dimensional spaces (images, under the action of the Weyl group, of the subspace V_6 defined in Section 4.8.6).

More explicitly, consider the matrix

$$\begin{pmatrix} 1 & 1 & 0 & 0 \\ 1 & -1 & 0 & 0 \\ 0 & 0 & 1 & 1 \\ 0 & 0 & 1 & -1 \end{pmatrix} \tag{4.119}$$

which maps the short roots of $\mathbf{F}_4$ to the long roots. The 6-space corresponding to a given long root r is spanned by the six basis vectors corresponding to the short roots having inner product 1 with r. This does not completely determine the automorphism, however: see Section 4.9 for more details.

There is no proper covering group of $F_4(q)$ except in the case $q = 2$, when there is a group $2{\cdot}F_4(2)$. This exceptional cover can be used to show that the

Ree group $^2F_4(2)$ has a subgroup of index 2, known as the Tits group. It is also of relevance to the Monster (see Section 5.8.1) on account of the chain of inclusions

$$2{\cdot}F_4(2) < 2^{2.2}E_6(2) < 2{\cdot}\mathbb{B} < \mathbb{M}. \tag{4.120}$$

4.8.11 An integral Albert algebra

Just as we have constructed integral forms of the quaternions and octonions in Section 4.4.2, so there is an interesting integral form of the exceptional Jordan algebra. The automorphism group of this algebra is the group $^3D_4(2){:}3$, though we shall not prove this here.

Consider first the 'monomial' subgroup of order $2^{12}.3$ generated by the maps (where the notation is as in (4.79))

$$
\begin{aligned}
(d, e, f \mid D, E, F) &\mapsto (d, e, f \mid iD, Ei, iFi) \text{ for } i = \pm i_t,\\
(d, e, f \mid D, E, F) &\mapsto (e, f, d \mid E, F, D),\\
(d, e, f \mid D, E, F) &\mapsto (d, f, e \mid \overline{D}, \overline{F}, \overline{E}).
\end{aligned}
\tag{4.121}
$$

Next consider the set of images under this group of the following three vectors (they are primitive idempotents in the sense defined in Section 4.8.8) in the algebra:

$$
\begin{aligned}
&(1, 0, 0 \mid 0, 0, 0),\\
&(0, \tfrac{1}{2}, \tfrac{1}{2} \mid \tfrac{1}{2}, 0, 0),\\
&(\tfrac{1}{2}, \tfrac{1}{4}, \tfrac{1}{4} \mid \tfrac{1}{4}, \tfrac{1}{4}s, \tfrac{1}{4}\overline{s}),
\end{aligned}
\tag{4.122}
$$

where $s = \tfrac{1}{2}(1 + i_0 + i_1 + \cdots + i_6)$. It is straightforward to check that there are $3 + 48 + 768 = 819$ images of these vectors. We call these 819 primitive idempotents *roots*, by analogy with Coxeter groups. Indeed, the group $^3D_4(2)$ is generated by the 819 'reflections' in the roots r, defined as the maps

$$x \mapsto x + 4(\mathrm{Tr}(x \circ r))r - 4x \circ r,$$

where $x \circ r$ as usual denotes the Jordan product. Note that this is the same map which was defined in (4.111).

It is also easy to see that the maps $i_t \mapsto i_{t+1}$ and $i_t \mapsto i_{2t}$ normalise the above monomial subgroup, and in fact extend it to a group

$$2^2.2^3.2^6.(7{:}3 \times S_3) \tag{4.123}$$

which is the normaliser in $^3D_4(2){:}3$ of the four-group generated by

$$
\begin{aligned}
(d, e, f \mid D, E, F) &\mapsto (d, e, f \mid D, -E, -F)\\
\text{and } (d, e, f \mid D, E, F) &\mapsto (d, e, f \mid -D, -E, F).
\end{aligned}
\tag{4.124}
$$

4.9 The large Ree groups

The large Ree groups $^2F_4(2^{2n+1})$ can be constructed by a similar but more complicated process to that described in Section 4.5 for constructing the small Ree groups $^2G_2(3^{2n+1})$. Instead of starting from an octonion algebra in characteristic 3, we use an Albert algebra in characteristic 2. The idea is to use 3^3:$SL_3(3)$, in the appropriate 26-dimensional representation over $\mathbb{F}_2$, to construct the Albert algebra and the outer automorphism of $F_4(2)$, in much the same way that we used $2^3 \cdot SL_3(2)$ to construct the octonion algebra and the outer automorphism of $G_2(3)$, in Sections 4.3.2 and 4.5.1.

Then we switch to a basis where the maximal torus is diagonal, to investigate the structure of the group. Since the first basis and the action of 3^3:$SL_3(3)$ are used only for motivation, in order to write down the correct formulae, and not in the proof of any properties of the Ree groups, we shall omit them here.

More details of my approach are given in [183]. A broadly similar, but somewhat different, description of the large Ree groups is given by Kris Coolsaet [35, 36, 37].

4.9.1 The outer automorphism of $F_4(2)$

The basis we shall use here is essentially the same as that given in Section 4.8.4 for the Albert algebra. The difference is that we want the algebra without the identity element, so we replace w_0, w_0' and w_0'' by

$$
\begin{aligned}
w_9 &= w_0 + 1, \\
w_9' &= w_0' + 1, \\
w_9'' &= w_0'' + 1.
\end{aligned}
\tag{4.125}
$$

Thus $w_9 + w_9' + w_9'' = 0$. Note, however, that this subspace (of trace 0 matrices) is not closed under multiplication. Thus we take a new multiplication by projecting onto this subspace, just as we did with the pure imaginary octonions in Section 4.5.2.

The outer automorphism of $F_4(2)$ described in Section 4.8.10 maps the stabilisers of the isotropic 1-spaces $\langle w_1 \rangle, \ldots, \langle w_8'' \rangle$ to the stabilisers of certain totally isotropic 6-spaces. Indeed, in the notation of Section 4.8.6, it maps the images of V_1 to the images of V_6. Moreover, as we saw in Section 4.8.6, the Albert algebra induces symplectic forms on these 6-spaces. Over general fields, these symplectic forms are determined up to scalar multiplication. Thus over $\mathbb{F}_2$ they are uniquely determined. For example, the symplectic form on the space $\langle w_1, w_1', w_1'', w_2, w_2', w_2'' \rangle$ may be defined by saying that the given basis is a symplectic basis, such that $\langle w_1, w_2 \rangle$, $\langle w_1', w_2' \rangle$ and $\langle w_1'', w_2'' \rangle$ are mutually orthogonal hyperbolic planes (see Section 3.4.4).

Just as in the case of $G_2(q)$ in characteristic 3, the outer automorphism of $F_4(q)$ in characteristic 2 may be specified by a vector space isomorphism between the 26-dimensional space W (the trace 0 part of the Albert algebra) and a suitable subquotient of its exterior square. Table 4.2 gives a particular choice of such an automorphism $*$, determined by (4.119), with $u.v$ standing for $u \wedge v$ modulo a certain 299-dimensional subspace. Effectively this means that the following relations hold:

$$w_1'.w_8' + w_2'.w_7' + w_3'.w_6' + w_4'.w_5' + w_1''.w_8'' + w_2''.w_7'' + w_3''.w_6'' + w_4''.w_5'' = 0,$$
$$w_2'.w_7'' + w_3'.w_6'' + w_5'.w_5'' + w_8'.w_1'' + w_4.w_9 = 0,$$
$$w_1'.w_5'' + w_2'.w_3'' + w_3'.w_2'' + w_4'.w_1'' + w_1.w_9 = 0,$$

and images under the Weyl group of $F_4(q)$. Note that the non-isotropic basis vectors w_9 and w_9', unlike the isotropic basis vectors, do not correspond to 6-spaces.

Table 4.2. The outer automorphism of $F_4(2)$

$$w_9^* = w_1.w_8 + w_1'.w_8' + w_1''.w_8'' + w_2.w_7 + w_2'.w_7' + w_2''.w_7''$$
$$w_9'^* = w_1.w_8 + w_1'.w_8' + w_1''.w_8'' + w_3.w_6 + w_3'.w_6' + w_3''.w_6''$$

$w_1^* = w_1.w_2 + w_1'.w_2' + w_1''.w_2''$	$w_8^* = w_7.w_8 + w_7'.w_8' + w_7''.w_8''$
$w_1'^* = w_1.w_4 + w_2'.w_3' + w_1''.w_5''$	$w_8'^* = w_5.w_8 + w_6'.w_7' + w_4''.w_8''$
$w_1''^* = w_1.w_3 + w_1'.w_3' + w_1''.w_3''$	$w_8''^* = w_6.w_8 + w_6'.w_8' + w_6''.w_8''$
$w_2^* = w_1.w_7 + w_3'.w_4' + w_3''.w_5''$	$w_7^* = w_2.w_8 + w_5'.w_6' + w_4''.w_6''$
$w_2'^* = w_1.w_5 + w_1'.w_4' + w_2''.w_3''$	$w_7'^* = w_4.w_8 + w_5'.w_8' + w_6''.w_7''$
$w_2''^* = w_1.w_6 + w_2'.w_4' + w_2''.w_5''$	$w_7''^* = w_3.w_8 + w_5'.w_7' + w_4''.w_7''$
$w_3^* = w_3.w_4 + w_3'.w_5' + w_1''.w_7''$	$w_6^* = w_5.w_6 + w_4'.w_6' + w_2''.w_8''$
$w_3'^* = w_2.w_3 + w_1'.w_5' + w_1''.w_4''$	$w_6'^* = w_6.w_7 + w_4'.w_8' + w_5''.w_8''$
$w_3''^* = w_2.w_4 + w_2'.w_5' + w_1''.w_6''$	$w_6''^* = w_5.w_7 + w_4'.w_7' + w_3''.w_8''$
$w_4^* = w_3.w_5 + w_1'.w_7' + w_3''.w_4''$	$w_5^* = w_4.w_6 + w_2'.w_8' + w_5''.w_6''$
$w_4'^* = w_2.w_6 + w_2'.w_6' + w_2''.w_6''$	$w_5'^* = w_3.w_7 + w_3'.w_7' + w_3''.w_7''$
$w_4''^* = w_4.w_7 + w_3'.w_8' + w_5''.w_7''$	$w_5''^* = w_2.w_5 + w_1'.w_6' + w_2''.w_4''$

For each $q = 2^{2n+1}$, the large Ree group $^2F_4(q)$ may be defined as the centraliser in $F_4(q)$ of the map $*$ defined by

$$* : \sum_{w \in \mathcal{B}} \lambda_w w \mapsto \sum_{w \in \mathcal{B}} (\lambda_w)^{2^{n+1}} w^* \tag{4.126}$$

where $\mathcal{B}$ is the given basis $\{w_1, \ldots, w_9'\}$ of the 26-space W.

4.9.2 Generators for the large Ree groups

To determine the maximal torus in $^2F_4(q)$, where $q = 2^{2n+1}$, work in the torus of $F_4(q)$ given in (4.96), and observe that the eigenvalues α, β, γ, δ on w_1, w_1', w_1'', w_2 map to $\alpha^{2^{n+1}}$, $\beta^{2^{n+1}}$, $\gamma^{2^{n+1}}$, $\delta^{2^{n+1}}$ on w_1^*, $w_1'^*$, $w_1''^*$, w_2^*. On the other hand, the eigenvalues on these symplectic 6-spaces are given by the action of the torus on the symplectic form, so are respectively $\alpha\delta$, $\alpha\beta^{-1}\gamma$, $\beta\gamma\delta^{-1}$ and $\alpha\delta^{-1}$. Solving these equations gives $\gamma = \alpha^{-1}\beta^{2^{n+1}+1}$ and $\delta = \alpha^{2^{n+1}-1}$. Thus we obtain from (4.96) two generators for the maximal torus of $^2F_4(2^{2n+1})$. Explicitly, this torus consists of the elements which fix w_9, w_9' and w_9'' and act on $w_1, w_1', \ldots, w_8', w_8''$ as follows (where the eigenvalue on w_{9-i} is the inverse of the eigenvalue on w_i, and similarly for w_i' and w_i'').

Vector	Eigenvalue	Vector	Eigenvalue
w_1	α	w_3	$\alpha^{-2^{n+1}-1}\beta^{2^{n+1}+2}$
w_1'	β	w_3'	$\alpha^{-2^{n+1}}\beta^{2^{n+1}+1}$
w_1''	$\alpha^{-1}\beta^{2^{n+1}+1}$	w_3''	$\alpha^{1-2^{n+1}}\beta$
w_2	$\alpha^{2^{n+1}-1}$	w_4	$\alpha^{-1}\beta^{2^{n+1}}$
w_2'	$\alpha^{2^{n+1}}\beta^{-1}$	w_4'	$\alpha^2\beta^{-1-2^{n+1}}$
w_2''	$\alpha^{2^{n+1}+1}\beta^{-2^{n+1}-1}$	w_4''	$\alpha^{-1}\beta$

The part of the Weyl group of F_4 which lies in $^2F_4(2^{2n+1})$ is a dihedral group of order 16, generated by the two coordinate permutations

$$r = (w_4w_5)(w_3w_6)(w_1'w_2')(w_3'w_4')(w_5'w_6')(w_7'w_8')(w_1''w_2'')(w_3''w_5'')(w_4''w_6'')(w_7''w_8''),$$
$$s = (w_9w_9'')(w_1w_1'')(w_4w_5'')(w_5w_4'')(w_8w_8'')(w_2w_3'')(w_3w_2'')(w_6w_7'')(w_7w_6'')$$
$$(w_2'w_3')(w_4'w_5')(w_6'w_7'). \tag{4.127}$$

In fact r and s are representatives of the two classes of involutions in $^2F_4(q)$. The following elements are also in $^2F_4(q)$, indeed in $^2F_4(2)$.

$$\begin{aligned}
t:\ & w_1'' \mapsto w_1'' + w_1, w_4 \mapsto w_4 + w_5'', w_4'' \mapsto w_4'' + w_5, w_8 \mapsto w_8 + w_8'', \\
& w_3' \mapsto w_3' + w_2', w_7' \mapsto w_7' + w_6', w_9 \mapsto w_9 + w_4', w_5' \mapsto w_5' + w_4' + w_9, \\
& w_3 \mapsto w_3 + w_3'' + w_2 + w_2'', w_3'' \mapsto w_3'' + w_2'', w_2 \mapsto w_2 + w_2'', \\
& w_7'' \mapsto w_7'' + w_7 + w_6'' + w_6, w_7 \mapsto w_7 + w_6, w_6'' \mapsto w_6'' + w_6, \\
x:\ & w_1' \mapsto w_1' + w_1'', w_2'' \mapsto w_2'' + w_1'' + w_1', w_2' \mapsto w_2' + w_1'' + w_1', \\
& w_3'' \mapsto w_3'' + w_3', w_4' \mapsto w_4' + w_5'' + w_3', w_5'' \mapsto w_5'' + w_3' + w_3'', \\
& w_4'' \mapsto w_4'' + w_5', w_6' \mapsto w_6' + w_6'' + w_5', w_6'' \mapsto w_6'' + w_5' + w_4'', \\
& w_7' \mapsto w_7' + w_7'', w_8'' \mapsto w_8'' + w_8' + w_7'', w_8' \mapsto w_8' + w_7'' + w_7', \\
& w_4 \mapsto w_4 + w_3, w_6 \mapsto w_6 + w_3 + w_5 + w_9, \\
& w_5 \mapsto w_5 + w_4 + w_3 + w_9, w_9' \mapsto w_9' + w_4.
\end{aligned} \tag{4.128}$$

4.9.3 Subgroups of the large Ree groups

The Borel subgroup has order $q^{12}.(q-1)^2$, and is generated by the maximal torus together with x, x^s, x^{sr}, x^{srs}, their squares, and t, t^r, t^{rs}, t^{rsr}. These twelve elements lie one in each of the twelve chief factors of order q. One of the maximal parabolic subgroups has shape $q.q^4.q.q^4{:}(\mathrm{Sz}(q) \times C_{q-1})$ and is obtained from the Borel subgroup by adjoining r. It can be generated by the maximal torus together with t, x, r and x^{srs}. A Levi complement $\mathrm{Sz}(q) \times C_{q-1}$ generated by the torus, r and x^{srs}, may be extended to a maximal subgroup

$Sz(q) \wr 2$ by adjoining srs. This maximal subgroup is the stabiliser of the vector w_9.

The other maximal parabolic has shape $q^2.[q^9]{:}GL_2(q)$, and is obtained by adjoining s to the Borel subgroup. A Levi complement $GL_2(q) \cong PSL_2(q) \times C_{q-1}$ generated by the torus, t and s, may be extended to $PSL_2(q) \wr 2$ by adjoining rsr. The latter subgroup is not however maximal, and may be extended to a maximal subgroup $Sp_4(q){:}2$ by adjoining r. This group therefore contains the full normaliser of the maximal torus, and is the stabiliser of the vector w_9'.

The complete list of maximal subgroups of $^2F_4(q)$, for $q = 2^{2n+1} \geqslant 8$, is given by Malle [129].

Theorem 4.5. *The maximal subgroups of $^2F_4(q)$, for $q = 2^{2n+1} \geqslant 8$, are as follows:*

   (i) the parabolic subgroup $q^1.q^4.q^1.q^4{:}(Sz(q) \times C_{q-1})$;
   (ii) the parabolic subgroup $q^2.[q^9]{:}GL_2(q)$;
  (iii) $SU_3(q){:}2$, the normaliser of a subgroup of order 3;
  (iv) $PGU_3(q){:}2$;
   (v) $Sz(q) \wr 2$;
  (vi) $Sp_4(q){:}2$;
 (vii) $^2F_4(q_0)$, where $q = q_0{}^r$ and r is prime;
(viii) $(q+1)^2{:}GL_2(3)$;
  (ix) $(q \pm \sqrt{2q} + 1)^2{:}4S_4$;
   (x) $(q^2 + q + 1 \pm \sqrt{2q}(q+1)){:}12$.

Other interesting subgroups which are not maximal are $SL_3(3)$ (which lies inside a maximal subgroup $SL_3(3){:}2$ of $^2F_4(2)'$, which itself has index 2 in $^2F_4(2)$) and $PSL_2(25)$ (which we shall see again in Section 5.9.3, in the construction of the sporadic Rudvalis group, which contains $^2F_4(2)$).

4.9.4 Simplicity of the large Ree groups

The large Ree groups are sometimes treated more geometrically than algebraically, as automorphism groups of 'generalised octagons'. These objects have 'points' which may be identified with the images of $\langle w_1 \rangle$ and 'lines' which may be identified with the images of $\langle w_1, w_1' \rangle$. Abstractly the points may be defined as 1-spaces $\langle v \rangle$ such that $v^* = v.a + b.c + d.e$ for suitable vectors a, b, c, d, e. Two points $\langle v \rangle$ and $\langle w \rangle$ are incident if $v^* = w.a + b.c + d.e$ for suitable vectors a, b, c, d, e.

The points fall into eight orbits under the Borel subgroup, represented by the eight images of $\langle w_1 \rangle$ under the Weyl group, namely the 1-spaces spanned by w_1, w_1'', w_2'', w_3, w_6, w_7'', w_8'' and w_8, and with sizes 1, q, q^3, q^4, q^6, q^7, q^9, and q^{10} respectively. Thus the total number of points is $(1+q)(1+q^3)(1+q^6)$, and it can also be shown that the point stabiliser is exactly the parabolic subgroup $q.q^4.q.q^4{:}(Sz(q) \times C_{q-1})$, of order $q^{12}(q-1)^2(q^2+1)$. Therefore

$$|{}^2F_4(q)| = q^{12}(q^6+1)(q^4-1)(q^3+1)(q-1). \qquad (4.129)$$

Since each line contains $q+1$ points, and each point lies in q^2+1 lines, it follows that the number of lines is $(1+q^2)(1+q^3)(1+q^6)$. The point-line incidence graph of the generalised octagon can then be summarised in the following two pictures. The first picture shows the orbits under the point stabiliser:

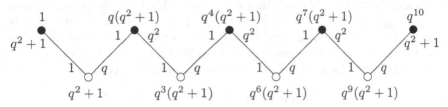

The second picture shows the orbits under the line stabiliser:

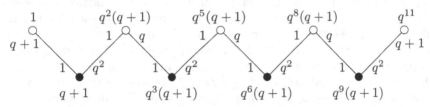

One can also prove simplicity of ${}^2F_4(q)$ for $q > 2$ by the usual method: the group acts primitively on the points of the generalised octagon, and the point stabiliser has a normal abelian subgroup containing an involution conjugate to x^2. Now if $q > 2$ then the conjugates of x^2 generate the group, and x^2 is a commutator, so Iwasawa's Lemma applies.

In fact ${}^2F_4(2)$ has a simple subgroup of index 2, generally known as the Tits group.

There are no non-trivial covers of any of the large Ree groups ${}^2F_4(2^{2n+1})$, and the only outer automorphisms are the field automorphisms, giving a cyclic outer automorphism group of order $2n+1$.

4.10 Trilinear forms and groups of type E_6

4.10.1 The determinant

The Albert algebra described in Section 4.8.1 can be used also to construct the groups of type E_6. These groups no longer preserve the algebra structure, or the inner product, but they do preserve a cubic form which can be interpreted as a type of determinant map. This is not quite the same cubic form as was defined in (4.83), but it is closely related. Indeed, up to an overall scalar, it agrees with the latter on the trace 0 matrices, as (4.95) shows. Of course, it is not obvious how to define a determinant even for matrices over a non-commutative ring, let alone a non-associative ring. However, for 3×3 Hermitian matrices over octonions there is a notion of determinant which makes sense, namely

$$\det(d, e, f \mid D, E, F) = def - dD\overline{D} - eE\overline{E} - fF\overline{F}$$
$$+ \mathrm{Re}\,(DEF) + \mathrm{Re}\,(FED). \qquad (4.130)$$

It is straightforward to calculate that the value of this determinant at a typical vector $\sum_{t=0}^{8}(\lambda_t w_t + \lambda_t' w_t' + \lambda_t'' w_t'')$ is as given in (4.95). The values of this determinant belong to the ground field $\mathbb{F}_q$, but beware that familiar identities like $\det(xy) = \det(x)\det(y)$ may not necessarily hold, or even make sense.

There is even a notion of *rank* for these matrices: clearly we want

$$\mathrm{rk}(x) = 3 \iff \det(x) \neq 0,$$
$$\mathrm{rk}(x) = 0 \iff x = 0, \qquad (4.131)$$

so all other matrices should have rank 1 or 2. The rank 1 matrices are essentially those whose rows are (left-octonion-)scalar multiples of each other. That is, if say $d \neq 0$, then the second row is $(\overline{F}, e, D) = d^{-1}(\overline{F}d, \overline{F}F, \overline{F}.E)$ and the third row is $(E, \overline{D}, f) = d^{-1}(Ed, EF, E\overline{E})$. (A slightly different definition is required if the diagonal of the matrix is zero. In fact, we take the matrices with $d = e = f = 0$ and $DE = EF = FD = 0$.) In some of the literature the vectors of rank 1, 2 and 3, or the 1-spaces they span, are called 'white', 'grey' and 'black' respectively.

It is possible to show (by direct, though tedious, calculations, similar to those in Sections 4.8.6 and 4.8.8) that the numbers of white, grey and black vectors are respectively $(q^9 - 1)(q^8 + q^4 + 1)$, $q^4(q^9 - 1)(q^8 + q^4 + 1)(q^5 - 1)$ and $q^{12}(q^9 - 1)(q^5 - 1)(q - 1)$. The last case covers $q - 1$ possibilities for the determinant, and it is easy to see that there are the same number of vectors of each non-zero determinant, so the number of vectors of determinant 1 is $q^{12}(q^9 - 1)(q^5 - 1)$. One of these is the identity matrix $(1, 1, 1 \mid 0, 0, 0)$, and if we fix this then we recover the Albert algebra. For the determinant gives rise to a trilinear form,

$$T(x, y, z) = \det(x + y + z) - \det(x + y) - \det(y + z) - \det(x + z)$$
$$+ \det x + \det y + \det z, \qquad (4.132)$$

and by substituting the identity matrix for one or two of the variables we obtain a bilinear and a linear form, and these three forms together define the algebra. (Details are left as an exercise: see Exercise 4.32.)

In other words, the stabiliser of a black vector is $F_4(q)$, and provided we can prove transitivity of our group on vectors of determinant 1 (which we shall not do here), we deduce that its order is $q^{12}(q^9 - 1)(q^5 - 1)|F_4(q)|$, that is

$$q^{36}(q^{12} - 1)(q^9 - 1)(q^8 - 1)(q^6 - 1)(q^5 - 1)(q^2 - 1). \qquad (4.133)$$

This group in general is not simple, as it may contain non-trivial scalars—this is so if and only if $\mathbb{F}_q$ contains a cube root ω of 1, for $\det(\omega x) = \omega^3 \det(x)$ which equals $\det(x)$ if and only if $\omega^3 = 1$. We define $E_6(q)$ to be the (in fact simple!) group obtained by factoring out the scalars of order 3, when $q \equiv 1 \bmod 3$.

By analogy with the linear groups, let $SE_6(q)$ denote the original matrix group, so that $SE_6(q) \cong 3 \cdot E_6(q)$ if $q \equiv 1 \bmod 3$ and $SE_6(q) \cong E_6(q)$ otherwise. Similarly, let $GE_6(q)$ denote the group of matrices which multiply the determinant by a scalar, so that $GE_6(q) \cong 3.(C_{(q-1)/3} \times E_6(q)).3$ if $q \equiv 1 \bmod 3$, and let $PGE_6(q)$ denote the quotient of $GE_6(q)$ by scalars. Thus $PGE_6(q) \cong E_6(q).3$ if $q \equiv 1 \bmod 3$, and $PGE_6(q) \cong E_6(q)$ otherwise.

4.10.2 Dickson's construction

It is not as well-known as it should be that the simple groups $E_6(q)$, and their triple covers when they exist, were first constructed by L. E. Dickson around 1901 (see [50], with corrections and simplifications in [51]). Incorrect statements on this point are frequent in the literature. Dickson constructed the 27-dimensional representation with respect to a basis which is essentially the same as the one we use in Section 4.10.3 below, and wrote down a large number of elements generating the group. He calculated the group order, as well as constructing a permutation representation of the simple group $E_6(q)$ on a set of $(q^9 - 1)(q^8 + q^4 + 1)/(q - 1)$ points (there are two such representations, related by a duality automorphism—see Sections 4.10.5 and 4.10.6 below), although he did not explicitly prove simplicity of the group.

Dickson's construction is more elementary than ours, but more or less equivalent. He takes 27 coordinates labelled x_i, y_i and $z_{ij} = -z_{ji}$ where i and j are distinct elements of the set $\{1, 2, 3, 4, 5, 6\}$, and defines his group as the stabiliser of a cubic form with 45 terms

$$\sum_{i \neq j} x_i y_j z_{ij} + \sum z_{ij} z_{kl} z_{mn}, \qquad (4.134)$$

where in the second summation $(ij \mid kl \mid mn)$ ranges over the 15 partitions of $\{1, 2, 3, 4, 5, 6\}$ into three pairs, ordered so that $\begin{pmatrix} 1 & 2 & 3 & 4 & 5 & 6 \\ i & j & k & l & m & n \end{pmatrix}$ is an even permutation.

It is a straightforward exercise (see Exercise 4.33) to show that this is the same cubic form as the determinant defined in (4.130) and (4.95) above, for example by taking $x_1, \ldots, x_6$ to correspond to $w_1, w_1', w_1'', w_2, w_2', w_2''$ and $y_1, \ldots, y_6$ to correspond to $w_7, \ldots, w_8''$ (in that order). Dickson enumerated the $(q^9 - 1)(q^8 + q^4 + 1)/(q - 1)$ white points, which can be defined abstractly as points spanned by vectors v with the property that $T(v, -, -)$ is a symmetric bilinear form with radical of dimension 17, where T is the symmetric trilinear form defined in (4.132). He then computed generators for the point stabiliser of shape $q^{16}\mathrm{Spin}_{10}^+(q).C_{q-1}$, and hence calculated the order of $E_6(q)$.

An alternative description of the cubic form may be obtained by defining the 27-space as the space of triples (x, y, z) of 3×3 matrices over $\mathbb{F}_q$, with the cubic form

$$\det x + \det y + \det z - \operatorname{Tr}(xyz). \qquad (4.135)$$

Again, this is easily seen to be a cubic form with 45 terms in 27 variables, and it is a straightforward exercise to show it is essentially the same cubic form.

4.10.3 The normaliser of a maximal torus

To exhibit the maximal split torus T we use the basis $\{w_0, \ldots, w_8''\}$ described in Section 4.8.4 for $F_4(q)$. With respect to this basis the determinant map takes a particularly nice form, as given in (4.95). This enables us to calculate the action of the torus, as follows.

Vector	Eigenvalue	Vector	Eigenvalue	Vector	Eigenvalue
w_0	λ	w_0'	μ	w_0''	$\lambda^{-1}\mu^{-1}$
w_1	α	w_1'	β	w_1''	γ
w_2	δ	w_2'	$\alpha\beta^{-1}\delta$	w_2''	$\alpha\gamma^{-1}\delta$
w_3	$\alpha^{-1}\beta\gamma\delta^{-1}\lambda^{-2}$	w_3'	$\gamma\delta^{-1}\lambda^{-2}$	w_3''	$\beta\delta^{-1}\lambda^{-2}$
w_4	$\beta^{-1}\gamma\lambda^{-1}\mu^{-1}$	w_4'	$\alpha\gamma^{-1}\lambda$	w_4''	$\alpha^{-1}\beta\mu$
w_5	$\beta\gamma^{-1}\mu$	w_5'	$\alpha^{-1}\gamma\lambda^{-1}\mu^{-1}$	w_5''	$\alpha\beta^{-1}\lambda$
w_6	$\alpha\beta^{-1}\gamma^{-1}\delta\lambda$	w_6'	$\gamma^{-1}\delta\lambda$	w_6''	$\beta^{-1}\delta\lambda$
w_7	$\delta^{-1}\lambda^{-1}$	w_7'	$\alpha^{-1}\beta\delta^{-1}\lambda^{-1}$	w_7''	$\alpha^{-1}\gamma\delta^{-1}\lambda^{-1}$
w_8	$\alpha^{-1}\lambda^{-1}$	w_8'	$\beta^{-1}\mu^{-1}$	w_8''	$\gamma^{-1}\lambda\mu$

$$(4.136)$$

If N is the normaliser of T, then $N/T \cong W(E_6) \cong \mathrm{GO}_6^-(2) \cong \mathrm{SO}_5(3)$, permuting the 27 coordinate 1-spaces transitively. (See Section 3.12.4 for proofs of these isomorphisms.)

We know that w_0 is a white vector, so by transitivity of the Weyl group, so are all the basis vectors.

4.10.4 Parabolic subgroups of $E_6(q)$

Up to conjugacy, all of the maximal parabolic subgroups contain the maximal split torus, and may be described as the stabilisers of certain subspaces spanned by subsets of the basis vectors. Specifically, let us define

$$
\begin{aligned}
W_1 &= \langle w_1 \rangle, \\
W_2 &= \langle w_1, w_1' \rangle, \\
W_3 &= \langle w_1, w_1', w_1'' \rangle, \\
W_6 &= \langle w_1, w_1', w_1'', w_2, w_2', w_2'' \rangle, \\
W_5 &= \langle w_1, w_1', w_1'', w_2, w_3 \rangle, \\
W_{10} &= \langle w_1, w_1', w_1'', w_2, w_2', w_3, w_3', w_0, w_4, w_5' \rangle.
\end{aligned}
\qquad (4.137)
$$

Then it turns out that the stabiliser in $SE_6(q)$ of W_i is a maximal subgroup. The precise structures of these subgroups depend on congruences of q modulo 3, 4 and 5. For simplicity, we describe the case when $q = 2^{2n+1}$, so that $q - 1$ is prime to 30. In this case the stabilisers of W_1, W_2, W_3, W_5, W_6 and W_{10} have shapes

$$q^{16}{:}(\Omega_{10}^+(q) \times C_{q-1}),$$
$$q^{5+20}{:}(\mathrm{SL}_2(q) \times \mathrm{SL}_5(q) \times C_{q-1}),$$
$$q^{2+9+18}{:}(\mathrm{SL}_3(q) \times \mathrm{SL}_3(q) \times \mathrm{SL}_2(q) \times C_{q-1}),$$
$$q^{5+20}{:}(\mathrm{SL}_2(q) \times \mathrm{SL}_5(q) \times C_{q-1}),$$
$$q^{1+20}{:}(\mathrm{SL}_6(q) \times C_{q-1}),$$
$$q^{16}{:}(\Omega_{10}^+(q) \times C_{q-1}). \tag{4.138}$$

Indeed, the stabilisers of W_1 and W_{10} are isomorphic, as are the stabilisers of W_2 and W_5. The stabiliser of W_6 is also the normaliser of a root subgroup (already defined in (4.104) above for $F_4(q)$) consisting of root elements

$$t_r(\lambda) : w_7 \mapsto w_7 - \lambda w_1, w_7' \mapsto w_7' - \lambda w_1', w_7'' \mapsto w_7'' - \lambda w_1'',$$
$$w_8 \mapsto w_8 + \lambda w_2, w_8' \mapsto w_8' + \lambda w_2', w_8'' \mapsto w_8'' + \lambda w_2''. \tag{4.139}$$

The stabiliser of a grey 1-space is a non-maximal subgroup of shape $q^{16}{:}2{\cdot}\Omega_9(q).C_{q-1}$. The sporadic subgroups $\mathrm{J}_3 < E_6(4)$ and $\mathrm{M}_{12} < E_6(5)$ are of particular interest [114], especially as one of these subgroups had previously been proved not to exist! For a great deal more on $E_6(q)$ and its subgroups, see the series of papers by Aschbacher [6, 7, 8, 9, 10].

4.10.5 The rank 3 action

The action of $E_6(q)$ on the $(q^8 + q^4 + 1)(q^9 - 1)/(q - 1)$ white points (i.e. the images of W_1) is particularly interesting as it has rank 3 (i.e. the point stabiliser has exactly 3 orbits on points). Two distinct white points either span a 2-space containing only white points, in which case the stabiliser of the two points is a group of shape $q^{15}.q^{10}.\mathrm{GL}_5(q)$, or they span a 2-space in which all the other points are grey, in which case the stabiliser of the two points is a group of shape $q^8.q^8.(2{\cdot})\Omega_8^+(q).(C_{q-1})^2$. (Here the factor 2 is omitted in characteristic 2.) Thus the non-trivial suborbit lengths of the action are $q(q^3 + 1)(q^8 - 1)/(q - 1)$ and $q^8(q^4 + 1)(q^5 - 1)/(q - 1)$ respectively. For example, the vectors $(1, 0, 0 \mid 0, 0, 0)$, $(0, 0, 0 \mid 0, x_1, 0)$ and $(0, 1, 0 \mid 0, 0, 0)$ all span white points, and $(1, 0, 0 \mid 0, x_1, 0)$ is white while $(1, 1, 0 \mid 0, 0, 0)$ is grey.

At this point it is fairly straightforward to prove simplicity of $E_6(q)$ using Iwasawa's Lemma in the usual way. The action on white points is primitive, and the point stabiliser has a normal abelian subgroup which contains root elements. It follows easily that root elements are commutators, and also that $SE_6(q)$ is generated by root elements. Hence the quotient of $SE_6(q)$ by any scalar matrices it contains is a simple group.

Now it turns out that the action of $E_6(q)$ on the images of W_{10} has exactly the same structure. Two distinct images of W_{10} intersect either in a 5-space (which is an image of W_5) or in a 1-space (an image of W_1). The respective stabilisers have the same shapes as above. Moreover, there is an outer automorphism of $E_6(q)$ which swaps these two actions of the group. To see this, notice that if we fix a white vector such as w_0, then the trilinear form T gives rise to a symmetric bilinear form $T(w_0, x, y)$. An easy calculation shows that this bilinear form has radical of dimension 17, spanned by w_0, w_i' and w_i''. Thus there is a 10-dimensional quotient space which supports this bilinear form. Therefore in the dual space we have a 10-dimensional subspace X_{10} supporting a corresponding symmetric bilinear form.

On the other hand, W_{10} supports a symmetric bilinear form, defined up to scalar multiplication by saying that the white vectors are isotropic and the grey vectors are non-isotropic. The outer automorphism of $SE_6(q)$ swaps the 27-dimensional representation with its dual, and identifies the images of W_{10} with the images of X_{10}, which in turn correspond to the images of W_1.

4.10.6 Covers and automorphisms

The method of construction shows that there is a triple cover $SE_6(q) \cong 3 \cdot E_6(q)$ whenever $q \equiv 1 \bmod 3$. In fact, there are no other covers, either generic or exceptional.

Besides field automorphisms, there is a 'diagonal' automorphism of order 3 whenever $q \equiv 1 \bmod 3$, obtained from linear maps which multiply the determinant by a non-cube, i.e. an element α of $\mathbb{F}_q^\times \setminus \mathbb{F}_q^{\times 3}$. In this case there are three orbits of $SE_6(q)$ on black 1-spaces, fused by the outer automorphism which multiplies the cubic form by α.

The only other outer automorphism is a 'duality' automorphism of order 2, which can be taken to centralise a maximal subgroup $F_4(q)$. This is analogous to the duality automorphism of $\mathrm{GL}_n(q)$ which can be taken to be the 'transpose–inverse' automorphism, which centralises $\mathrm{GO}_n^\varepsilon(q)$ for some ε (or $\mathrm{Sp}_n(q)$ if both n and q are even). If the field has order q^2, then it has an automorphism $x \mapsto x^q$ of order 2, and multiplying this duality automorphism of $E_6(q)$ by a field automorphism of order 2 gives a twisted duality automorphism which centralises a subgroup called $^2E_6(q)$ of $E_6(q^2)$ (see Section 4.11). These are analogous to the unitary groups inside the general linear groups.

4.11 Twisted groups of type $^2E_6$

The duality on the groups of type E_6 can also be expressed by the natural inner product on the Albert algebra. Twisting this in the usual way to make a Hermitian form over the field of order q^2, we can consider the subgroup of $E_6(q^2)$ which preserves this Hermitian form. This subgroup, modulo scalars, is denoted $^2E_6(q)$, and bears the same relationship to $E_6(q)$ as $\mathrm{PSU}_n(q)$ bears to $\mathrm{PSL}_n(q)$. There is a triple cover $3{\cdot}^2E_6(q)$ whenever $q \equiv 2 \bmod 3$, and the order of the covering group is

$$q^{36}(q^{12} - 1)(q^9 + 1)(q^8 - 1)(q^6 - 1)(q^5 + 1)(q^2 - 1).$$

More concretely, take the basis $\{w_0, \ldots, w_8''\}$ of the 27-space defined for $E_6(q^2)$, and define the Hermitian form such that this basis is orthonormal. Now $^2E_6(q)$ is defined to be the quotient by any scalars of the group which preserves both the determinant (4.130) and the Hermitian form (or inner product).

Now a scalar $\lambda \in \mathbb{F}_{q^2}$ preserves the trilinear form if and only if $\lambda^3 = 1$, and preserves the Hermitian form if and only if $\lambda^{q+1} = 1$, so the group contains nontrivial scalars (necessarily of order 3) exactly when $3 \mid (q + 1)$.

The stabiliser of $(1, 1, 1 \mid 0, 0, 0)$ is the subgroup of $F_4(q^2)$ which preserves the Hermitian form. But $F_4(q^2)$ already preserves a symmetric bilinear form over $\mathbb{F}_{q^2}$, and it follows that the required stabiliser is $F_4(q)$. If we could count the images of $(1, 1, 1 \mid 0, 0, 0)$ we would find that the total number of them is $q^{12}(q^9 + 1)(q^5 + 1)(q - 1)$, and we would deduce the order of $^2E_6(q)$ as given above.

The maximal torus is obtained from the maximal torus of $E_6(q^2)$ by fixing the Hermitian form. Since the form is the standard one with respect to a basis of eigenvectors of the maximal torus, it follows at once that the eigenvalues λ satisfy $\lambda^{q+1} = 1$, so that the torus has structure $(C_{q+1})^6$. Similar arguments give the structures of the maximal parabolic subgroups. As in the case of $E_6(q)$, the precise structures depend on congruences of q modulo 3, 4, and 5.

The automorphism groups and covers behave very much as in the untwisted groups of type E_6. If $q = p^a$ then the field automorphisms (of the field of order p^{2a}) form a cyclic group of order $2a$. If $q \equiv 2 \bmod 3$ there is a diagonal automorphism of order 3 and a triple cover.

There is just one exceptional cover, namely the fourfold cover $2^{2{\cdot}2}E_6(2)$ of $^2E_6(2)$. This group lies inside the Monster (see Section 5.8.1), where its normaliser is a maximal subgroup of shape $2^{2{\cdot}2}E_6(2){:}S_3$. Indeed there is a chain of subgroups

$$2{\cdot}F_4(2) < 2^{2{\cdot}2}E_6(2) < 2{\cdot}\mathbb{B} < \mathbb{M}.$$

4.12 Groups of type E_7 and E_8

The groups of type E_7 and E_8 cannot easily be defined in a 'classical' way. For E_8, there is no representation smaller than the Lie algebra, of dimension 248, so the only sensible way of describing the group is in terms of this Lie algebra. For E_7, in addition to the Lie algebra, of dimension 133, there is a 56-dimensional representation (which also lies inside the Lie algebra of type E_8). This is a representation of $E_7(q)$ if q is a power of 2, and a representation of a double cover $2{\cdot}E_7(q)$ if q is odd. It may be defined as the group of automorphisms of a suitable quartic form. See for example [11, 17, 39, 135] for more details.

4.12.1 Lie algebras

For completeness we give here a very brief introduction to the Lie algebra of type E_8. A *Lie algebra* is a (non-associative) algebra whose (bilinear) product, traditionally written $[a, b]$ or $[ab]$, satisfies $[aa] = 0$ for all a (and therefore $[ab] = -[ba]$) and

$$[[ab]c] + [[bc]a] + [[ca]b] = 0. \tag{4.140}$$

This last condition is called the *Jacobi identity*. The canonical example is the algebra of $n \times n$ matrices, which can be made into a Lie algebra by defining the multiplication $[AB] = AB - BA$: it is obvious that $[AA] = 0$, and it takes only a few moments to verify the Jacobi identity (see Exercise 4.35).

Simple Lie algebras over the complex numbers are completely classified, and parametrised by the Dynkin diagrams A_n, B_n, C_n, D_n, G_2, F_4, E_6, E_7, E_8 (see Section 2.8.3). Since the families A_n, B_n, C_n, D_n correspond to classical groups, and

$$G_2 < F_4 < E_6 < E_7 < E_8, \tag{4.141}$$

the only Lie algebra we really need to understand is E_8.

First we need to understand its Weyl group: the details we need are in Section 3.12.4. Also note the following important properties of the root system. The roots corresponding to the nodes of the Dynkin diagram are called *fundamental* roots, and the roots which are positive integer linear combinations of the fundamental roots are called *positive* roots. The negative of a positive root is called a *negative* root. The fact that every pair of fundamental roots forms an obtuse angle or a right angle implies (after some work) that every root is either positive or negative.

Observe that there are 240 roots r, coming in 120 pairs $(r, -r)$. To build the Lie algebra, of dimension 248, take 240 basis vectors e_r corresponding to the roots r, and an 8-space H spanned by the E_8 lattice, but with coefficients in the desired field F. We may take this 8-space H (known as a *Cartan subalgebra*) to be spanned by vectors h_r, where r is one of the eight fundamental roots, corresponding to one of the nodes of the Dynkin diagram. (Such a basis for the Lie algebra is known as a *Chevalley basis*.)

The general form of the multiplication in the algebra is that the e_r are simultaneous eigenvectors for multiplication by every h_s, and that $[e_r e_s]$ is a scalar multiple of e_{r+s} (interpreted as 0 if $r + s$ is not a root, except that $[e_r e_{-r}] = h_r$). More precisely,

$$\begin{aligned} [h_r h_s] &= 0, \\ [h_r e_s] &= (r.s)e_s \end{aligned} \tag{4.142}$$

where $r.s$ denotes the inner product in the E_8-lattice, scaled so that $r.r = 2$. Now given roots r and s, at most one of $r + s$ and $-r + s$ can be a root. [Beware that this is only true for root systems of type A_n, D_n or E_n: the so-called *simply-laced* types.] In the former case we have $[e_r e_s] = \pm e_{r+s}$ and in the latter case $[e_r e_s] = 0$.

To specify the precise form of the multiplication we need to specify the signs. Some signs may be chosen arbitrarily and then the rest are determined by certain relations arising from the Jacobi identity. Details may be found in Carter's book [21].

The group $E_8(q)$ may be defined as the automorphism group of the Lie algebra. In the next section we give generators for the group and some of its subgroups (modulo the above sign problem, which is not solved here).

4.12.2 Subgroups of $E_8(q)$

It is easy to see from our partial definition of the Lie algebra that there is a subgroup of diagonal automorphisms isomorphic to a direct product of 8 cyclic groups of order $q - 1$, since if $r_1, \ldots, r_8$ is a fundamental set of roots, we may multiply $e_{r_1}, \ldots, e_{r_8}$ independently by scalars $\lambda_1, \ldots, \lambda_8$, and then the relations imply that if $\sum_{i=1}^{8} a_i r_i$ is a root s, say, then e_s is multiplied by $\prod_{i=1}^{8}(\lambda_i)^{a_i}$, and that h_r is fixed for all r. This subgroup is the *maximal split torus* T, and its normaliser N has the property that N/T is isomorphic to the Weyl group of type E_8, acting in the natural way on the 8-space spanned by the h_r, and permuting the 240 1-spaces spanned by the e_r. (It is not obviously a split extension, except in characteristic 2, as the sign problem may prevent splitting.)

Consider the map

$$\mathrm{ad}\, e_r : x \mapsto [e_r x]. \tag{4.143}$$

Ignoring scalars for the moment, this map takes e_{-r} to h_r, and h_r to e_r, and e_r to 0; any other e_s gets mapped to 0 after at most two steps. Therefore $(\mathrm{ad}\, e_r)^3$ is the zero map. (In characteristic 2 we have $(\mathrm{ad}\, e_r)^2 = 0$.) Define the *root elements* $x_r(\lambda)$ by

$$x_r(\lambda) = 1 + \lambda \mathrm{ad}\, e_r + \tfrac{1}{2}\lambda^2(\mathrm{ad}\, e_r)^2 \tag{4.144}$$

if the characteristic is not 2, and by

$$x_r(\lambda) = 1 + \lambda \mathrm{ad}\, e_r \tag{4.145}$$

if the characteristic is 2. It is straightforward to show that these maps preserve the Lie algebra (once the signs have been fixed!). The *root subgroups*

$$X_r = \langle x_r(\lambda) \mid \lambda \in F \rangle$$

are isomorphic to the additive group of the field F, since for any root r and any $\lambda, \mu \in F$, we have $x_r(\lambda)x_r(\mu) = x_r(\lambda + \mu)$.

Many interesting subgroups are generated by suitable sets of root subgroups. These include the *maximal parabolic* subgroups, and the *maximal rank* subgroups. The former are analogous to the stabilisers of totally isotropic subspaces in the classical groups, while the latter are analogous to the stabilisers of non-singular subspaces.

The set of all X_r as r ranges over positive roots generates the Sylow p-subgroup of $E_8(q)$, which has order q^{120}. This is normalised by the diagonal subgroup, of order $(q-1)^8$, to give the *Borel subgroup* B.

If r is any root, then the two root subgroups X_r and X_{-r} together generate a group isomorphic to $\mathrm{SL}_2(q)$, whose Lie algebra is the 3-space $\langle e_r, e_{-r}, h_r \rangle$ on which the group acts as $\Omega_3(q)$ (at least if the characteristic is not 2). If s is any root orthogonal to r, then both X_r and X_{-r} centralise e_s. The other 112 roots come in 56 pairs (e_t, e_{r+t}) on which the group acts as $\mathrm{SL}_2(q)$ in its natural representation.

The *parabolic* subgroups are obtained by adjoining some of the negative root subgroups X_{-r} to the Borel subgroup. We may choose any subset of the fundamental roots for this purpose: a set of all but one of the fundamental roots gives rise to a *maximal parabolic subgroup*. Their shapes can be read off from the Dynkin diagram, although there are subtleties concerning the precise isomorphism types of these subgroups.

For example, the maximal parabolic obtained by leaving out the end node of the long arm of the diagram has shape $q^1.q^{56}{:}(C_{q-1} \times E_7(q))$ for q even, and $q^1.q^{56}{:}2^{\cdot}(C_{(q-1)/2} \times E_7(q)).2$ for q odd. The next one along has shape $q^2.q^{27}.q^{54}{:}(C_{q-1} \times \mathrm{SL}_2(q) \times E_6(q))$ in the simplest case, when q is an odd power of 2. When $q \equiv 1 \bmod 6$, the shape is $q^2.q^{27}.q^{54}{:}6^{\cdot}(C_{(q-1)/6} \times \mathrm{PSL}_2(q) \times E_6(q)).6$.

It is not particularly easy to calculate the group order directly from the definition. In fact

$$|E_8(q)| = q^{120}(q^{30} - 1)(q^{24} - 1)(q^{20} - 1)(q^{18} - 1) \\ (q^{14} - 1)(q^{12} - 1)(q^8 - 1)(q^2 - 1). \tag{4.146}$$

At this point simplicity of $E_8(q)$ can be proved using Iwasawa's Lemma in the usual way. The action on the root subgroups can be shown to be primitive, and the root elements generate $E_8(q)$. Moreover, the root subgroups lie in the derived subgroup, so $E_8(q)$ is perfect. Hence $E_8(q)$ is simple.

To describe the maximal rank subgroups, we extend the Dynkin diagram so that the long arm of the diagram is extended by one further node. (Clearly there is a unique vector which has these specified inner products with the fundamental roots, and it is easy to show that this vector is also a root.) Then the maximal rank subgroups can be described by deleting one node of the extended diagram. They are generated by the corresponding root subgroups X_r and X_{-r} as r ranges over the remaining roots in the diagram.

For example, deleting the node adjacent to the extending node gives a group which in characteristic 2 has shape $\mathrm{SL}_2(q) \times E_7(q)$, and in other characteristics has shape $\mathrm{SL}_2(q) \circ 2{\cdot}E_7(q) \cong 2{\cdot}(\mathrm{PSL}_2(q) \times E_7(q))$. The latter subgroup is not itself maximal, but its normaliser $2{\cdot}(\mathrm{PSL}_2(q) \times E_7(q)).2$ is.

There are a few more exotic subgroups, such as $2^{5+10}{\cdot}\mathrm{GL}_5(2) < E_8(q)$ for every odd q (it is maximal if and only if q is an odd prime) and $5^3{:}\mathrm{SL}_3(5) < E_8(q)$ for all q prime to 5 (again it is maximal just when q is a prime distinct from 5).

The group $E_8(q)$ has no proper covering groups, and the only outer automorphisms are the field automorphisms.

4.12.3 $E_7(q)$

Most of the important properties of $E_7(q)$ can be read off from the embedding in $E_8(q)$ of $\mathrm{SL}_2(q) \times E_7(q)$ (in characteristic 2) or $2{\cdot}(\mathrm{PSL}_2(q) \times E_7(q)).2$ (in odd characteristic). For example we see the Lie algebra, of dimension 133, spanned by the h_r as r runs over the seven fundamental roots of $\mathbf{E_7}$ and the e_r as r runs over the 126 roots of $\mathbf{E_7}$.

We also see the 56-dimensional representation of $2{\cdot}E_7(q)$ (or $E_7(q)$ in characteristic 2): the 112 roots of $\mathbf{E_8}$ which are neither in $\mathbf{E_7}$ nor orthogonal to it fall into 56 pairs $(t, r+t)$, where $\pm r$ are the two roots orthogonal to $\mathbf{E_7}$. The 112-dimensional representation of $2{\cdot}(\mathrm{PSL}_2(q) \times E_7(q))$ is the tensor product of the 2-dimensional representation of $\mathrm{SL}_2(q)$ with the 56-dimensional representation of $2{\cdot}E_7(q)$, so the latter can be extracted.

In particular note that in the case q odd there is both a double cover of $E_7(q)$ and a diagonal automorphism of order 2. As noted above, the double cover has a representation of degree 56, which is considerably smaller than the Lie algebra, of dimension 133.

The order of $E_7(q)$ in characteristic 2 is

$$q^{63}(q^{18} - 1)(q^{14} - 1)(q^{12} - 1)(q^{10} - 1)(q^8 - 1)(q^6 - 1)(q^2 - 1).$$

In odd characteristic it is half this.

Further reading

Much of the material in this chapter is not available in book form anywhere, certainly not in one place. For the standard treatment of exceptional groups via Lie algebras, see Carter's 'Simple groups of Lie type' [21]. This book is essential reading for anyone seriously interested in finite simple groups. For the approach using algebraic groups, see 'Linear algebraic groups' by Humphreys [86] or 'An introduction to algebraic geometry and algebraic groups' by Geck [65].

A more abstract treatment of octonions, Jordan algebras and exceptional groups is given by Springer and Veldkamp [158], who deal thoroughly with the groups of type G_2, ${}^3D_4$, F_4 and E_6, although they do not always treat the exceptional characteristics 2 and 3. A detailed description of rings of 'integral' octonions (in characteristic 0) may be found in the recent book 'On quaternions and octonions' by Conway and Smith [32]. For a thorough treatment of Jordan algebras in all their glory, I recommend the immensely readable book 'A taste of Jordan algebras' by Kevin McCrimmon [133].

Adams's book 'Lectures on Lie groups' [1], although devoted to Lie groups over the complex numbers, also reveals a lot of the structure of finite groups of Lie type. The book 'Orthogonal decompositions and integral lattices' by Kostrikin and Tiep [115] deals with various aspects of Lie algebras of relevance to finite groups.

The Suzuki groups are treated in a number of more general texts, for example Huppert and Blackburn's 'Finite groups. III' [87] (which also includes a discussion of the small Ree groups, but without proofs), and Taylor's book 'The geometry of classical groups' [162], as well as the book 'Die Suzukigruppen unde ihre Geometrien' by Lüneburg [125]. More details of my approach to the Ree groups may be found in [182, 183, 184], and Coolsaet's approach is described in [35, 36, 37].

Exercises

4.1. Verify that the map $*$ defined in (4.2) commutes with the coordinate permutation (2,3,4,5,6), i.e. with the symplectic transformation

$$e_1 \mapsto f_1, \; f_1 \mapsto e_1 + f_1 + f_2, \; e_2 \mapsto f_2, \; f_2 \mapsto e_2 + f_1.$$

4.2. Verify (see Section 4.2.2) that if $x, y, z \in \mathbb{F}_{2^{2n+1}}$, then $z = xy + x^{2^{n+1}+2} + y^{2^{n+1}}$ implies $z^{2^{n+1}+1} + xyz^{2^{n+1}} + y^{2^{n+1}+2} + x^{2^{n+1}}z^2 = 0$.

4.3. Show that if x, y and z are three mutually orthogonal purely imaginary quaternions of norm 1, then $xy = \pm z$.

4.4. Prove that every automorphism of the real quaternion algebra $\mathbb{H} = \mathbb{R}[i, j, k]$ is an inner automorphism $\alpha_q : x \mapsto q^{-1}xq$.

4.5. Show that in a Moufang loop $(xy)x = x(yx)$. Show also that $(xx)y = x(xy)$ and $y(xx) = (yx)x$. Deduce that any 2-generator Moufang loop is a group.

4.6. Verify that the Moufang identities $(xy)(zx) = (x(yz))x$ and $x(y(xz)) = ((xy)x)z$ and $((yx)z)x = y(x(zx))$ hold in the octonion algebra generated by $i_0, \ldots, i_6$ over any field.

[Hint: use the reduction to the case $y = i_0$, $z = i_1$, sketched in Section 4.3.2.]

4.7. Calculate the multiplication table of the octonion algebra with respect to the basis $\{1, i, j, ij, l, il, jl, (ij)l\}$ defined in Section 4.3.3, and relabel the basis vectors to show that this is essentially the same table as (4.18).

4.8. Show that the monomial subgroup $2^3\mathrm{GL}_3(2)$ of $G_2(q)$ is a non-split extension.

[Hint: any subgroup $\mathrm{GL}_3(2)$ would have to be generated by a Sylow 7-normaliser (of shape 7:3) together with a subgroup S_3 of index 2 in the Sylow 3-normaliser (of shape $2 \times S_3$), but both possibilities generate the whole of $2^3\mathrm{GL}_3(2)$.]

4.9. Prove that if q is odd then $G_2(q)$ is simple.

4.10. Show that the maps r and s defined in (4.30) preserve the multiplication table of the split octonion algebra given in (4.27).

4.11. Mimic the calculation of the order of $^3D_4(q)$ in Section 4.6 to calculate the order of $G_2(q)$ for arbitrary q.

4.12. Show that in the real octonion algebra (see Section 4.4.2) with

$$\omega_t = \tfrac{1}{2}(-1 + i_t + i_{t+1} + i_{t+3}),$$

$(\overline{\omega}_0\overline{\omega}_1)i_1 = \tfrac{1}{2}(-1 - i_0 + i_2 + i_5)$ and deduce that $\tfrac{1}{2}(i_5 + i_6)$ is in the ring generated by the ω_t and i_t.

Hence show that the algebra generated by the i_t and ω_t is not discrete.

4.13. Show that the octonions 1, $\tfrac{1}{2}(-1 - i_0 + i_4 + i_5)$, i_0, $\tfrac{1}{2}(-i_0 - i_1 - i_5 - i_6)$, i_1, $\tfrac{1}{2}(-i_1 - i_2 - i_3 + i_6)$, i_2 and $\tfrac{1}{2}(-i_0 + i_3 - i_4 + i_6)$ form the fundamental roots of an E_8 lattice, and verify that products of the fundamental roots are again roots.

4.14. Verify that the map e defined in (4.38) in Section 4.4.2 is an automorphism of the integral octonion algebra and that it fuses the four orbits of the monomial group on the 126 vectors of norm 1.

4.15. Compute directly the kernel of the natural map $G_2(\mathbb{Z}) \to G_2(q)$ (see Section 4.4.3) for odd q.

4.16. Verify the congruences in (4.49).

4.17. In the notation of Section 4.5, use the relations (4.49) to show that if v is a vector satisfying $v^* \equiv v \wedge w \bmod W_0$, where v^* is defined by (4.46), then

(i) either v is a scalar multiple of v_1 or v is a scalar multiple of a vector $v_7 + \sum_{i=1}^{6} \alpha_i v_i$;

(ii) by using the lower triangular matrices (4.50) show that we may assume $\alpha_6 = \alpha_5 = \alpha_4 = 0$;

(iii) use (4.49) again to show that if $\alpha_6 = \alpha_5 = \alpha_4 = 0$ then $\alpha_3 = \alpha_2 = \alpha_1 = 0$;

(iv) deduce that there are precisely $(q-1)(q^3+1)$ non-zero vectors v such that $v^* \equiv v \wedge w \bmod W_0$.

4.18. Show that an involution in $^2G_2(q)$ fixes exactly $q+1$ of the q^3+1 points, and that each point is fixed by exactly q^2 involutions. Deduce that the number of involutions in $^2G_2(q)$ is $q^4 - q^3 + q^2$ and that the centraliser of an involution has order $q(q^2-1)$.

[Hint: it is necessary to show that all the involutions in $^2G_2(q)$ are conjugate.]

4.19. Complete the proof of the order formula (4.67) for $^3D_4(q)$ by classifying completely the isotropic vectors v satisfying $v * v = 0$ in the twisted octonion algebra.

4.20. Complete the proof of simplicity of $^3D_4(q)$ by proving that the number $(q^8+q^4+1)(q^6-1)$ is not divisible by any of the integers $1+q^3+q^4$, $1+q^7+q^8$ or $1+q^3+q^4+q^7+q^8$.

4.21. Prove that if (α, β, γ) is an isotopy of the octonion algebra over the field $\mathbb{F}_q$ of characteristic 2, then the homomorphism $(\alpha, \beta, \gamma) \mapsto \gamma$ is an isomorphism of the group of isotopies with $\Omega_8^+(q)$.

4.22. Show that in any associative algebra, the Jordan product $a \circ b$ defined by $a \circ b = \frac{1}{2}(ab+ba)$ satisfies the Jordan identity $((a \circ a) \circ b) \circ a = (a \circ a) \circ (a \circ b)$.

4.23. Verify that the Albert algebra as defined in Section 4.8.1 is closed under Jordan multiplication. More specifically, show that

$$(d, e, f \mid D, E, F) \circ (p, q, r \mid P, Q, R) = (x, y, z \mid X, Y, Z),$$

where $x = dp + \operatorname{Re}(F\overline{R} + \overline{E}Q)$ and

$$X = \tfrac{1}{2}(e+f)P + \tfrac{1}{2}(q+r)D + \tfrac{1}{2}(\overline{ER} + \overline{QF}).$$

4.24. Show that if T is a symmetric trilinear form, and $C(x) = T(x, x, x)$, then

$$6T(x, y, z) = C(x+y+z) - C(x+y) - C(y+z) - C(z+x)$$
$$+ C(x) + C(y) + C(z).$$

Show also that

$$24T(x, y, z) = C(x+y+z) + C(x-y-z)$$
$$+ C(-x+y-z) + C(-x-y+z).$$

4.25. Show that the map $(d, e, f \mid D, E, F) \mapsto (d, e, f \mid uD, Eu, \overline{u}F\overline{u})$, where u is an octonion of norm 1, preserves the Jordan product.

4.26. Compute the cubic form $C(d, e, f \mid D, E, F) = def - dD\overline{D} - eE\overline{E} - fF\overline{F} + \mathrm{Re}\,(DEF + DFE)$ where $d = \lambda_0$, $e = \lambda_0'$, $f = \lambda_0''$, $D = \sum_{i=1}^{8} \lambda_i x_i$, $E = \sum_{i=1}^{8} \lambda_i' x_i$ and $F = \sum_{i=1}^{8} \lambda_i'' x_i$, and hence verify the formula (4.95).

4.27. Show that the short root element of $F_4(q)$ defined in (4.105) preserves the Jordan product.

4.28. Show that there are at most two automorphisms of the Albert algebra which fix all the vectors $(1, 0, 0 \mid 0, 0, 0)$ and $(0, e, -e \mid D, 0, 0)$ (see Section 4.8.8).

4.29. Show that the stabiliser in $F_4(q)$ (q odd) of the decomposition of 1 as the sum of three orthogonal primitive idempotents

$$(1, 1, 1 \mid 0, 0, 0) = (1, 0, 0 \mid 0, 0, 0) + (0, 1, 0 \mid 0, 0, 0) + (0, 0, 1 \mid 0, 0, 0)$$

is the group $2 \cdot \mathrm{P\Omega}_8^+(q){:}S_3$ constructed in Section 4.8.3.

4.30. Determine the order of $F_4(q)$ (q odd) by counting decompositions of 1 as a sum of three orthogonal primitive idempotents, and proving transitivity of $F_4(q)$ on them.

4.31. Show that (if q is prime to 6) the determinant map defined for $E_6(q)$ satisfies

$$\det x = \tfrac{1}{3}\mathrm{Tr}(x \circ x \circ x) - \tfrac{1}{2}\mathrm{Tr}(x)\mathrm{Tr}(x \circ x) + \tfrac{1}{6}\mathrm{Tr}(x)^3,$$

where $\circ$ is the multiplication in the Albert algebra.

4.32. Reconstruct the Albert algebra from the determinant defined in Section 4.10.1, using $(1, 1, 1 \mid 0, 0, 0)$ as the identity element. Deduce that the stabiliser in $E_6(q)$ of a black vector is $F_4(q)$.

4.33. Prove that Dickson's cubic form defined in (4.134) is equivalent to the determinant defined in Section 4.10.1.

Do the same for the form defined in (4.135).

4.34. Show that the number of white vectors for $E_6(q)$ is $(q^9 - 1)(q^8 + q^4 + 1)$.

4.35. Verify the Jacobi identity for the Lie bracket $[AB] = AB - BA$ of matrices.

4.36. Define a multiplication on the 3-dimensional space of pure imaginary quaternions by taking the imaginary part of the usual quaternion product. Show that this makes the space into a Lie algebra.

4.37. A *derivation* of an algebra A with multiplication $*$ over a field F is a map $\delta : A \to A$ which satisfies

$$\delta(a * b) = \delta(a) * b + a * \delta(b).$$

Show that if γ and δ are derivations of A, then so is $\gamma\delta - \delta\gamma$.

Deduce that the set of derivations of A forms a Lie algebra under the Lie bracket $[\gamma, \delta] = \gamma\delta - \delta\gamma$.

4.38. Let $\mathbb{H} = \mathbb{R}[i, j, k]$ be the algebra of real quaternions, and let

$$\delta_1 : a + bi + cj + dk \mapsto ck - dj,$$
$$\delta_2 : a + bi + cj + dk \mapsto di - bk,$$
$$\delta_3 : a + bi + cj + dk \mapsto bj - ci.$$

Show that δ_1, δ_2 and δ_3 are derivations of $\mathbb{H}$ and that $\delta_3 = [\delta_1, \delta_2]$.

Show also that the algebra of derivations is spanned as a vector space by δ_1, δ_2, δ_3.

4.39. (i) Define the linear map β_0 on the real octonions by

$$\beta_0 : 1, i_0, i_1, i_3 \mapsto 0,$$
$$i_2 \mapsto -i_6 \mapsto -i_2, i_4 \mapsto i_5 \mapsto -i_4.$$

Show that β_0 is a derivation.
(ii) Define γ_0 and δ_0 similarly:

$$\gamma_0 : 1, i_0, i_2, i_6 \mapsto 0,$$
$$i_4 \mapsto -i_5 \mapsto -i_4, i_1 \mapsto i_3 \mapsto -i_1,$$
$$\delta_0 : 1, i_0, i_4, i_5 \mapsto 0,$$
$$i_1 \mapsto -i_3 \mapsto -i_1, i_2 \mapsto i_6 \mapsto -i_2.$$

Show that $\beta_0 + \gamma_0 + \delta_0 = 0$ and that the Lie bracket of any two of β_0, γ_0, δ_0 is 0.
(iii) Now define β_t, γ_t and δ_t similarly, by adding t to all the subscripts. Show that these maps span the algebra of derivations of the octonion algebra, and deduce that the derivation algebra has dimension 14.
(iv) Show that $[\beta_0, \beta_1] = \delta_3$, and compute also $[\beta_0, \gamma_1]$, $[\gamma_0, \beta_1]$, and $[\gamma_0, \gamma_1]$.
(v) Using the symmetry $\beta_t \mapsto \gamma_{2t} \mapsto \delta_{4t} \mapsto \beta_t$ write down the multiplication table of the derivation algebra with respect to a suitable basis.

[This Lie algebra is in fact the compact real form of the simple algebra of type G_2.]

4.40. Show that the algebra of derivations of the Albert algebra has dimension 78.

4.41. If L is a Lie algebra, and $x \in L$, define the map $\operatorname{ad} x : L \to L$ by $\operatorname{ad} x : y \mapsto [x, y]$. Use the Jacobi identity to show that $\operatorname{ad} x$ is a derivation on L. [It is called an *inner derivation*.]

Define the *centre* of L to be

$$Z(L) = \{x \in L \mid [xy] = 0 \text{ for all } y \in L\}.$$

Show that $Z(L)$ is an ideal in L and that the algebra of inner derivations is isomorphic to $L/Z(L)$.

5

The sporadic groups

5.1 Introduction

In this chapter we introduce the 26 sporadic simple groups. These are in many ways the most interesting of the finite simple groups, but are also the most difficult to construct. It is not possible here to provide complete proofs in all cases, but merely to indicate the general lines such proofs might take. Roughly speaking, proofs are given as far as the middle of Section 5.7, which deals with the Fischer groups. Section 5.2 deals with the large Mathieu groups M_{24}, M_{23} and M_{22}, and then the small Mathieu groups M_{12} and M_{11} are treated in Section 5.3. These groups have been known since Mathieu's papers [130, 131, 132] of the 1860s and 1870s. In all these cases it is possible to give complete constructions in a reasonably small number of pages, computing the group orders and proving simplicity, as well as exhibiting a number of important subgroups. Other facts are stated with varying degrees of justification.

In Sections 5.4 and 5.5 we construct the Leech lattice and the Conway groups Co_1, Co_2 and Co_3, together with the McLaughlin group McL and the Higman–Sims group HS. Although the latter two groups had been discovered earlier, their most natural home is with the Conway groups, and they are both contained in both Co_2 and Co_3. The Conway groups were discovered in around 1968. Again, our treatment includes more-or-less complete proofs for the basic facts. Analogous constructions in Section 5.6 of various complex and quaternionic versions of the Leech lattice lead to constructions of the sporadic groups of Suzuki (Suz) and Hall–Janko (J_2), as well as the exceptional double cover of $G_2(4)$. A new octonionic construction of the Leech lattice is also described.

The three Fischer groups Fi_{22}, Fi_{23} and Fi'_{24} are described in Section 5.7 in terms of their actions on graphs with 3510, 31671 and 306936 vertices, respectively. Parker's loop is introduced both as a means of elucidating the structure of the largest Fischer group Fi'_{24}, and as a key ingredient in the construction of the Monster, which is described in Section 5.8.1. This leads into brief descriptions of the remaining 'Monstrous' groups: the Baby Monster

R.A. Wilson, *The Finite Simple Groups*,
Graduate Texts in Mathematics 251,

(𝔹), Thompson's group (Th), the Harada–Norton group (HN), and the Held group (He).

Finally we describe the six 'pariahs' in Section 5.9: the remaining three Janko groups J_1, J_3 and J_4, and the simple groups of O'Nan (O'N), Rudvalis (Ru) and Lyons (Ly). It should be noted that the only apparent relationship between these groups is that J_1 is a subgroup of O'N. In particular, the three Janko groups are unrelated to each other. The behaviour of these six groups is so bizarre that any attempt to describe them ends up looking like a disconnected sequence of unrelated facts. I make no apology for this—it is simply the nature of the subject.

5.2 The large Mathieu groups

5.2.1 The hexacode

To construct the Mathieu groups we start by defining the *hexacode*. [A (linear) *code* is just a subspace of a finite vector space $\mathbb{F}_q{}^n$.] The hexacode is the 3-dimensional subspace of $\mathbb{F}_4{}^6$ spanned by the vectors $(\omega, \overline{\omega}, \overline{\omega}, \omega, \overline{\omega}, \omega)$, $(\overline{\omega}, \omega, \omega, \overline{\omega}, \overline{\omega}, \omega)$ and $(\overline{\omega}, \omega, \overline{\omega}, \omega, \omega, \overline{\omega})$, where $\mathbb{F}_4 = \{0, 1, \omega, \overline{\omega}\}$ is the field of order 4. Since the sum of these three vectors is $(\omega, \overline{\omega}, \omega, \overline{\omega}, \omega, \overline{\omega})$, there is an obvious symmetry group $3 \times S_4$ generated by scalar multiplications and the coordinate permutations $(1, 2)(3, 4)$, $(1, 3, 5)(2, 4, 6)$ and $(1, 3)(2, 4)$. The non-zero vectors fall into four orbits under this group, as follows:

Orbit representative	Length of orbit	
$(1, 1, 1, 1, 0, 0)$	9	
$(\omega, \overline{\omega}, \omega, \overline{\omega}, \omega, \overline{\omega})$	12	(5.1)
$(1, 1, \omega, \omega, \overline{\omega}, \overline{\omega})$	6	
$(0, 1, 0, 1, \omega, \overline{\omega})$	36	

The group of automorphisms of this code is defined to be the set of monomial permutations of the coordinates which fix the code as a set. It is immediate that any diagonal symmetry is a scalar, as it maps each of $(1, 1, 1, 1, 0, 0)$ and $(0, 0, 1, 1, 1, 1)$ to a scalar multiple of itself. Modulo scalars, we can extend the group S_4 of permutations to A_6 by adjoining the map

$$(x_1, \ldots, x_6) \mapsto (\omega x_1, \overline{\omega} x_2, x_3, x_6, x_4, x_5). \tag{5.2}$$

On the other hand, no automorphism induces an odd permutation, for if so, then looking at the images of $(1, 1, 1, 1, 0, 0)$ and $(0, 0, 1, 1, 1, 1)$ we deduce that the coordinate permutation (5,6) is an automorphism, but $(0, 1, 0, 1, \overline{\omega}, \omega)$ is not in the hexacode. However, odd permutations are allowed provided they are always followed by the field automorphism $\omega \mapsto \overline{\omega}$. This gives rise to a group $3 \cdot S_6$ of *semi-automorphisms* of the hexacode. (Compare the construction of $3 \cdot A_6$ and $3 \cdot S_6$ in Section 2.7.3.)

5.2.2 The binary Golay code

We next construct a set of 24 points, labelled (i, x) where i is an integer from 1 to 6 (corresponding to one of the six coordinates of the hexacode) and $x \in \mathbb{F}_4$. Let the hexacode act on this set in the 'obvious' way, by addition: a hexacode word $(x_1, \ldots, x_6)$ maps (i, x) to $(i, x + x_i)$. Similarly the group $3 \cdot S_6$ of semi-automorphisms acts on the set in the 'obvious' way: if the group element maps 1 in the ith coordinate to λ in the jth coordinate, then it maps (i, x) to $(j, \lambda x)$. These 24 points are generally arranged in a 6×4 array with columns labelled 1 to 6 and rows labelled $0, 1, \omega, \overline{\omega}$ (called the MOG, or Miracle Octad Generator, by Curtis, who first used such an array as a practical tool for calculating in the Golay code [41]) where these symmetries can be conveniently visualised. For example, the group $2^6{:}3 \cdot S_6$ may be generated by the following permutations of 24 points (where cycles of length 5 are represented by arrows, and it is understood that the head of the arrow maps back to the start).

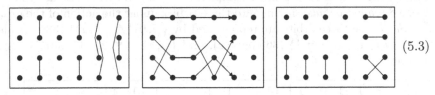

$$(5.3)$$

The largest Mathieu group M_{24} may be viewed as a permutation group on these 24 points, containing the group $2^6{:}3 \cdot S_6$ just constructed, as a maximal subgroup. To effect this construction we shall define the *(extended binary) Golay code* as a set of binary vectors of length 24, with coordinates indexed by the 24 points (i, x). For convenience, we identify each vector with its support (that is, the set of points where it has coordinate 1). The group M_{24} will then be defined as the group of permutations which preserve this set of vectors.

From each hexacode word $(x_1, \ldots, x_6)$ we derive 64 words in the Golay code, as follows. Each coordinate x_i corresponds to a set of points (i, y) such that $\sum y = x_i$. Thus x_i corresponds either to an odd-order set, $\{(i, x_i)\}$ or its complement $\{(i, x) \mid x \neq x_i\}$, or to an even-order set, $\{(i, x_i)\} \triangle \{(i, 0)\}$ (where $\triangle$ as usual denotes symmetric difference of sets) or its complement $\{(i, x) \mid 0 \neq x \neq x_i \text{ or } 0 = x = x_i\}$. In pictures, each x_i has two odd and two even interpretations:

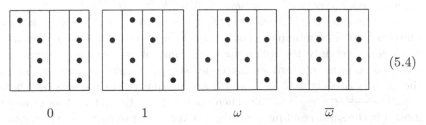

$$(5.4)$$

$$0 \qquad\qquad 1 \qquad\qquad \omega \qquad\qquad \overline{\omega}$$

Now impose the further conditions that all 6 coordinates have the same parity, and that this parity is equal to the parity of the number of $(i, 0)$ in the set, i.e. the parity of the top row. Thus we may choose the set corresponding to the first coordinate in 4 ways, and the next four coordinates in 2 ways each, and the last set is determined. As an example, here are three of the 64 Golay code words corresponding to the hexacode word $(0, 1, 0, 1, \omega, \overline{\omega})$.

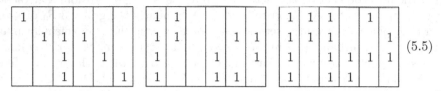

$$(5.5)$$

It is almost immediate to check from this definition that the Golay code is closed under vector addition (symmetric difference of sets), and so forms a subspace of dimension 12 of $\mathbb{F}_2^{24}$. Moreover, the whole set (corresponding to the vector with 24 coordinates 1) satisfies the conditions to be in the Golay code, so the code is closed under complementation. In total there are 759 sets of size 8 (called *octads*), falling into three orbits under the action of the group $2^6{:}3{\cdot}S_6$. These three orbits are represented by $\{(i, 0)\} \triangle \{(0, x)\}$ (384 octads), $\{(i, x) \mid i \leqslant 4, x = 0 \text{ or } 1\}$ (360 octads), and $\{(i, x) \mid i \leqslant 2\}$ (15 octads), or in pictures:

$$
\begin{array}{|ccccc|} \hline
  & 1 & 1 & 1 & 1 & 1 \\
1 & & & & & \\
1 & & & & & \\
1 & & & & & \\ \hline
\end{array}
\begin{array}{|cccc|}\hline
1 & 1 & 1 & 1 \\
1 & 1 & 1 & 1 \\
 & & & \\
 & & & \\ \hline
\end{array}
\begin{array}{|cc|}\hline
1 & 1 \\
1 & 1 \\
1 & 1 \\
1 & 1 \\ \hline
\end{array}
\qquad (5.6)
$$

Hence, by complementation, there are 759 sets of size 16 in the code. All the remaining sets have size 12: there are 2576 of them, lying in three orbits, of lengths $576 + 720 + 1280$, under $2^6{:}3{\cdot}S_6$. In pictures:

$$
\begin{array}{|cccccc|}\hline
1 & 1 & 1 & 1 & 1 & 1 \\
 & & & & & \\
1 & & 1 & & 1 & \\
 & 1 & & 1 & & 1 \\ \hline
\end{array}
\begin{array}{|ccccc|}\hline
1 & 1 & 1 & & & \\
 & 1 & 1 & 1 & & \\
 & 1 & 1 & 1 & & \\
 & 1 & 1 & 1 & & \\ \hline
\end{array}
\begin{array}{|ccccc|}\hline
1 & 1 & 1 & & & 1 \\
1 & 1 & 1 & & & 1 \\
 & & & 1 & & 1 \\
 & & & 1 & & 1 \\ \hline
\end{array}
\quad (5.7)
$$

Thus the code has *weight distribution* $0^1 8^{759} 12^{2576} 16^{759} 24^1$.

Now consider the 2^{12} cosets of the Golay code in $\mathbb{F}_2^{24}$, and look for coset representatives of minimal weight (i.e. with as few non-zero coordinates as possible). Certainly the difference (i.e. sum) of two representatives for the same coset is an element of the code, so has weight at least 8. Therefore the vectors of weight 0, 1, 2, and 3 are unique in their cosets, so there are $1 + 24 + \frac{24.23}{2} + \frac{24.23.22}{2.3} = 2325$ such cosets. Two distinct vectors of weight at most 4 in the same coset must be disjoint vectors of weight 4, so there can be at most 6 such vectors in each coset. Therefore there are at least $\frac{24.23.22.21}{4.3.2.6} = 1771$ such cosets. But now we have accounted for at least $2325 + 1771 = 4096 = 2^{12}$

cosets. In particular, every vector of weight 4 determines a partition of the 24 points into 6 sets of size 4. These partitions are called *sextets*, consisting of six *tetrads*.

Now we can explicitly calculate these sextets, and we find that there are just four orbits under the group $2^6{:}3{\cdot}S_6$. The first orbit, of size 1, is the sextet consisting of the 6 columns of the MOG. The second orbit, of size 90, consists of sextets defined by two points in one column and two points in another. The third orbit, of size 240, consists of sextets defined by three points in one column and one in another column. The fourth orbit, of size 1440, consists of sextets defined by two points in one column and two other points in two other columns. This accounts for all 1771 sextets. In pictures, representatives of the three non-trivial orbits are

1	1	3	3	5	5
1	1	3	3	5	5
2	2	4	4	6	6
2	2	4	4	6	6

1	2	3	3	3	3
2	1	4	4	4	4
2	1	5	5	5	5
2	1	6	6	6	6

1	1	1	2	5	6
1	2	2	2	4	3
3	5	6	4	3	6
4	6	5	3	5	4

$$(5.8)$$

where the six tetrads of each sextet are labelled by the numbers 1 up to 6.

5.2.3 The group M_{24}

By checking the action on a basis of the Golay code, it is easy to verify that the element

$$\alpha = \qquad\qquad\qquad\qquad\qquad\qquad (5.9)$$

fixes the code. Moreover, α fuses the four orbits of $2^6{:}3{\cdot}S_6$ on sextets. Since this group is the full sextet stabilizer (see Exercise 5.2), it follows that the order of M_{24} is

$$|M_{24}| = 1771.2^6.3.6! = 244823040 = 2^{10}.3^3.5.7.11.23. \qquad (5.10)$$

It follows almost immediately from the transitivity of M_{24} on sextets that it is 5-transitive on points. For we can take the sextet defined by the first four points to any sextet, and then the sextet stabiliser is transitive on its six tetrads. Moreover, fixing the tetrad we have a full S_4 of permutations of the tetrad. Finally, the pointwise stabiliser of this tetrad is a group $2^4{:}A_5$ which is visibly transitive on the remaining 20 points.

Consequently, every set of five points is contained in an octad, but since $\binom{24}{5} = 759 \binom{8}{5}$, this octad is unique. (Alternatively, this follows from the fact that the sum of two octads is in the code, so the octads cannot intersect in more than four points.) This combinatorial property is the defining property of a *Steiner system* $S(5,8,24)$. More generally, a Steiner system $S(t,k,v)$ is a system of special k-subsets (called *blocks*) of a set of size v, with the property that every set of size t is contained in a unique block. Counting the number of subsets of size t in two different ways shows that the number of blocks in an $S(t,k,v)$ is $\binom{v}{t} \Big/ \binom{k}{t}$.

It is easy to see that the (extended binary) Golay code gives rise to a Steiner system $S(5,8,24)$, but the converse is not obvious. It follows however from the fact that, up to relabelling the points, there is a unique $S(5,8,24)$. The proof of uniqueness also gives an alternative proof that M_{24} is 5-transitive. We now sketch this proof.

5.2.4 Uniqueness of the Steiner system $S(5,8,24)$

If $\{1,2,3,4,5,6,7,8\}$ is an octad, the Steiner system property implies that the number of octads containing $\{1,\ldots,i\}$ is $\binom{24-i}{5-i} \Big/ \binom{8-i}{5-i}$ for $0 \leqslant i \leqslant 4$ and is 1 for $5 \leqslant i \leqslant 8$. These numbers are 759, 253, 77, 21, 5, 1, 1, 1, 1 for $i = 0,\ldots,8$, respectively. Hence the number of octads not containing the point 1 is $759 - 253 = 506$, and so on. We complete the *Leech triangle* (see Figure 5.1), in which the ith entry in the jth row is the number of octads intersecting $\{1,\ldots,j-1\}$ in exactly $\{1,\ldots,i-1\}$, using this property that each entry is the sum of the two nearest entries in the row below (cf. Pascal's triangle). In particular we see from the bottom row of the triangle that two

```
                                759
                         506          253
                     330      176          77
                  210     120      56         21
               130     80     40      16      5
            78     52     28     12      4     1
         46     32     20     8      4     0     1
      30     16     16     4      4     0     0     1
   30     0     16     0     4     0     0     0     1
```

Fig. 5.1. The Leech triangle

distinct octads intersect in 0, 2 or 4 points. It follows easily that every set of 4 points determines a *sextet*, with the properties described above, and that every octad is either the sum of two tetrads of the sextet, or cuts across these tetrads with intersections of sizes $(3,1^5)$ or $(2^4,0^2)$. Moreover, in the $(3,1^5)$ case, we may choose the three points in one column in 6×4 ways, and two more points in the first two available columns in 4×4 ways, so there are 384 such octads, and therefore there are 360 of type $(2^4, 0^2)$.

Now we might as well arrange our first sextet to consist of the six columns of a MOG array, and choose an octad of shape $(3, 1^5)$ to be the first column plus the top row. Indeed, we can arrange the points so that the sextet containing the tetrad consisting of the last four points in the top row also contains the tetrads consisting of the last four points in the other three rows. These now correspond to hexacode words $(0, 0, 1, 1, 1, 1)$ and its scalar multiples, and the sextet is the second one displayed in (5.8). Moreover, every tetrad consisting of three points in one column and one point in another determines a sextet which cuts across the columns in the same pattern, which we may abbreviate as $((3, 1^5)^2, (1^4)^4)$. Since each such sextet contains two tetrads of this type, it follows that there are 240 such sextets, which between them contain all octads, as unions of two tetrads.

Now if two octads intersect in four points, then (using the sextet determined by their intersection) their symmetric difference is also an octad. Therefore it is sufficient to determine the 64 octads of shape $(3, 1^5)$ with three points in the first column, as then the rest of the octads are determined. Adding the first column to all such octads we obtain a set of 64 'hexads' consisting of one point in each column. Labelling the rows $0, 1, \omega, \overline{\omega}$ we may identify these hexads with vectors in $\mathbb{F}_4{}^6$, forming a certain (not necessarily linear) 'code'.

The Steiner system property implies that any two of these 64 words must differ in at least four coordinates. We shall now show that, up to reordering the columns, and the points in each column, these words are the words in the hexacode, and therefore the Steiner system is the one already constructed.

For the first three coordinates determine five points and so a unique octad—in other words they determine the remaining three coordinates. We have already assumed without loss of generality that the words $(0, 0, 1, 1, 1, 1)$, $(0, 0, \omega, \omega, \omega, \omega)$ and $(0, 0, \overline{\omega}, \overline{\omega}, \overline{\omega}, \overline{\omega})$ are in the code. The word beginning $(0, 1, 0, \ldots)$ completes with $1, \omega, \overline{\omega}$ in some order, so we may assume it is $(0, 1, 0, 1, \omega, \overline{\omega})$. Then, swapping ω and $\overline{\omega}$ in the second column if necessary, we may assume the code contains $(0, \omega, 0, \omega, \overline{\omega}, 1)$ and $(0, \overline{\omega}, 0, \overline{\omega}, 1, \omega)$.

Next, $(0, 1, 1, \ldots)$ completes with $0, \omega, \overline{\omega}$ in some order, and the only possibility which differs in at least four places from all words already chosen is $(0, 1, 1, 0, \overline{\omega}, \omega)$. Similarly, $(0, \omega, \omega, \ldots)$ completes to $(0, \omega, \omega, 0, 1, \overline{\omega})$ and $(0, \overline{\omega}, \overline{\omega}, \ldots)$ completes to $(0, \overline{\omega}, \overline{\omega}, 0, \omega, 1)$. Continuing in this way we find unique completions for all words $(0, x, y, \ldots)$.

Then $(1, 0, 0, \ldots)$ completes with $1, \omega, \overline{\omega}$ in the opposite cyclic ordering from $(0, 1, 0, 1, \omega, \overline{\omega})$, so, relabelling the first column if necessary, we may assume the code contains $(1, 0, 0, 1, \overline{\omega}, \omega)$, and similarly $(\omega, 0, 0, \omega, 1, \overline{\omega})$ and $(\overline{\omega}, 0, 0, \overline{\omega}, \omega, 1)$. After this it is easy to see that all the remaining triples $(x, y, z, \ldots)$ have completions which are uniquely determined by the choices already made.

5.2.5 Simplicity of M_{24}

To prove that M_{24} is simple we can use Iwasawa's Lemma (see Theorem 3.1) as with the classical and exceptional groups. It is easy to see that the action of M_{24} on the 1771 sextets is primitive, since the sextet stabiliser $2^6{:}3S_6$ has orbits of lengths $1 + 90 + 240 + 1440$ on the 1771 sextets. Moreover, this group has a normal abelian subgroup of order 2^6, whose elements are easily seen to be commutators of elements of $2^6{:}3{\cdot}S_6$. The latter group is generated by conjugates of the third element of (5.3), and to generate M_{24} we need only adjoin the element α defined in (5.9). Since both of these elements are in the normal 2^6 of the stabiliser of the first sextet of (5.8), we have verified all the hypotheses of Iwasawa's Lemma, and therefore conclude that M_{24} is simple.

5.2.6 Subgroups of M_{24}

The stabiliser of a point is by definition the group M_{23}, which therefore has order $|M_{24}|/24 = 10200960$, while the stabiliser of two points is the group M_{22} of order $|M_{23}|/23 = 443520$. These three groups M_{22}, M_{23} and M_{24} are collectively known as the large Mathieu groups, and were first described by Mathieu in his 1873 paper [132]. The stabiliser of three points similarly has order $|M_{22}|/22 = 20160$, and is sometimes called M_{21}, but turns out to be isomorphic to $\mathrm{PSL}_3(4)$ (see Exercise 5.6). This isomorphism can be seen by showing that the Golay code structure gives rise to a projective plane of order 4 on the 21 remaining points. Of course, the multiple transitivity of M_{24} shows that these groups extend to subgroups $M_{22}{:}2$ and $\mathrm{PSL}_3(4){:}S_3$, both of which are in fact maximal subgroups of M_{24}.

The stabiliser of an octad has order $|M_{24}|/759 = 322560$. Now if all 8 points of the octad are fixed, then we may assume these are the first two columns of the MOG, so by looking inside the sextet stabiliser we see that there is only an elementary abelian group of order 16 left. Modulo this, the permutation action on the octad has order 20160, and is therefore A_8. Thus the octad stabiliser is $2^4 A_8$. Indeed, by fixing one of the 16 points outside the octad, we see a subgroup A_8, so the extension splits (i.e. the octad stabiliser is a semidirect product $2^4{:}A_8$). Moreover, the 16 points outside the octad now have a vector space structure on them, and we see the isomorphism $A_8 \cong L_4(2)$ again (compare Section 3.12.1).

It is easy to show (see Exercise 5.3) that M_{24} acts transitively on the Golay code words of weight 12 (the *dodecads*), so that the stabiliser of one of them is a group of order $|M_{24}|/2576 = 95040$, and is the group M_{12}. The fact that M_{12} is a subgroup of M_{24} was not known to Mathieu, and was first discovered by Frobenius. In fact, it is not maximal, as the complement of a dodecad is another dodecad, and since M_{24} is transitive on dodecads, we have a subgroup $M_{12}{:}2$ (which is, in fact, maximal) fixing the pair of complementary dodecads.

Fixing a point in one of these dodecads we obtain the smallest Mathieu group, M_{11} of order $|M_{12}|/12 = 7920$. (See Section 5.3 for alternative definitions of M_{12} and M_{11}.)

There are just three more classes of maximal subgroups of M_{24} (for a proof, see Curtis [42], and for a complete list of the maximal subgroups of all the Mathieu groups, see Table 5.1). One is the stabiliser of a set of three mutually disjoint octads (such as the three bricks of the MOG: these *bricks* are the left-most 8 points, the rightmost 8 points, and the middle 8 points), and has the shape $2^6{:}(\mathrm{GL}_3(2) \times S_3)$, with the quotient S_3 acting as permutations of the three octads. Another is the group $\mathrm{PSL}_2(23)$, which was known to Mathieu, and was in a sense the basis of his original construction. (He first constructed M_{23} and then extended the maximal subgroup $23{:}11$ to $\mathrm{PSL}_2(23)$ to generate M_{24}. It is not easy to prove that M_{24} exists from this definition, and it appears that Mathieu's construction did not entirely convince his contemporaries. Even 30 years later it was possible for Miller to publish a paper [139] purporting to prove that M_{24} did not exist—although, to be fair, he quickly realised his mistake and retracted the claim. The first really convincing construction was that of Witt [187] in 1938—his paper is well worth reading, even today, for those who read German.)

Another way to see the subgroup $\mathrm{PSL}_2(23)$ is to construct the Golay code by taking the 24 points to be the points of the projective line $\mathbb{F}_{23} \cup \{\infty\}$, and defining the code to be spanned by the set $\{x^2 \mid x \in \mathbb{F}_{23}\}$ and its images under $t \mapsto t+1$, and complementation. It takes a little bit of work to show that this defines a code with the same weight distribution as the Golay code, and then the uniqueness of the Steiner system $S(5,8,24)$ implies it is isomorphic to the Golay code (see Exercise 5.7). The group $\mathrm{PSL}_2(23)$ of symmetries generated by $t \mapsto t+1$, $t \mapsto 2t$, and $t \mapsto -1/t$ can be extended to M_{24} by adjoining the map $t \mapsto t^3/9$ for t a quadratic residue or 0 (i.e. for t a square in $\mathbb{F}_{23}$), and $t \mapsto 9t^3$ for t a non-residue or ∞, i.e. the permutation

$$(1,18,4,2,6)(8,16,13,9,12)(5,21,20,10,7)(11,19,22,14,17).$$

A correspondence with the MOG is given in the Atlas [28] by labelling the points of the MOG with the points of the projective line as follows:

$$
\begin{array}{|cccccc|}
\hline
0 & \infty & 1 & 11 & 2 & 22 \\
19 & 3 & 20 & 4 & 10 & 18 \\
15 & 6 & 14 & 16 & 17 & 8 \\
5 & 9 & 21 & 13 & 7 & 12 \\
\hline
\end{array}
\tag{5.11}
$$

The last, and smallest, maximal subgroup of M_{24} is a subgroup $\mathrm{PSL}_2(7)$ of order 168. As this contains a subgroup S_4 which commutes with an element of order 2 in M_{24}, this leads to a way of generating M_{24} in a nice symmetric way with seven elements of order 2, as in Section 5.2.7.

5.2.7 A presentation of M_{24}

Label these seven elements of order 2 as $t_0, \ldots, t_6$ in such a way that $\mathrm{PSL}_2(7) \cong \mathrm{PSL}_3(2)$ acts on them as described in Section 3.3.5, that is, according to the permutations $(0,1,2,3,4,5,6)$, $(1,2,4)(3,6,5)$ and $(1,2)(3,6)$. Calculations in M_{24} reveal that $(t_0)^{t_1}$ commutes with t_3, and that $(t_0 t_3)^3$ is equal to the permutation $(2,4)(5,6)$ in $PSL_3(2)$, and that $(1,2,4)(3,6,5)t_3$ has order 11. It turns out that these relations are sufficient to define M_{24}. For more details, see [45]. (Compare the Coxeter presentation of S_n given in Section 2.8.1.)

5.2.8 The group M_{23}

We defined M_{23} as the stabiliser of a point in M_{24}, so M_{23} has order $10200960 = 2^7.3^2.5.7.11.23$. Since M_{24} is 5-transitive, it is clear that M_{23} is 4-transitive. The Steiner system $S(5,8,24)$ gives rise to a Steiner system $S(4,7,23)$, by taking all the octads containing a given point, and then removing this point from the set. The number of resulting 'heptads' in this Steiner system is $\binom{23}{4} / \binom{7}{4} = 253$.

Similarly, if we remove one (fixed) coordinate from all the 2^{12} codewords in the Golay code, we obtain a code, called the (unextended) *binary Golay code*, with weight distribution

$$0^1 7^{253} 8^{506} 11^{1288} 12^{1288} 15^{506} 16^{253} 23^1. \qquad (5.12)$$

In particular, the minimum non-zero weight in this code is 7, which means that any 'errors' in at most 3 coordinates of a codeword can be 'corrected' uniquely to restore the original codeword. Now the number of such errors is: 1 trivial error (in no coordinates), 23 errors in one coordinate, $\binom{23}{2} = 253$ errors in two coordinates, and $\binom{23}{3} = 1771$ errors in three coordinates, making $1 + 23 + 253 + 1771 = 2048 = 2^{11}$ in total. These errors, applied to all 2^{12} codewords, exactly account for all 2^{23} vectors in the underlying space, so the code is called *perfect*.

Simplicity of M_{23} can be proved by applying Iwasawa's Lemma either to the action on triples, in which case the stabiliser has shape $2^4{:}3{:}S_5$, or to the action on heptads, in which case the stabiliser is $2^4{:}A_7$ (see Exercise 5.11).

It turns out that all of the maximal subgroups of M_{23} can be obtained by fixing a point in the action of one of the maximal subgroups of M_{24}. For example, the subgroup $M_{22}{:}2$ has two orbits, of lengths 2 and 22 on the 24 points. Fixing one of the first two points gives a maximal subgroup M_{22} of M_{23}, while fixing one of the other points gives a maximal subgroup $PSL_3(4){:}2$. Similarly, the octad stabiliser $2^4{:}A_8$ in M_{24} has orbits of lengths 8 and 16, giving rise to maximal subgroups $2^4{:}A_7$ and A_8 respectively in M_{23}. From the sextet stabiliser $2^6{:}3{\cdot}S_6$ in M_{24} we obtain the subgroup $2^4{:}3{:}S_5$ in M_{23}, and from $M_{12}{:}2$ we obtain M_{11}. Finally, the point stabiliser in $PSL_2(23)$ is $23{:}11$, which Mathieu proved is maximal in M_{23}.

5.2.9 The group M_{22}

We defined M_{22} as the stabiliser of a point in M_{23}, so M_{22} has order $443520 = 2^7.3^2.5.7.11$. By the same arguments as in Section 5.2.8 we see that M_{22} is 3-transitive on 22 points, and preserves a Steiner system $S(3, 6, 22)$. The number of 'hexads' in this Steiner system is $\binom{22}{3} \Big/ \binom{6}{3} = 77$.

Since M_{24} is 2-transitive, it contains elements interchanging the two points fixed by M_{22}. These elements must normalise M_{22}, and clearly cannot centralise it. Therefore they extend it to a group of shape $M_{22}{:}2$, which is in fact the full automorphism group of M_{22}. It is also the automorphism group of the Steiner system $S(3, 6, 22)$.

The Leech triangle (Fig. 5.1) shows that each hexad is disjoint from 16 others. Indeed, the hexad stabiliser $2^4{:}A_6$ has orbits of lengths $1 + 16 + 60$ on the 77 hexads, with representatives as follows.

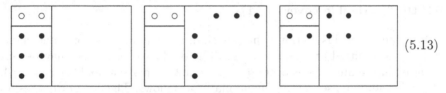

$$(5.13)$$

We may identify the 16 points on the right hand side of the MOG with the 16 points of an affine 4-space over $\mathbb{F}_2$, in natural bijection with the 16 hexads of the second orbit. The underlying vector space has a natural symplectic form defined by the action of $A_6 \cong \mathrm{Sp}_4(2)'$, so there are 15 totally isotropic vector subspaces of dimension 2. The 60 hexads in the third orbit correspond to the 4.15 cosets of these subspaces. (These are called *affine* subspaces.)

Now construct a graph on the 77 hexads by joining two hexads by an edge in the graph if they are disjoint as sets. It is easy to see that the edges are defined in terms of $2^4{:}A_6$ as follows: an affine 2-space is joined to the four points it contains, and to the twelve 2-spaces which are disjoint from it but not parallel to it. (Here, *parallel* means that they are cosets of the same vector subspace.) We summarise the structure of this graph in the following picture.

$$\overset{}{\underbrace{1}}\overset{16}{\underset{1}{\rule{3em}{0.4pt}}}\overset{}{\underbrace{16}}\overset{15}{\underset{4}{\rule{3em}{0.4pt}}}\overset{12}{\underbrace{60}} \qquad (5.14)$$

The automorphism group of this graph is $M_{22}{:}2$, since the point stabiliser cannot be bigger than $2^4{:}\mathrm{Sp}_4(2) \cong 2^4{:}S_6$.

At the same time we can prove simplicity of M_{22} using Iwasawa's Lemma. The action of M_{22} on the 77 hexads is obviously faithful and primitive, and it is easy to check that the group is generated by conjugates of the normal abelian subgroup 2^4 of the hexad stabiliser. Moreover, it is obvious that this 2^4 lies in the derived subgroup. Therefore M_{22} is simple.

Again it is true that every maximal subgroup of M_{22} is obtained from a maximal subgroup of M_{24} by fixing two points. From $PSL_3(4){:}S_3$, we obtain $PSL_3(4)$ and $2^4{:}S_5$; from $2^4{:}A_8$ we obtain $2^4{:}A_6$, $2^3{:}GL_3(2)$ and two conjugacy classes of A_7; from $M_{12}{:}2$ we obtain $PSL_2(11)$ and $A_6{\cdot}2$ (the stabiliser in M_{12} of an ordered pair of points, and sometimes called M_{10}). Incidentally, Mathieu already showed that $PSL_2(11)$ is a subgroup of M_{22}.

The group M_{22} however has some more interesting properties which are not immediately obvious from its containment in M_{24}. For example it has a covering group $12{\cdot}M_{22}$ in which the centre is a cyclic group of order 12. Factoring this group by the normal subgroup of order 2 gives a group $6{\cdot}M_{22}$ which is a subgroup of the largest Janko group J_4 (see Section 5.9.6). Factoring by a further normal subgroup of order 2 gives a triple cover $3{\cdot}M_{22}$, which is a subgroup of $SU_6(2) \cong 3{\cdot}PSU_6(2)$. Thus $3{\cdot}M_{22}$ can be generated by some 6×6 unitary matrices over $\mathbb{F}_4$.

5.2.10 The double cover of M_{22}

(The material in this section is harder than surrounding sections, and might sensibly be omitted at a first reading.) The double cover of M_{22} may be constructed in ten dimensions over $\mathbb{Q}(\sqrt{-7})$. First we make a double cover $2^5{:}A_6$ of the hexad stabiliser, acting monomially as follows. The 10 coordinates are indexed by the splittings of a set of 6 points into two triples, and A_6 permutes the coordinates naturally. (Alternatively, index the coordinates by the points of the projective line over $\mathbb{F}_9$, and use the natural action of $PSL_2(9)$.) Each sign-change is defined by a syntheme $(ab \mid cd \mid ef)$, which negates the four coordinates $(ace \mid bdf)$, $(bce \mid adf)$, $(ade \mid bcf)$, $(acf \mid bde)$. (Or in the alternative description, by one of the images under $PSL_2(9)$ of the set $\{\pm 1 \pm i\}$ of non-squares.)

To extend this group $2^5{:}A_6$ to $2{\cdot}M_{22}$ we adjoin an outer automorphism of the permutation group A_6, extending it to $M_{10} \cong A_6{\cdot}2$ (modulo scalars). A suitable extending element is

$$g = \frac{1}{4} \begin{pmatrix} -\lambda & -1 & -1 & -1 & -1 & -1 & -1 & -1 & -1 & -1 \\ 1 & \lambda & 1 & 1 & 1 & -1 & -1 & 1 & -1 & -1 \\ 1 & -1 & 1 & -1 & -1 & 1 & -1 & 1 & \lambda & 1 \\ 1 & -1 & -1 & 1 & 1 & 1 & \lambda & -1 & -1 & 1 \\ 1 & -1 & -1 & 1 & -1 & -1 & 1 & 1 & 1 & \lambda \\ 1 & 1 & -1 & -1 & \lambda & 1 & 1 & 1 & -1 & -1 \\ 1 & 1 & \lambda & 1 & -1 & 1 & -1 & -1 & 1 & -1 \\ 1 & -1 & 1 & -1 & 1 & \lambda & 1 & -1 & 1 & -1 \\ 1 & 1 & 1 & \lambda & -1 & -1 & 1 & -1 & -1 & 1 \\ 1 & 1 & -1 & -1 & 1 & -1 & -1 & \lambda & 1 & 1 \end{pmatrix}$$

where $\lambda = \sqrt{-7}$ and the coordinates are written in the following order: $\infty, 0, 1, -1, i, 1 + i, -1 + i, -i, 1 - i, -1 - i$. To show that this is a double cover of M_{22}, we reconstruct the graph on the hexads. Observe first that the matrix g given above maps the standard coordinate frame in which the vectors have shape $(4, 0^9)$ to one in which the vectors have the shape $(\lambda, 1^9)$. Moreover, there are 16 such coordinate frames obtained by applying the sign-changes. Conjugating such a sign-change by the matrix g gives an element such as

$$h = \frac{1}{4} \begin{pmatrix} 2 & -2 & -\beta & -\beta & -\beta & 0 & 0 & -\beta & 0 & 0 \\ -2 & 2 & -\beta & -\beta & -\beta & 0 & 0 & -\beta & 0 & 0 \\ -\gamma & -\gamma & -1 & -1 & 1 & -\gamma & \gamma & 1 & -\gamma & \gamma \\ -\gamma & -\gamma & -1 & -1 & 1 & \gamma & -\gamma & 1 & \gamma & -\gamma \\ -\gamma & -\gamma & 1 & 1 & -1 & -\gamma & -\gamma & -1 & \gamma & \gamma \\ 0 & 0 & -\beta & \beta & -\beta & 2 & 0 & \beta & 0 & 2 \\ 0 & 0 & \beta & -\beta & -\beta & 0 & 2 & \beta & 2 & 0 \\ -\gamma & -\gamma & 1 & 1 & -1 & \gamma & \gamma & -1 & -\gamma & -\gamma \\ 0 & 0 & -\beta & \beta & \beta & 0 & 2 & -\beta & 2 & 0 \\ 0 & 0 & \beta & -\beta & \beta & 2 & 0 & -\beta & 0 & 2 \end{pmatrix},$$

where $\beta = \frac{1}{2}(-1 + \sqrt{-7})$ and $\gamma = \frac{1}{2}(-1 - \sqrt{-7})$. This gives another type of coordinate frame in which there are four vectors of shape $(1^4, \gamma^6)$, and six vectors of shape $(\beta^4, 0^4, 2^2)$. There are 60 images of this coordinate frame under the monomial group $2^5{:}A_6$. Now we obtain the graph on 77 points by joining two of these coordinate frames if no vector in one frame is perpendicular to any vector in the other.

We need to check two things: (i) that this set of 77 coordinate frames is invariant under g, and (ii) that the graph defined here is the same as the graph on the 77 hexads of the Steiner system $S(3, 6, 22)$. To verify (i), check first that g normalises A_6. Then we need only check the images of one representative of each A_6 orbit. These orbits have lengths $1 + 1 + 15 + 15 + 45$ and only the orbit of length 45 remains to be checked. To verify (ii), it is sufficient to verify that the two non-trivial orbits of edges under $2^4{:}A_6$ are the same in both graphs. Thus we need only label the vertices and check one edge in each orbit. (See Exercise 5.10.)

Finally, observe that the 10-dimensional representation of $2{\cdot}M_{22}$ becomes orthogonal when reduced modulo 7. Since $\Omega_{10}^+(7) \cong P\Omega_{10}^+(7) \times 2$, we have $2{\cdot}M_{22} < \Omega_{10}^-(7) \cong 2{\cdot}P\Omega_{10}^-(7)$. Therefore, we can lift to the double cover of the orthogonal group, that is the spin group (see Section 3.9.2). But in this spin group, the central involution of the orthogonal group becomes an element of order 4. Therefore we obtain a proper quadruple cover $4{\cdot}M_{22}$.

5.3 The small Mathieu groups

5.3.1 The group M_{12}

In Section 5.2.6 we defined M_{12} as the stabiliser of a dodecad in the extended binary Golay code. In particular, the order of M_{12} is $95040 = 2^6.3^3.5.11$. In this code, an octad can intersect a dodecad in either 2, 4, or 6 points. A straightforward counting argument (or explicit enumeration: see Exercise 5.4) shows that for a fixed dodecad, there are 132 octads intersecting it in 2 points, 132 intersecting it in 6 points, and 495 intersecting it in 4 points. In particular the intersections of size 6 form a set of 132 blocks (called *hexads*), with the property (inherited from the Steiner system $S(5,8,24)$) that every set of 5 points is in exactly one of these blocks. In other words, M_{12} preserves a Steiner system $S(5,6,12)$.

Now the pointwise stabiliser of a hexad also fixes pointwise the octad of $S(5,8,24)$ which it is contained in, and it is easy to check that no such group element fixes our given dodecad. In other words, the pointwise stabiliser of a hexad in M_{12} is the trivial group. Since there are just 132 hexads, and $|M_{12}| = 132|S_6|$, it follows that M_{12} is transitive on the hexads, and that the setwise stabiliser of a hexad is S_6 acting naturally on the six points of the hexad. Hence M_{12} is 5-transitive on the 12 points.

These blocks do not give rise to a binary code in the same way as the blocks of the Steiner system $S(5,8,24)$, as two blocks can intersect in 0, 2, 3, or 4 points. However, we can attach signs to the points in the blocks in such a way that we obtain a ternary code (known as the *extended ternary Golay code*), that is a vector space over $\mathbb{F}_3$ (see Section 5.3.5).

5.3.2 The Steiner system $S(5,6,12)$

Before we do this, however, let us study the Steiner system itself. First define the *tetracode* to be the linear code of length 4 and dimension 2 over $\mathbb{F}_3 = \{0,+,-\}$ whose non-zero codewords are

$$\pm(0+++),$$
$$\pm(+0+-),$$
$$\pm(+-0+),$$
$$\pm(++-0). \tag{5.15}$$

The MINIMOG is then a 4×3 array whose rows are labelled by the elements $0, +, -$ of $\mathbb{F}_3$ and whose columns are labelled by the coordinates 1, 2, 3, 4 (or, better, ∞, 0, 1, 2) of the tetracode. Words in the code act on the MINIMOG via column-wise addition, so that for example the words $(0+++)$ and $(+0+-)$ act as follows:

$$\tag{5.16}$$

The automorphism group of the code is a group $GL_2(3) \cong 2{\cdot}S_4$ which acts on the MINIMOG with generators

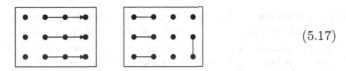

$$(5.17)$$

Together these generate a group of shape $3^2{:}\mathrm{GL}_2(3)$.

Then the 132 hexads of the Steiner system $S(5, 6, 12)$ can be described as the sets of six points with column distribution $(3^2 0^2)$, $(3^1 1^3)$, $(2^3 0^1)$ or $(2^2 1^2)$ whose 'odd men out' form part of a tetracode word. Here the 'odd man out' is either the unique point of the column which is in the hexad, or the unique point of the column which is not in the hexad, if either of these is defined. For example, the set

$$(5.18)$$

has odd men out $(+0+-)$ in the four columns. Since each tetracode word is determined by any two of its coordinates, the odd men out always determine a unique tetracode word. Thus there are 6 hexads with point distribution $(3, 3, 0, 0)$ across the columns, 36 with distribution $(3, 1, 1, 1)$, 36 with distribution $(2, 2, 2, 0)$ and 54 with distribution $(2, 2, 1, 1)$. In pictures, we may take representatives of the orbits under $3^3{:}\mathrm{GL}_2(3)$ as follows.

$$(5.19)$$

5.3.3 Uniqueness of $S(5, 6, 12)$

In the same way that we proved in Section 5.2.4 that there is a unique Steiner system of type $S(5, 8, 24)$, up to isomorphism, we can show that there is a unique Steiner system of type $S(5, 6, 12)$, up to isomorphism. First create a Leech triangle for such a Steiner system:

$$
\begin{array}{ccccccccccccc}
 & & & & & & 132 & & & & & & \\
 & & & & & 66 & & 66 & & & & & \\
 & & & & 30 & & 36 & & 30 & & & & \\
 & & & 12 & & 18 & & 18 & & 12 & & & \\
 & & 4 & & 8 & & 10 & & 8 & & 4 & & \\
 & 1 & & 3 & & 5 & & 5 & & 3 & & 1 & \\
1 & & 0 & & 3 & & 2 & & 3 & & 0 & & 1 \\
\end{array}
\qquad (5.20)
$$

In particular, since two hexads cannot intersect in exactly one point, the complement of a hexad is a hexad. Now pick any four points, and consider the hexads containing these four points: by the defining property of the Steiner system, any fifth point determines the sixth point of the hexad, so there are exactly four such hexads. We can arrange them in a 4×3 MINIMOG array such that the first four points form the top row, and the pairs which complete it to a hexad are the remaining points in the four columns, thus:

$$
\begin{array}{|cccc|}
\hline
0 & 0 & 0 & 0 \\
1 & 2 & 3 & 4 \\
1 & 2 & 3 & 4 \\
\hline
\end{array}
\tag{5.21}
$$

Since the complements of hexads are also hexads, we already know two hexads containing the four points labelled 1 and 2, so the other two each contain two of the points labelled 0. In other words they determine a splitting of the top row into two pairs: we can arrange that this splitting is between the first two and the last two points. Similarly for the points labelled 1 and 3. In particular, the union of any two columns is a hexad.

Now any set of three points lies in exactly $9.8/3.2 = 12$ hexads. For example, taking the three points in the first column of the array, we may re-arrange the points so that the hexads are given by pairs of numbers in one of the three arrays:

$$
\begin{array}{|cccc|}
\hline
0 & 1 & 1 & 1 \\
0 & 2 & 2 & 2 \\
0 & 3 & 3 & 3 \\
\hline
\end{array}
\quad
\begin{array}{|cccc|}
\hline
0 & 1 & 2 & 3 \\
0 & 3 & 1 & 2 \\
0 & 2 & 3 & 1 \\
\hline
\end{array}
\quad
\begin{array}{|cccc|}
\hline
0 & 1 & 2 & 3 \\
0 & 2 & 3 & 1 \\
0 & 3 & 1 & 2 \\
\hline
\end{array}
\tag{5.22}
$$

Similarly, if we take the three points in the second column then without loss of generality the hexads are given by

$$
\begin{array}{|cccc|}
\hline
1 & 0 & 1 & 1 \\
2 & 0 & 2 & 3 \\
3 & 0 & 3 & 2 \\
\hline
\end{array}
\quad
\begin{array}{|cccc|}
\hline
1 & 0 & 3 & 2 \\
2 & 0 & 1 & 1 \\
3 & 0 & 2 & 3 \\
\hline
\end{array}
\quad
\begin{array}{|cccc|}
\hline
1 & 0 & 2 & 3 \\
2 & 0 & 3 & 2 \\
3 & 0 & 1 & 1 \\
\hline
\end{array}
\tag{5.23}
$$

After this, there is a unique possibility for the hexads containing the third column, and similarly for the fourth column. We have shown that the hexads with column distribution 3111 are the same as in the Steiner system already constructed. Similarly, the hexads with column distribution 222 are just their complements.

It remains to show that the remaining hexads are the ones with column distribution 2211 which we exhibited before. Now if we take two points in each of two columns, then we already have a hexad containing these four points, together with two points in either one of the remaining columns. Therefore the last hexad is determined. This concludes the proof that there is, up to permutations, a unique Steiner system $S(5, 6, 12)$.

5.3.4 Simplicity of M_{12}

We have made extensive use of the subgroup $3^2{:}GL_2(3)$, which has index 220 and acts transitively but imprimitively (with four blocks of size 3) on the 12 points. Such a partition of the twelve points into four triples, any two of which form a hexad of the Steiner system, is called a *quartet*. (In the Atlas [28], they are called *linked threes*.) In constrast to the situation in M_{24}, a quartet is not determined by one of the triples, but it is determined by two of the triples.

Straightforward calculation shows that the quartet stabiliser has orbits of lengths $1 + 12 + 27 + 72 + 108$ on the 220 quartets, and hence the action is primitive. For reference, here are representatives of the four non-trivial orbits:

$$
\begin{array}{|cccc|}\hline 0&1&1&1\\ 0&2&2&2\\ 0&3&3&3\\\hline\end{array}
\quad
\begin{array}{|cccc|}\hline 1&0&3&2\\ 0&1&2&3\\ 0&1&2&3\\\hline\end{array}
\quad
\begin{array}{|cccc|}\hline 0&0&2&3\\ 0&2&1&3\\ 1&3&2&1\\\hline\end{array}
\quad
\begin{array}{|cccc|}\hline 1&0&1&1\\ 0&2&3&3\\ 0&3&2&2\\\hline\end{array}
\tag{5.24}
$$

It is clear that the elements of the normal 3^2 in $3^2{:}GL_2(3)$ are commutators, and it is not hard to show that M_{12} is generated by conjugates of these elements. In particular, M_{12} is perfect, and hence, by Iwasawa's Lemma (Theorem 3.1) is simple.

5.3.5 The ternary Golay code

Since M_{12} contains involutions with cycle type (2^6), which lift to elements of order 4 in the Schur double cover $2 \cdot A_{12}$ (see Section 2.7.2), we see that there is a proper double cover $2 \cdot M_{12}$ of M_{12}. More concretely, we shall see $2 \cdot M_{12}$ as the automorphism group of the extended ternary Golay code.

This code can be obtained by attaching signs to the points in the hexads, turning them into vectors over $\mathbb{F}_3$. The signs have the property that the sum of the entries in the four positions corresponding to any tetracode word is independent of the word, and is equal to minus the sum of the entries in any column. One can check that for each hexad there is a unique choice of signs (up to an overall sign change), and that the resulting 264 vectors of weight 6 in $\mathbb{F}_3{}^{12}$ are mutually orthogonal with respect to the usual inner product. Hence they span a totally isotropic space of dimension 6, called the (extended) *ternary Golay code*.

Indeed, the code consists of all vectors satisfying the above linear conditions, and one can easily verify that it contains 440 vectors of weight 9, and 24 vectors of weight 12. The latter fall into three orbits of lengths 6, 9 and 9 under the action of $3^2{:}2S_4$, with representatives

$$
\begin{array}{|cccc|}\hline +&+&-&-\\ +&+&-&-\\ +&+&-&-\\\hline\end{array}
\quad\text{and}\quad
\pm\begin{array}{|cccc|}\hline -&-&-&-\\ +&+&+&+\\ +&+&+&+\\\hline\end{array}
\tag{5.25}
$$

We now check that the element

$$(5.26)$$

preserves the code, and therefore the Steiner system. Since the codewords of weight 6 map two-to-one onto the hexads of the Steiner system $S(5,6,12)$, and the only automorphism of the code which fixes all the hexads is -1, it follows that the group generated by these elements is a double cover $2 \cdot M_{12}$ of M_{12}, with central element acting as -1 on all twelve coordinates.

The stabiliser of one of the 24 vectors of weight 12 in the code is a subgroup isomorphic to M_{11}, permuting the 12 coordinates transitively. If we change basis by negating (say) the top row of the MINIMOG, we can choose the fixed vector of M_{11} to be $(+^{12})$, so that M_{11} acts as a group of pure permutations (with no sign changes). This basis will be useful for example in the construction of the sporadic Suzuki group (see Sections 5.6.10 and 5.6.11).

It is easy to see that the Golay code gives rise to a Steiner system $S(5,6,12)$, by taking the codewords of weight 6 and ignoring the signs. We have already proved that this Steiner system is unique up to isomorphism. To obtain an explicit isomorphism between the construction in this section and the one given in Section 5.2.6 we may embed the MINIMOG in the MOG, by swinging the first column of the MINIMOG upwards and to the left thus:

$$
\begin{array}{cccc}
1 & 2 & 3 & 4 \\
5 & 6 & 7 & 8 \\
9 & 10 & 11 & 12
\end{array}
\quad \mapsto \quad
\begin{array}{cccc}
9 & 5 & 1 & \\
& 2 & 3 & 4 \\
& 6 & 7 & 8 \\
& 10 & 11 & 12
\end{array}
\qquad (5.27)
$$

The generators of M_{12} given in (5.16), (5.17) and (5.26) above (ignoring signs) extend to elements of M_{24} as follows.

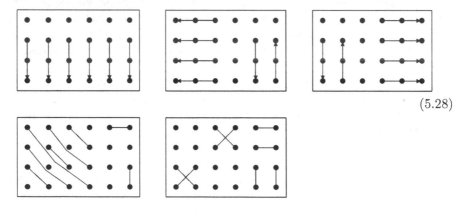

$$(5.28)$$

5.3.6 The outer automorphism of M_{12}

As we have seen above, the group M_{12} acts on the 24 points of the MOG with two orbits of size 12, which are 'dodecads' in the Golay code. Since M_{24} is transitive on dodecads, there is an element of M_{24} which interchanges these two orbits. Such an element cannot centralise M_{12}, so acts as an outer automorphism of M_{12}. Inside M_{24} the outer automorphism of M_{12} may be realised by the following permutation.

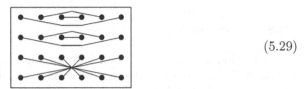

$$(5.29)$$

We have already seen that if we fix two points in one dodecad, then there are exactly two octads containing these two points and no others in the dodecad. These octads define a splitting of the complementary dodecad into two hexads of an $S(5,6,12)$. In particular, the point stabiliser M_{11} in one dodecad does not fix a point in the other dodecad, so (since it contains elements of order 11) acts transitively, and indeed 2-transitively. (We shall see in a moment that it is actually 3-transitive.) This tells us that we have two classes of M_{11} inside M_{12}, interchanged by the outer automorphism. (Compare the outer automorphism of S_6 described in Section 2.4.2, interchanging two classes of subgroups S_5.)

5.3.7 Subgroups of M_{12}

We already know that any element of M_{12} which fixes five of the 12 points is trivial, so a transposition in the hexad-stabiliser S_6 must act fixed-point-freely on the complementary hexad. Therefore the actions of S_6 on the two hexads are related by the outer automorphism of S_6. There is also an element of M_{12} which interchanges the hexad with its complement, and therefore realises the outer automorphism of S_6. This subgroup $S_6.2$ has orbit lengths 2 and 10 on the points of the other dodecad, so the stabiliser of a pair of points also has the shape $S_6.2$.

The stabiliser of a triple of points has order $|M_{12}|.3!/12.11.10 = 2^4.3^3$ and structure $3^2{:}2S_4$. The image of this under the outer automorphism of M_{12} fixes a quartet, that is, a partition of the 12 coordinates into four sets of size 3, with the property that the union of any two of them is a hexad of the Steiner system (e.g. the columns of the MINIMOG). Notice that the point-wise stabiliser of three points has structure $3^2{:}Q_8$ and acts transitively on the complementary dodecad. Another way of saying this is that the point stabiliser M_{11} in one dodecad acts 3-transitively on the complementary dodecad.

The stabiliser of a set of four points has order $|M_{12}|.4!/12.11.10.9 = 2^6.3$ and has structure $2^{1+4}S_3$. In this case there is a unique octad of $S(5,8,24)$

intersecting the dodecad in exactly these four points, so this gives rise to a well-defined set of four points in the complementary dodecad as well. Thus there is only one conjugacy class of subgroups of this type.

There is in fact a maximal subgroup $\mathrm{PSL}_2(11)$, acting on the 12 points in the same way that it acts on the projective line (see Section 3.3.5). This extends to a subgroup $\mathrm{PGL}_2(11)$ in $\mathrm{M}_{12}{:}2$. There are just three other classes of maximal subgroups, of shapes $2 \times S_5$, $4^2{:}D_{12}$ and $A_4 \times S_3$. (See Table 5.1 for the full list of maximal subgroups.)

5.3.8 The group M_{11}

The definition of M_{11} as the point stabiliser in M_{12} shows that it has order $7920 = 2^4.3^2.5.11$ and acts 4-transitively on a set of 11 points, and preserves a Steiner system $S(4,5,11)$. The usual calculation (see Section 5.2.3) shows that the number of pentads in this Steiner system is $\binom{11}{4} \Big/ \binom{5}{4} = 66$.

Since the stabiliser in M_{12} of a triple of points is an affine group $3^2{:}\mathrm{GL}_2(3)$, it follows that the stabiliser in M_{11} of a pair of points is $3^2{:}SD_{16}$, where SD_{16} is the Sylow 2-subgroup of $\mathrm{GL}_2(3)$, known as the *semidihedral* group of order 16, with the presentation

$$\langle a, b \mid a^8 = b^2 = 1, a^b = a^3 \rangle. \tag{5.30}$$

As the Sylow 3-subgroup of M_{11} has order 9, this group $3^2{:}SD_{16}$ is its normaliser, and so all elements of order 3 in M_{11} are conjugate.

Now it is easy to see that the elements of order 3 generate a 3-transitive subgroup. Moreover, the pointwise stabiliser of three points is Q_8, whose normaliser in M_{11} is $Q_8.S_3 \cong \mathrm{GL}_2(3)$, so the elements of order 3 also generate this Q_8. Therefore they generate a 4-transitive subgroup, which has order at least $11.10.9.8. = 7920$ so is the whole of M_{11}. Clearly these elements of order 3 are commutators, so M_{11} is perfect. By 4-transitivity, the action on pairs is primitive. Hence by Iwasawa's Lemma (Theorem 3.1), M_{11} is simple.

Inside M_{24} we see M_{11} with orbits of sizes 1, 11, and 12. In particular we see that M_{11} has a transitive action on 12 points, which we proved in Section 5.3.7 is actually 3-transitive. It follows that these two faithful representations on 11 and 12 points are primitive. The point stabilisers are maximal subgroups $A_6\cdot 2$ and $\mathrm{PSL}_2(11)$ respectively. Fixing a pair of the 11 points gives, as we have seen, a soluble maximal subgroup $3^2{:}SD_{16}$ of order 144, which has orbits of lengths 2 and 9 on the 11 points and acts transitively on the 12 points. Similarly, fixing a pair of the 12 points gives a maximal subgroup S_5, which has orbit lengths 2 and 10 on the 12 points, and 5 and 6 on the 11 points. The only other maximal subgroups are the involution centralisers $2\cdot S_4 \cong \mathrm{GL}_2(3)$ which have orbit lengths 3 and 8 on the 11 points and 4 and 8 on the 12 points. (The complete list of maximal subgroups for all the Mathieu groups is given in Table 5.1.)

Table 5.1. Maximal subgroups of the Mathieu groups

Group	M_{24}	M_{23}	M_{22}	M_{12}	M_{11}
cocode subgroups	M_{23} $M_{22}{:}2$ $PSL_3(4){:}S_3$ $2^6{:}3S_6$	M_{22} $PSL_3(4){:}2$ $2^4{:}(3 \times A_5){:}2$	$PSL_3(4)$ $2^4{:}S_5$	M_{11} $A_6{\cdot}2^2$ $3^2{:}2S_4$ $2^{1+4}S_3$ $4^2 D_{12}$	$A_6{\cdot}2$ $3^2{:}SD_{16}$ $2S_4$
code subgroups	$2^4{:}A_8$ $M_{12}{:}2$ $2^6{:}(GL_3(2) \times S_3)$	$2^4{:}A_7$ A_8 M_{11}	$2^4{:}A_6$ A_7 A_7 $2^3{:}GL_3(2)$ $A_6{\cdot}2$ $PSL_2(11)$	M_{11} $A_6{\cdot}2^2$ $3^2{:}2S_4$	$PSL_2(11)$ S_5
others	$PSL_2(23)$ $PSL_2(7)$	$23{:}11$		$PSL_2(11)$ $2 \times S_5$ $A_4 \times S_3$	

5.4 The Leech lattice and the Conway group

5.4.1 The Leech lattice

We construct the biggest Conway group $2{\cdot}Co_1$ (a double cover of the simple group Co_1) as a group of 24×24 real matrices. At the same time we construct a *lattice* (i.e. a discrete set of vectors closed under addition and subtraction) which is invariant under the group, in order to determine the order of the group and other properties. The process is analagous to the construction of M_{24} and the Golay code, using the subgroup $2^6{:}3{\cdot}S_6$ derived from the hexacode. Here we use instead a group $2^{12}{:}M_{24}$ derived from the Golay code to help with the construction, which is based on Conway's original construction [25].

There is an obvious action of M_{24} on 24-space, in which it permutes the coordinate vectors naturally. There is also an action of the Golay code itself, whereby a codeword acts by negating all the coordinates corresponding to a 1 in the word. Thus an octad acts by negating an 8-space, for example. Taking both groups together we obtain a group $2^{12}{:}M_{24}$ acting monomially on 24-space.

The Leech lattice is named after John Leech [116] although it was apparently first discovered by Witt in 1940. It may be defined as the $\mathbb{Z}$-linear combinations of the $1104 + 97152 + 98304 = 196560$ images under the group $2^{12}{:}M_{24}$ of the following vectors

$$
\begin{array}{|cc|c|c|}
\hline
4\ \ 4 & & & \\
\hline
\end{array}
\quad
\begin{array}{|cc|c|c|}
\hline
2\ \ 2 & & & \\
2\ \ 2 & & & \\
2\ \ 2 & & & \\
2\ \ 2 & & & \\
\hline
\end{array}
\quad
\begin{array}{|cc|cc|cc|}
\hline
-3\ \ 1 & 1\ \ 1 & 1\ \ 1 \\
1\ \ 1 & 1\ \ 1 & 1\ \ 1 \\
1\ \ 1 & 1\ \ 1 & 1\ \ 1 \\
1\ \ 1 & 1\ \ 1 & 1\ \ 1 \\
\hline
\end{array}
\qquad (5.31)
$$

More helpfully, it may be defined as the set of all integral vectors $(x_1, \ldots, x_{24})$ (i.e. $x_i \in \mathbb{Z}$) such that either all the x_i are even or they are all odd, and congruent modulo 2 to $\frac{1}{4}$ of the sum of the coordinates, and additionally the residue classes modulo 4 are in the Golay code. Thus

$$x_i \equiv m \bmod 2,$$

$$\sum_{i=1}^{24} x_i \equiv 4m \bmod 8, \text{ and}$$

for each k, the set $\{i \mid x_i \equiv k \bmod 4\}$ is in the Golay code.   (5.32)

   To show that these two definitions are equivalent we must first show that all the spanning vectors in the first definition satisfy the congruence conditions in the second definition. This is a straightforward exercise (see Exercise 5.13). Conversely, suppose that x is a vector satisfying the conditions of the second definition (5.32). If the coordinates of x are odd, subtract the vector $(-3, 1^{23})$ to get a new vector x with even coordinates, still satisfying the conditions. If now x has some coordinates not divisible by 4, we can subtract some octad vectors (i.e. vectors of shape $(2^8, 0^{16})$) until all the coordinates are divisible by 4, and the new vector x still satisfies the conditions. Now the sum of the coordinates is congruent to 0 modulo 8, so the vector x is a sum of vectors of the shape $(\pm 4, \pm 4, 0^{22})$. Hence x is in the lattice spanned by the vectors given in the first definition (5.31).

   Now it is easy to see that the 196560 vectors listed in (5.31) are the only vectors of smallest norm in the lattice. (We scale the usual norm $\sum_{i=1}^{24} x_i^2$ by dividing by 8, so that these vectors have norm (i.e. squared length) 4. This is the smallest scale on which all inner products of vectors in the lattice are integers.) Similarly, the vectors of norm 6 fall into four orbits under the group $2^{12}{:}M_{24}$, with lengths $2576.2^{11} = 5275648$, $\binom{24}{3}.2^{12} = 8290304$, $759.16.2^8 = 3108864$, and $24.2^{12} = 98304$ (making 16773120 in all), and representatives as follows.

2	2	2			
			2	2	2
			2	2	2
			2	2	2

-3	-3	-3	1	1	1
1	1	1	1	1	1
1	1	1	1	1	1
1	1	1	1	1	1

(5.33)

2	-2	4			
2	2				
2	2				
2	2				

5	1	1	1	1	1
1	1	1	1	1	1
1	1	1	1	1	1
1	1	1	1	1	1

   The 398034000 vectors of norm 8 similarly fall into nine orbits under the group $2^{12}{:}M_{24}$. (See Exercise 5.14.)

Let Λ denote the Leech lattice. As an abelian group under addition, Λ is isomorphic to $\mathbb{Z}^{24}$, so that $\Lambda/2\Lambda \cong \mathbb{F}_2^{24}$, a vector space of dimension 24 over $\mathbb{F}_2$. In particular there are just 2^{24} congruence classes mod 2Λ of vectors in Λ. These are just the cosets of 2Λ in Λ. If x and y are in the same congruence class, then $x \pm y$ both lie in 2Λ, so have norm 0 or at least 16. Therefore the sum of the norms of x and y must be at least 16, unless $x = \pm y$. In particular, the vectors of norms 0, 4, and 6 are congruent only to their negatives, while two vectors of norm 8 can be congruent only if they are either negatives of each other, or perpendicular to each other. Since perpendicular vectors are linearly independent, we have accounted for at least

$$1 + 196560/2 + 16773120/2 + 398034000/48 = 16777216$$
$$= 2^{24} \tag{5.34}$$

congruence classes.

5.4.2 The Conway group Co_1

Thus all the vectors of norm 8 in the Leech lattice fall into congruence classes of 48 pairs of mutually perpendicular vectors (which we call *coordinate frames* or *crosses*, or sometimes *double bases*), and therefore there are exactly $398034000/48 = 8292375$ such crosses. Since the stabiliser of a cross is just $2^{12}{:}M_{24}$ (as is clear from the second definition (5.32) of the Leech lattice), we only need to prove transitivity on the crosses in order to compute the order of the automorphism group as $8292375.2^{12}.|M_{24}|$. Clearly this automorphism group has a centre of order 2 generated by the scalar -1. The quotient by the centre is therefore a group Co_1 of order

$$|Co_1| = 4\,157\,776\,806\,543\,360\,000 = 2^{21}.3^9.5^4.7^2.11.13.23. \tag{5.35}$$

To prove transitivity on the crosses it is sufficient to exhibit an element which fuses the orbits of the monomial group $2^{12}{:}M_{24}$. Alternatively, a non-constructive proof can be obtained by showing that any 24-dimensional *even (integral) lattice* (i.e. a lattice such that all norms are even integers, and therefore all inner products are integers) containing the given numbers of vectors of all norms up to and including 8, is isomorphic to the Leech lattice. In effect this gives us a third definition of the Leech lattice, which we record formally in the following theorem.

Theorem 5.1. *If Λ is a 24-dimensional even (integral) lattice containing no vectors of norm 2, 196560 vectors of norm 4, 16773120 vectors of norm 6 and 398034000 vectors of norm 8, then Λ is isomorphic to the Leech lattice.*

Proof. The same counting argument as above (5.34) shows that in any such lattice the vectors of norm 8 form coordinate frames. Writing the lattice with respect to a basis such that one such coordinate frame consists of the vectors of shape $(\pm 8, 0^{23})$, we know that $(\pm 8, \pm 8, 0^{22})$ is in twice the lattice, so that $(\pm 4, \pm 4, 0^{22})$ belongs to the lattice. Since all inner products are integral (after dividing by 8), it follows that for any vector in the lattice, all its coordinates are integers, and are congruent, to m say, modulo 2. Hence all vectors of shape $(\pm 4, \pm 4, \pm 4, \pm 4, 0, \ldots, 0)$ belong to the lattice and also form coordinate frames. These vectors therefore determine splittings of the 24 coordinates into sextets, or equivalently, every set of five coordinates determines an octad. Thus we obtain the structure of a Steiner system $S(5, 8, 24)$ on the 24 coordinates. Using the fact that this Steiner system is essentially unique (see Section 5.2.3) we may label the vectors of our coordinate frame so that it is the same Steiner system as before. Moreover, vectors of shape $(\pm 2^8, 0^{16})$ with 2s on an octad are also in the lattice (with signs yet to be determined). Since there are no vectors of norm less than 4, and there are 196560 vectors of norm 4, there must be some vectors of norm 4 with odd coordinates. Changing signs on some coordinates if necessary, we may assume this vector is $(-3, 1^{23})$. Therefore the octad vectors have an even number of minus signs, and we have the same Leech lattice as we had before.

5.4.3 Simplicity of Co_1

We can prove that Co_1 is simple using Iwasawa's Lemma again. To do this we must first show that the group acts primitively on the crosses. This is an easy exercise (see Exercise 5.14). We have already shown that the stabiliser of a cross in $2 \cdot Co_1$ is $2^{12}{:}M_{24}$. Next we show that $2 \cdot Co_1$ is generated by conjugates of the normal abelian subgroup 2^{12} of the cross stabiliser, by first finding an element of this form in $2^{12}{:}M_{24} \setminus 2^{12}$, so that by the simplicity of M_{24} the whole group $2^{12}{:}M_{24}$ is generated by such elements. A suitable element is the first involution displayed in (5.3), which acts as sign-changes on the coordinate frame defined by the first vector displayed in (5.37), and is therefore in the normal 2^{12} subgroup of the stabiliser $2^{12}{:}M_{24}$ of this coordinate frame.

Since $2^{12}{:}M_{24}$ is maximal in $2 \cdot Co_1$, there are conjugates of the 2^{12} not contained in this copy of $2^{12}{:}M_{24}$, so $2 \cdot Co_1$ is generated by these conjugates, as required. Finally, we easily see that the 2^{12} is generated by commutators already in $2^{12}{:}M_{24}$. Hence we have all the ingredients of Iwasawa's Lemma (Theorem 3.1), and conclude that Co_1 is simple.

5.4.4 The small Conway groups

Now that we have shown, in Section 5.4.2, that $2 \cdot Co_1$ is transitive on crosses, it follows easily that it is transitive on vectors of norm 4, and on vectors of norm 6. For the cross stabiliser is transitive on norm 4 vectors having inner products ± 4 or 0 with each vector of the cross: thus for example $(4, 4, 0^{22})$ with respect to the standard cross, and $(2^8, 0^{16})$ with respect to the cross containing $(4^4, 0^{20})$, and $(3, -1^7, 1^{16})$ with respect to the cross containing $(5, -3, -3, 1^{21})$. Therefore the three orbits of $2^{12}{:}M_{24}$ on vectors of norm 4 are fused into a single orbit under $2 \cdot \mathrm{Co}_1$. The stabiliser of a vector of norm 4 is denoted Co_2, and has order

$$\begin{aligned}
|\mathrm{Co}_2| &= |\mathrm{Co}_1|/98280 \\
&= 42\,305\,421\,312\,000 \\
&= 2^{18}.3^6.5^3.7.11.23.
\end{aligned} \tag{5.36}$$

We shall study this group in more detail in Section 5.5.4.

Similarly, the cross stabiliser is transitive on norm 6 vectors having inner products ± 2 or 0 with each vector of the cross. One example of such a vector is $(2^{12}, 0^{12})$ with respect to the standard cross. Representatives for the other three orbits under the monomial group are displayed in the MOG array below. The top row contains a representative vector of a suitable cross, and the bottom row the vector of norm 6.

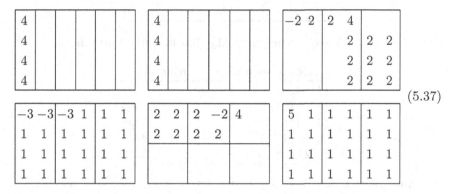

$$(5.37)$$

It follows that $2 \cdot \mathrm{Co}_1$ is transitive on the vectors of norm 6. The stabiliser of a vector of norm 6 is denoted Co_3, and has order

$$\begin{aligned}
|\mathrm{Co}_3| &= |\mathrm{Co}_1|/8386560 \\
&= 495\,766\,656\,000 \\
&= 2^{10}.3^7.5^3.7.11.23.
\end{aligned} \tag{5.38}$$

We shall study this group in more detail in Section 5.5.3.

We can proceed further, to define the McLaughlin group and the Higman–Sims group, and determine their orders, by proving transitivity of Co_2 or Co_3 on suitable sets of vectors. In the McLaughlin group case we need to prove that Co_3, the stabiliser of a vector v of norm 6, is transitive on the 552 vectors of norm 4 which have inner product -3 with v. If $v = (-2^{12}, 0^{12})$ then the monomial group $2 \times M_{12}$ fixes v and has orbits of lengths $24 + 264 + 264$ on these vectors, with representatives $(1^{12}, -3, 1^{11})$, $(2^6, 0^6, 2^2, 0^{10})$ and $(3, -1, 1^{10}, 1^6, -1^6)$ respectively. On the other hand, if $v = (-5, -1^{23})$, then the monomial group M_{23} has orbits of lengths $23 + 23 + 253 + 253$, with representatives $(4, 4, 0^{22})$, $(1, -3, 1^{22})$, $(2, 2^7, 0^{16})$ and $(3, -1^7, 1^{16})$ respectively. The only way for both these sets of orbits to fuse into orbits for Co_3 is as a single orbit of length 552.

Thus the stabiliser in Co_3 of such a vector is a subgroup of index 552 in Co_3, so has order $|Co_3|/552 = 898128000 = 2^7.3^6.5^3.7.11$. This is the McLaughlin group, which we study in more detail in Section 5.5.2.

Similarly, in the Higman–Sims case we need to prove that Co_3 is transitive on the 11178 vectors of norm 4 which have inner product -2 with v. When $v = (-2^{12}, 0^{12})$, the monomial group $2 \times M_{12}$ has six orbits on these vectors, as follows:

Orbit representative	Orbit length	
$(-3, 1^{11}, 1^{12})$	24	
$(4^2, 0^{10}, 0^{12})$	66	
$(2^5, -2, 0^6, 2^2, 0^{10})$	1584	(5.39)
$(1^{10}, -1^2, -1^6, -3, 1^5)$	1584	
$(2^4, 0^8, 2^4, 0^8)$	3960	
$(3, -1^3, 1^8, -1^4, 1^8)$	3960	

But when $v = (-5, -1^{23})$, the group M_{23} has five orbits, as follows.

Orbit representative	Orbit length	
$(4, -4, 0^{22})$	23	
$(0, 2^8, 0^{15})$	506	
$(3, -1^{11}, 1^{12})$	1288	(5.40)
$(1, 3, -1^{15}, 1^7)$	4048	
$(2, 2^5, -2^2, 0^{16})$	5313	

Again it is easy to see that Co_3 must act transitively on these vectors.

Thus the stabiliser in Co_3 of such a vector is a subgroup of index 11178 in Co_3, so has order $|Co_3|/11178 = 44352000 = 2^9.3^2.5^3.7.11$. This is the Higman–Sims group, which we study in more detail in Section 5.5.1.

5.4.5 The Leech lattice modulo 2

Reducing the Leech lattice modulo 2, as in Section 5.4.1, gives rise to a 24-dimensional representation of $2 \cdot Co_1$ over $\mathbb{F}_2$, in which the central involution acts trivially. Thus we obtain a 24-dimensional representation of the simple group Co_1 over $\mathbb{F}_2$. Moreover, the real norm, divided by 2 and reduced modulo 2, gives rise to a non-degenerate quadratic form, of plus type. Thus we have an embedding of Co_1 in $\Omega_{24}^+(2)$. We shall use this embedding in the construction of the Monster in Section 5.8.1 below.

Many of the maximal subgroups of the Conway group are best seen as stabilisers of sublattices, or of subspaces of the lattice modulo 2. We have already seen that there are just three orbits of Co_1 on subspaces of dimension 1 in $\Lambda/2\Lambda$: a single orbit on the 8386560 vectors of norm 1, coming from the vectors of norm 6 in the Leech lattice, and two orbits of lengths 98280 and 8292375 on the isotropic vectors in $\Lambda/2\Lambda$, coming from the vectors of norms 4 and 8 in Λ respectively. We shall call these vectors of type 3, 2 and 4 respectively. (More generally, the *type* of a Leech lattice vector is half its norm.) The corresponding stabilisers are the maximal subgroups Co_3, Co_2, and $2^{11}{:}\mathrm{M}_{24}$ respectively.

Next we enumerate the subspaces of dimension 2. Those subspaces which contain vectors of type 4 are easy to classify, as we can work inside the monomial subgroup $2^{12}{:}\mathrm{M}_{24}$ of $2{\cdot}\mathrm{Co}_1$ to write down all possibilities. The other cases require more work, as it is not obvious how many orbits the group has on subspaces of any given type. The results are listed in Table 5.2.

For each of the first six rows of Table 5.2, we need to prove transitivity of $2{\cdot}\mathrm{Co}_1$ on certain configurations of vectors, or equivalently to prove transitivity of Co_2 or Co_3 on certain sets of vectors. We have already proved the cases which lead to the McLaughlin and Higman–Sims groups, and we leave the rest as exercises (see Exercises 5.15, 5.16, 5.17). More details can be found in [178].

Table 5.2. Orbits of Co_1 on 2-spaces of $\Lambda/2\Lambda$

Type	Representatives	Centraliser
222	$(4,-4,0^{22}),(0,4,-4,0^{21})$	$\mathrm{PSU}_6(2)$
223	$(4,4,0^{22}),(-3,1^{23})$	McL
233	$(4,-4,0^{22}),(5,1^{23})$	HS
233*	$(2^7,-2,4,0^{15}),(0^{22},4,4)$	$\mathrm{PSU}_4(3){:}2^2$
333	$(5,1^{23}),(0^{12},2^{12})$	$3^5{:}\mathrm{M}_{11}$
333*	$(5,1^{23}),(-1^8,5,1^{15})$	$\mathrm{PSU}_3(5)$
224	$(8,0^{23}),(4,4,0^{22})$	$2^{10}{:}\mathrm{M}_{22}{:}2$
234	$(8,0^{23}),(-3,1^{23})$	M_{23}
244	$(8,0^{23}),((2^8,0^{16})$	$2^{1+8}_{+}{:}A_8$
334	$(8,0^{23}),(2^7,-2,4,0^{15})$	$2^4{\cdot}A_8$
334*	$(8,0^{23}),(2^{12},0^{12})$	$2 \times \mathrm{M}_{12}$
344	$(8,0^{23}),(-3^3,1^{21})$	$\mathrm{PSL}_3(4){:}S_3$
444	$(8,0^{23}),(4^4,0^{20})$	$2^{4+12}{:}3{\cdot}S_6$
444	$(8,0^{23}),(2^{11},-2,4,0^{11})$	$\mathrm{M}_{12}{:}2$
444	$(8,0^{23}),(2^8,4^2,0^{14})$	$[2^{11}]{:}\mathrm{PSL}_3(2)$

Note: a $*$ in the type symbol indicates that it is impossible to choose three Leech lattice vectors of these types adding to 0 and mapping modulo 2 to the three vectors in the given 2-space.

Note in particular the appearance of the group $\mathrm{PSU}_6(2)$ in this table. In this sense we may regard $\mathrm{PSU}_6(2)$ as a 'Conway group': we shall see in Section 5.7, especially Section 5.7.1, that it also has an incarnation as a 'Fischer group'. Other groups which arise here are HS:2 and $\mathrm{PSU}_4(3){:}D_8$, which are maximal in Co_2. The 333* case is particularly interesting: if u, v, w are type 3 representatives for the three fixed cosets of 2Λ, then three of the vectors $u \perp v \perp w$ have type 2 and one has type 3. Thus there is a group $\mathrm{PSU}_3(5){.}S_3$

permuting u, v, w (up to signs), contained in the group Co_3 which fixes this last vector.

Two interesting local maximal subgroups of Co_1, of shape $3^{1+4}{:}\text{Sp}_4(3){:}2$ and $5^{1+2}{:}\text{GL}_2(5)$, also arise as stabilisers of sublattices. All these sublattices are examples of $\mathcal{S}$-lattices, which were completely classified by Curtis [43], as part of his work on maximal subgroups of Co_1. He defined an $\mathcal{S}$-lattice as a sublattice spanned by vectors of type 2 and 3, containing no vector which is congruent modulo 2Λ to a type 4 vector, with the additional property that if v, w are in the lattice, and are congruent modulo 2Λ, then $\frac{1}{2}(v - w)$ is also in the lattice. This somewhat technical definition ensures that the sublattice stabiliser is not obviously contained in $2^{12}{:}\text{M}_{24}$.

The full list of maximal subgroups is obtained in [174], building on the work of Curtis and others [44, 138]. (But beware of errors in virtually all the published lists: the corrected list is given in Table 5.3.) We consider some of these subgroups in more detail in the following sections.

5.5 Sublattice groups

In this section we shall study the groups Co_2, Co_3, McL and HS in more detail. We begin with the smallest, the Higman–Sims group HS, and work our way up.

5.5.1 The Higman–Sims group HS

This group was defined in Section 5.4.4 as the stabiliser of two type 3 vectors v and w with inner product -4 (so that their sum has type 2). We showed that $2{\cdot}\text{Co}_1$ is transitive on such pairs of vectors, so that HS is well-defined up to isomorphism, and has order $44352000 = 2^9.3^2.5^3.7.11$. Moreover, if we take $v = (5, 1, 1^{22})$ and $w = (-1, -5, -1^{22})$ then we see immediately that there is an involution in the monomial group $\text{M}_{22}{:}2$ which interchanges them. Therefore HS extends to HS:2, which is in fact the full automorphism group of HS.

We can also use this monomial group $\text{M}_{22}{:}2$ to recover the original definition used by Donald Higman and Charles Sims [75]. Consider the type 2 vectors which have inner product 3 with v and -3 with w. It is straightforward to show that there are just 100 such vectors, falling into three orbits under the permutation group M_{22}: the single vector $(4, 4, 0^{22})$, 22 vectors of shape $(1, 1, -3, 1^{21})$ and 77 vectors of shape $(2, 2, 2^6, 0^{16})$. They are mapped to their negatives by the outer automorphism of M_{22}. The number of orbits of the point stabiliser is called the *rank*, so this action of HS on 100 points has rank 3.

Now $|\text{HS}| = 100|\text{M}_{22}|$, so HS acts transitively on these 100 vectors. Thus if we construct a graph on 100 vertices, joining two vertices when the corresponding vectors have inner product 1, then we obtain a regular graph of

Table 5.3. Maximal subgroups of the Conway groups

Co_1	Co_2	Co_3	McL	HS
Co_2	McL	McL:2		
Co_3	HS:2	HS		
$2^{11}{:}\mathrm{M}_{24}$	$2^{10}{:}\mathrm{M}_{22}{:}2$	$2 \times \mathrm{M}_{12}$	M_{22}	M_{22}
	M_{23}	M_{23}	M_{22}	M_{11}
			M_{11}	M_{11}
$\mathrm{PSU}_6(2){:}S_3$	$\mathrm{PSU}_6(2){:}2$			
	$\mathrm{PSU}_4(3).D_8$	$\mathrm{PSU}_4(3).2^2$	$\mathrm{PSU}_4(3)$	
		$\mathrm{PSU}_3(5){:}S_3$	$\mathrm{PSU}_3(5)$	$\mathrm{PSU}_3(5){:}2$
				$\mathrm{PSU}_3(5){:}2$
		$\mathrm{PSL}_3(4).D_{12}$	$\mathrm{PSL}_3(4){:}2$	$\mathrm{PSL}_3(4){:}2$
				S_8
$3^{1+4}{:}\mathrm{Sp}_4(3){:}2$	$3^{1+4}{:}2^{1+4}.S_5$	$3^{1+4}{:}4S_6$		
$5^{1+2}{:}\mathrm{GL}_2(5)$	$5^{1+2}{:}4S_4$			
			$5^{1+2}{:}3{:}8$	
		$2^4{\cdot}A_8$	$2^4{:}A_7$	$2^4{\cdot}S_6$
			$2^4{:}A_7$	
$3{\cdot}\mathrm{Suz}{:}2$				
$(A_4 \times G_2(4)){:}2$				
$(A_5 \times \mathrm{J}_2){:}2$				
$(A_6 \times \mathrm{PSU}_3(3)){:}2$				
$(A_7 \times \mathrm{PSL}_3(2)){:}2$				
$A_9 \times S_3$		$\mathrm{PSL}_2(8){:}3 \times S_3$		
$2^{2+12}{:}(A_8 \times S_3)$				
$2^{4+12}{\cdot}(S_3 \times 3{\cdot}S_6)$	$2^{4+10}.(S_3 \times S_5)$			$4^3{:}\mathrm{GL}_3(2)$
		$A_4 \times S_5$		
		$2^2[2^7.3^2]S_3$		
$3^6{:}2{\cdot}\mathrm{M}_{12}$		$3^5{:}(2 \times \mathrm{M}_{11})$	$3^4{:}A_6{\cdot}2$	
$3^{3+4}{:}2(S_4 \times S_4)$				
$2^{1+8}{\cdot}\Omega_8^+(2)$	$2^{1+8}\mathrm{Sp}_6(2)$	$2{\cdot}\mathrm{Sp}_6(2)$	$2{\cdot}A_8$	$2 \times A_6{\cdot}2^2$
	$2^{1+6+4}A_8$			$4{\cdot}2^4.S_5$
$3^2{\cdot}\mathrm{PSU}_4(3).D_8$				
$(D_{10} \times (A_5 \times A_5)).2$				
$5^3{:}(4 \times A_5).2$				$5{:}4 \times A_5$
$5^2{:}4{\cdot}A_5$				
$7^2{:}(3 \times 2{\cdot}A_4)$				

degree 22. The 100 vertices may be labelled by $*$, together with the 22 points and 77 hexads of the Steiner system $S(3, 6, 22)$. Translating the inner product condition gives the edges in the graph: a hexad is joined to the 6 points it contains, and the 16 hexads disjoint from it.

$$
\begin{array}{ccccccc}
 & 22 & & 1 & 21 & 6 & 16 \\
\boxed{1} & \!\!\!\!\rule[2pt]{2em}{0.5pt}\!\!\!\! & \boxed{22} & \!\!\!\!\rule[2pt]{2em}{0.5pt}\!\!\!\! & & \!\!\!\!\rule[2pt]{2em}{0.5pt}\!\!\!\! & \boxed{77}
\end{array}
\tag{5.41}
$$

(Observe the similarity with the graph on 77 points acted on by M_{22}, described in Section 5.2.9.) Higman and Sims constructed the group in this way, extending the stabiliser $M_{21} \cong PSL_3(4)$ of two adjacent vertices to the edge stabiliser $M_{21}{:}2$ in order to generate the group HS.

The group HS was also constructed independently by Graham Higman [76] as the automorphism group of a kind of geometry on 176 points and 176 'quadrics', which were identified with certain sets of 50 points. It was not immediately obvious at the time that this gave the same group as the Higman–Sims construction, but this is clear from the Leech lattice approach. The Higman geometry can be seen by looking at the type 2 vectors which have inner product 2 with v and -3 with w, or vice versa. The former fall naturally into 176 pairs, such that the sum of any pair is $-w$. We shall call these pairs 'points'. The latter similarly fall into 176 pairs with sum $-v$. We shall call these pairs 'quadrics'.

Explicitly, using the permutation group M_{22} as before, the 'points' are the pairs

$$
\begin{array}{|cc|cc|}
\hline
2 & 2 & & \\
\hline
2 & 2 & & \\
2 & 2 & & \\
2 & 2 & & \\
\hline
\end{array}
\qquad
\begin{array}{|cc|cc|cc|}
\hline
1 & 3 & -1 & 1 & 1 & 1 \\
\hline
1 & -1 & -1 & 1 & 1 & 1 \\
1 & -1 & -1 & 1 & 1 & 1 \\
1 & -1 & -1 & 1 & 1 & 1 \\
\hline
\end{array}
\tag{5.42}
$$

and similarly the 'quadrics' are the pairs

$$
\begin{array}{|cc|cc|}
\hline
-2 & & -2 & \\
\hline
-2 & & -2 & \\
-2 & & -2 & \\
-2 & & -2 & \\
\hline
\end{array}
\qquad
\begin{array}{|cc|cc|cc|}
\hline
-3 & -1 & 1 & -1 & -1 & -1 \\
\hline
1 & -1 & 1 & -1 & -1 & -1 \\
1 & -1 & 1 & -1 & -1 & -1 \\
1 & -1 & 1 & -1 & -1 & -1 \\
\hline
\end{array}
\tag{5.43}
$$

Thus the points and quadrics correspond to the two orbits of M_{22} on 'heptads': that is, octads of $S(5,8,24)$ which contain exactly one of the two fixed points of M_{22}. Two heptads in different orbits intersect in either 0, 2 or 4 points. If the intersection has size 0 or 4 then the inner products in the Leech lattice are $\{0, 0, -2, -2\}$, while if the intersection has size 2, the inner products are all -1. For a fixed heptad there are 50 of the former type and 126 of the latter.

Many of the maximal subgroups of HS are visible inside $2^{12}{:}M_{24}$, by choosing the fixed vectors of HS appropriately. For example, choosing the pair $\{(5, 1^{11}, 1^{12}), (-2, -2^{11}, 0^{12})\}$ exhibits (two classes of) M_{11}, while the pair

$$
\begin{array}{|ccc|cc|}
\hline
-2 & 2 & 4 & & \\
\hline
2 & 2 & & & \\
2 & 2 & & & \\
2 & 2 & & & \\
\hline
\end{array}
\qquad
\begin{array}{|ccc|cc|}
\hline
-2 & 2 & -4 & & \\
\hline
-2 & -2 & & & \\
-2 & -2 & & & \\
-2 & -2 & & & \\
\hline
\end{array}
\tag{5.44}
$$

exhibits $2^4 {\cdot} S_6$. Similarly, the pair

-2	2	4			
2	2				
2	2				
2	2				

2	-2				
-2	-2	4			
-2	-2				
-2	-2				

$$(5.45)$$

exhibits $4^3{:}L_3(2)$, while

		2	2	2	
2	2	2			
2	2	2			
2	2	2			

		-2	2	2	
-2	-2	-2			
-2	-2	-2			
-2	-2	-2			

$$(5.46)$$

exhibits $2 \times A_6 {\cdot} 2^2$. The pair $\{(5, 1^{15}, 1^8), (-5, -1^{15}, 1^8)\}$ exhibits a subgroup A_8 which has index 2 in a maximal subgroup S_8. The odd permutations here interchange the standard coordinate frame with the one determined by $(0, 0^{15}, -6, 2^7)$.

It is possible, but not easy, to prove simplicity of the Higman–Sims group using Iwasawa's Lemma. Instead, consider the primitive action of HS on 100 points, in which the point stabiliser is M_{22}, which we already know is simple. Therefore any non-trivial proper normal subgroup N of HS is transitive on the 100 points and intersects M_{22} trivially. Thus N has order 100, so has a characteristic subgroup of order 25, which is also normal in HS. This contradiction proves that HS is simple.

One interesting facet of the Higman–Sims group which is not immediately obvious from the Leech lattice viewpoint is the existence of a proper double cover $2{\cdot}$HS. To prove this, we first show that in the permutation representation on 100 points there is an involution with no fixed points. Then these involutions lift to elements of order 4 in the Schur double cover of A_{100}, so we have an embedding of $2{\cdot}$HS in $2{\cdot}A_{100}$. Take the copy of HS which fixes the pair of vectors in (5.44). Then the 100 vectors fall into four orbits under $2^4 S_6$, with lengths 2, 6, 32 and 60, and representatives

2	2				
2	2				
2	2				
2	2				

			4		
4					

$$(5.47)$$

1	1	3	-1	-1	-1
1	1	-1	-1	-1	-1
1	1	-1	-1	-1	-1
1	1	-1	-1	-1	-1

		2	2		
		2	2		
2	2				
2	2				

It is then easy to see that the involution in $2^{12}{:}M_{24}$ which negates the top row and the third column of the MOG, and then swaps the first two columns and swaps the last two columns, lies in HS and permutes these 100 vectors fixed-point-freely.

Other interesting facts about the Higman–Sims group are that the double cover $2{\cdot}$HS has a pair of mutually dual 28-dimensional representations over $\mathbb{F}_5$, related to its embedding in the 56-dimensional representation of $2{\cdot}E_7(5)$ (see Section 4.12.3). These are all properties shared by the Rudvalis group (see Section 5.9.3), but as far as I know there is no evidence that this is anything more than a coincidence!

The group $2{\cdot}$HS is also contained in the Harada–Norton group, which we shall meet later on, in Section 5.8.8.

5.5.2 The McLaughlin group McL

The McLaughlin group was defined in Section 5.4.4 as the stabiliser of two type 2 vectors v and w with inner product -1 (so that $v + w$ has type 3). We proved that $2{\cdot}\mathrm{Co}_1$ is transitive on such (ordered) pairs of vectors, so that McL is well-defined, and has order $898128000 = 2^7.3^6.5^3.7.11$. Moreover, this shows that McL has an outer automorphism swapping the vectors v and w.

Let us begin by proving simplicity of McL, along the same lines as the proof of simplicity of the Higman–Sims group in Section 5.5.1. If we take the fixed type 2 vectors v and w to be

4	4				

1	−3	1	1	1	1
1	1	1	1	1	1
1	1	1	1	1	1
1	1	1	1	1	1

$$(5.48)$$

then we see a subgroup M_{22} of permutations. Now straightforward calculations show that there are 2025 vectors of type 2 which are perpendicular to v and have inner product 2 with w, falling into four orbits of lengths $1 + 330 + 462 + 1232$ under M_{22}, with representatives as follows:

4	−4				

		2	2		
		2	2		
		2	2		
		2	2		

$$(5.49)$$

2	−2				
2	−2				
2	2				
2	2				

1	−1	3	−1	−1	−1
−1	1	1	1	1	1
−1	1	1	1	1	1
−1	1	1	1	1	1

Therefore McL acts primitively on these 2025 points, and since $|\text{McL}| = 2025|\text{M}_{22}|$ the point stabiliser is M_{22}, which we already know is simple.

So any non-trivial normal subgroup of McL is regular on these $2025 = 3^4.5^2$ points. But any group of order $3^4.5^2$ has a characteristic subgroup of order 5 or 5^2, which would thus be normal in McL but not regular. This contradiction implies that McL is simple.

We also showed that for a given type 3 vector $v + w$ there are exactly 276 pairs of type 2 vectors with sum $v + w$. Now consider all the other pairs $\{v', w'\}$ of type 2 vectors whose sum is $v + w$. Straightforward calculation with inner products shows that the inner product of v' with v is either 1 or 2. So we might as well assume it is 2, and enumerate the 275 such vectors v'. With v and w as above, there are three orbits of v' under M_{22}, with lengths 22, 77, and 176, and representatives as follows:

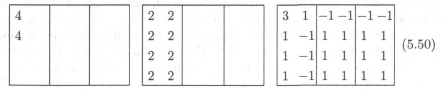

$$(5.50)$$

On the other hand, if we take the fixed type 2 vectors v and w to be

$$(5.51)$$

the vectors fall into four orbits, of lengths $2 + 56 + 105 + 112$, under the permutation group $\text{L}_3(4){:}2$, as follows:

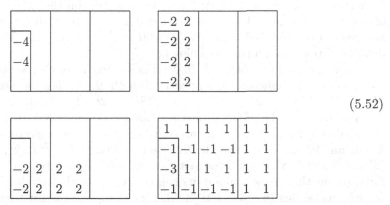

$$(5.52)$$

Thus McL acts transitively on these 275 vectors.

Now construct a graph with 275 vertices corresponding to these vectors, joining two vertices by an edge when the vectors have inner product 1. We obtain a regular graph of degree 112, which may be summarised in the following diagram.

$$(5.53)$$

It can be shown that the vertex stabiliser is $PSU_4(3) \cong P\Omega_6^-(3)$. Certainly we know it has the right order, since $|McL| = 275|PSU_4(3)|$. Moreover, the 112 neighbours of the fixed vertex correspond to the isotropic 1-spaces in the orthogonal 6-space, joined in the graph if they are perpendicular to each other. McLaughlin constructed his group originally [134] by explicitly constructing the graph in terms of $PSU_4(3)$ concepts, and proving that the automorphism group of the graph is transitive on the vertices.

Most of the maximal subgroups of the McLaughlin group can be seen as stabilisers of particular vectors in the Leech lattice representation. We have already seen such subgroups $PSU_4(3)$ and $PSL_3(4):2$, as well as M_{22}. There is another conjugacy class of subgroups M_{22}, obtained by interchanging the roles of the two fixed type 2 vectors v and w. Similarly we may take $v = (-3, 1^{11}, 1^{12})$ and $w = (3, -1^{11}, 1^{12})$ to reveal a maximal subgroup M_{11}, and $v = (-3, 1^7, 1^{16})$ and $w = (-2, -2^7, 0^{16})$ or vice versa to reveal two conjugacy classes of maximal subgroups $2^4:A_7$. The full list of maximal subgroups, calculated by Finkelstein [60] is given in Table 5.3.

One other important fact about the McLaughlin group which we cannot prove here is that it has a triple cover $3\cdot McL$, which is a subgroup of the Lyons group Ly (see Section 5.9.5). This triple cover has a 45-dimensional unitary representation over $\mathbb{F}_{25}$.

5.5.3 The group Co_3

In Section 5.4.4 we showed that $2\cdot Co_1$ is transitive on the 16773120 vectors of norm 6 in the Leech lattice, and the stabiliser of any such vector therefore has order $|Co_1|/8386560 = 495\,766\,656\,000 = 2^{10}.3^7.5^3.7.11.23$. This group is denoted Co_3 and is simple (see below).

From Table 5.2 we obtain the following subgroups of Co_3 as stabilisers of vectors in the 23-dimensional representation obtained from the Leech lattice: $McL:2$, HS, $PSU_4(3):2^2$, $3^5:(M_{11} \times 2)$, $PSU_3(5):2$, M_{23}, $2^4\cdot A_8$, $2 \times M_{12}$ and $PSL_3(4):D_{12}$. The last four groups in this list are obtained from stabilisers of coordinate frames (crosses): for simplicity, take the standard coordinate frame, and let the vector fixed by Co_3 be respectively $(5, 1^{23})$, $(2^7, -2, 4, 0^{15})$, $(2^{12}, 0^{12})$ and $(-3^3, 1^{21})$. In the last case we only see a subgroup $PSL_3(4):S_3$ directly, but there is also an automorphism swapping the standard cross with the one containing $(5, -3, -3, 1^{21})$. All nine groups in the list turn out to be maximal except $PSU_3(5):2$, which is contained in a maximal subgroup $McL:2$. There is also another (non-conjugate) subgroup $PSU_3(5)$ whose normaliser is a maximal subgroup $PSU_3(5):S_3$. This group is the stabiliser of a particular type of 3-dimensional sublattice in the Leech latttice (see Section 5.4.5).

The largest subgroup of Co_3 is McL:2, which is obtained by fixing a pair of type 2 vectors whose sum is equal to the type 3 vector fixed by Co_3. There are just 276 such pairs, giving rise to a transitive permutation representation of Co_3 on 276 points. Indeed, we have already seen that the point stabiliser McL:2 is transitive on the remaining 275 points, so Co_3 acts 2-transitively on 276 points.

These 276 pairs of vectors form an interesting combinatorial structure known as a 2-*graph*. This is a hypergraph in which the hyperedges contain three vertices, with the additional property that on any set of four vertices there is an even number of hyperedges. The easiest way to think of this hypergraph is that the hyperedges containing a given vertex are just the edges of the McLaughlin graph (augmented by the fixed vertex).

To prove simplicity of Co_3, note that the action on 276 points is 2-transitive, so primitive. Moreover, the point stabiliser is McL:2, so any proper non-trivial normal subgroup N of Co_3 intersects it either trivially or in McL. In the former case, N is regular of order $276 = 23.2^2.3$, so N has a characteristic subgroup of order 23, which is normal in Co_3 (contradiction). In the latter case, N has index 2 in Co_3, and splits the 552 vectors listed above into 276 'left-hand' vectors and 276 'right-hand' vectors. Two vectors would be on the same side if and only if their inner product was some (specified) fixed value. But we have already seen in the action of McL on 275 points that this does not happen, since the contruction of the graph (5.53) uses the fact that two 'left-hand' vectors can have either of the two possible inner products.

The maximal subgroups, determined by Finkelstein [60], are listed in Table 5.3.

5.5.4 The group Co_2

The second Conway group is defined as the stabiliser of a type 2 vector in the Leech lattice. In Section 5.4.4 we showed that $2^{\cdot}Co_1$ is transitive on such vectors, and deduced that Co_2 has order $42\,305\,421\,312\,000 = 2^{18}.3^6.5^3.7.11.23$.

By taking the fixed vector of Co_2 to be $(4, -4, 0^{22})$ we see a subgroup $2^{10}{:}M_{22}{:}2$ of index 46575 fixing the standard cross, and therefore fixing the pair $\pm(4, 4, 0^{22})$ of minimal vectors orthogonal to $(4, -4, 0^{22})$. The 46575 points permuted by Co_2 may therefore be identified with the pairs of minimal vectors orthogonal to the fixed vector of Co_2. These are easily seen to be as follows.

Orbit representative	Orbit length	
$\pm(4, 4, 0^{22})$	1	
$\pm(0, 0, 4^2, 0^{20})$	$22.21 = 462$	
$\pm(2, 2, 2^6, 0^{16})$	$77.2^5 = 2464$	(5.54)
$\pm(0, 0, 2^8, 0^{14})$	$330.2^6 = 21120$	
$\pm(1, 1, -3, 1^{21})$	$22.2^{10} = 22528$	

It is easy to deduce that the action of Co_2 on 46575 points is primitive (and therefore the subgroup $2^{10}{:}M_{22}{:}2$ is maximal). Moreover, easy calculations show that $2^{10}{:}M_{22}{:}2$ (and therefore also Co_2) is generated by conjugates of the normal subgroup 2^{10}, and it is clear that this subgroup lies in the derived group. Therefore Co_2 is simple by Iwasawa's Lemma.

Taking the fixed vector of Co_2 as $(2^8, 0^{16})$ we see a subgroup $2^5.2^4.A_8$ which however is not maximal: it contains a central involution negating the last 16 coordinates, and the full centraliser of this involution has shape $2^{1+8}Sp_6(2)$. Taking the fixed vector as $(-3, 1^{23})$ we see the maximal subgroup M_{23}.

Other maximal subgroups may be obtained by fixing other vectors. For example, consider the type 2 vectors which have inner product -2 with the fixed vector. If our fixed vector is $(4, 4, 0^{22})$, then we see that the 4600 such vectors are the images under $2^{10}{:}M_{22}{:}2$ of the following vectors.

Orbit representative	Orbit length
$(-4, 0, 4, 0^{21})$	88
$(-2, -2, 2^6, 0^{16})$	$77.32 = 2464$
$(-3, -1, -1^7, 1^{15})$	$2^{11} = 2048$

(5.55)

On the other hand if the fixed vector is $(-3, 1^{23})$ then they are the images under M_{23} of the following vectors.

Orbit representative	Orbit length
$(4, -4, 0^{22})$	23
$(1, -3, 1^{22})$	23
$(0, -2^8, 0^{15})$	506
$(-3, -1^8, 1^{15})$	506
$(2, 2, -2^6, 0^{16})$	$253.7 = 1771$
$(-1, 3, -1^6, 1^{16})$	$253.7 = 1771$

(5.56)

But no subsum of $88 + 2464 + 2048$ is equal to any subsum of $23 + 23 + 506 + 506 + 1771 + 1771$, so Co_2 is transitive on these 4600 vectors. As these vectors come in pairs whose difference is the fixed vector of Co_2, we see that Co_2 has a subgroup of index 2300. This is in fact isomorphic to $PSU_6(2){:}2$, and is the largest proper subgroup.

In this action of Co_2 on these 2300 points the point stabiliser has orbits $1 + 891 + 1408$. It is possible to construct Co_2 from first principles, using the action of $PSU_6(2){:}2$ on the 891 totally isotropic 3-spaces in the unitary 6-space, and the action by conjugation on 1408 conjugates of a subgroup $PSU_4(3){:}2^2$. (Compare the constructions of the Higman–Sims and McLaughlin groups described above.)

Next consider the type 2 vectors with inner product -1 with the fixed vector (i.e. the type 2 vectors whose sum with the fixed vector has type 3). There are 47104 such vectors, and a similar argument to the above shows that Co_2 is transitive on these vectors, with point stabiliser McL of order $|Co_2|/47104 = 898128000 = 2^7.3^6.5^3.7.11$. Indeed, this follows directly from Section 5.4.4, where we showed that McL is the stabiliser in $2{\cdot}Co_1$ of such a configuration of vectors.

Similarly there are 476928 pairs of type 3 vectors whose difference is the fixed vector of Co_2. Again the group acts transitively on these vectors, with stabiliser HS, the Higman–Sims group, of order 44352000. The two vectors of the pair can be interchanged, giving a maximal subgroup HS:2 of Co_2. A full list of maximal subgroups, determined in [173], is given in Table 5.3.

5.6 The Suzuki chain

We turn now to the description of another family of subgroups of $2 \cdot Co_1$, which act irreducibly on the Leech lattice and cannot be described as stabilisers of sublattices.

It turns out that there is cyclic subgroup of order 3 in Co_1, generated by an element in class $3D$ (in Atlas [28] notation), whose normaliser in Co_1 is $A_9 \times S_3$. If we take a descending chain of subgroups $A_9 > A_8 > \cdots > A_4 > A_3$ we obtain a corresponding increasing chain of centralisers

$$S_3 < S_4 < \mathrm{PSL}_3(2) < \mathrm{PSU}_3(3) < \mathrm{J}_2 < G_2(4) < 3 \cdot \mathrm{Suz}, \qquad (5.57)$$

where J_2 is the second Janko group (also known as the Hall–Janko group), of order 604800, and Suz is the Suzuki sporadic group, of order $448\,345\,497\,600$. The corresponding normalisers are respectively

$$
\begin{aligned}
&A_9 \times S_3, \\
&A_8 \times S_4, \\
&(A_7 \times \mathrm{PSL}_3(2)){:}2, \\
&(A_6 \times \mathrm{PSU}_3(3)){:}2, \\
&(A_5 \times \mathrm{J}_2){:}2, \\
&(A_4 \times G_2(4)){:}2, \\
&3 \cdot \mathrm{Suz}{:}2,
\end{aligned}
\qquad (5.58)
$$

all except the second of which are maximal subgroups of Co_1. The group $A_8 \times S_4$ is contained in a maximal subgroup of shape $2^{2+12}{:}(S_3 \times A_8)$. The double covers of these groups, in $2 \cdot Co_1$, have the shapes

$$
\begin{aligned}
&2 \cdot A_9 \times S_3, \\
&2 \cdot A_8 \times S_4, \\
&(2 \cdot A_7 \times \mathrm{PSL}_3(2)){:}2, \\
&(2 \cdot A_6 \times \mathrm{PSU}_3(3)){:}2, \\
&(2 \cdot A_5 \circ 2 \cdot \mathrm{J}_2){:}2, \\
&(2 \cdot A_4 \circ 2 \cdot G_2(4)){:}2, \\
&6 \cdot \mathrm{Suz}{:}2.
\end{aligned}
\qquad (5.59)
$$

Although we cannot prove these facts here (some will be proved later on), they provide motivation for our constructions of the sporadic groups Suz and J_2. The central element of order 3 in $6 \cdot \mathrm{Suz}$ acts on the Leech lattice without fixed points, and therefore we can identify it with scalar multiplication by $\omega = e^{2\pi i/3} \in \mathbb{C}$. This gives the Leech lattice a 12-dimensional complex structure, and the automorphism group of this new structure is 6·Suz. Similarly, in the group $2 \cdot A_5 \circ 2 \cdot J_2$ the normal subgroup $2 \cdot A_5$ consists entirely of fixed-point-free elements. However, it is non-abelian, so cannot be embedded in $\mathbb{C}$. It turns out that it can be embedded in the quaternions, and gives the Leech lattice a 3-dimensional quaternionic structure. We discuss this case first, and then the 6-dimensional quaternionic lattice for $2 \cdot G_2(4)$, before moving on to the Suzuki group. Finally in Section 5.6.12 we show how to use the octonions to unify all these constructions.

5.6.1 The Hall–Janko group J_2

There are two important constructions of the Hall–Janko group J_2. Historically, the first is Marshall Hall's original construction as a permutation group on 100 points. In fact J_2:2, the automorphism group of J_2, is the automorphism group of a graph on $1+36+63$ points. The second (which we treat first) is a construction of the double cover $2 \cdot J_2$ as a quaternionic reflection group in 3 dimensions. There is a corresponding 'lattice' which can be interpreted as a version of the Leech lattice, thereby giving an embedding of $(A_5 \times J_2)$:2 in the Conway group Co_1.

5.6.2 The icosians

The icosian ring is a remarkable ring of quaternions, discovered by Hamilton, and used by him in the design of his 'icosian game'. This game was based on finding Hamilton cycles on a dodecahedron and is thereby responsible for Hamilton's name being attached to the concept of a cycle which visits every vertex of a graph. The ring, which we denote $\mathbb{I}$, is generated under addition and multiplication by the quaternions i and $\frac{1}{2}(1 + \sigma i + \tau k)$, where $\sigma = \frac{1}{2}(1 - \sqrt{5})$ and $\tau = \frac{1}{2}(1 + \sqrt{5})$. The group of units of norm 1 is isomorphic to $\mathrm{SL}_2(5)$, and consists of the images, under sign-changes and even permutations of $\{1, i, j, k\}$, of the quaternions

$$1 \qquad \text{(8 of these)},$$
$$\tfrac{1}{2}(1 + i + j + k) \quad \text{(16 of these)},$$
$$\tfrac{1}{2}(i + \sigma j + \tau k) \quad \text{(96 of these)},$$

making 120 units in all. (This group $\mathrm{SL}_2(5) \cong 2 \cdot A_5$ is sometimes known as the *binary icosahedral group*, since it is a double cover of the rotation group of the icosahedron.)

The purely imaginary units of norm 1 mark the 30 midpoints of the edges of a regular dodecahedron (or icosahedron). The full group of units is $2 \cdot A_5 \times \mathbb{Z}$, generated by the units of norm 1 together with σ. There are 600 icosians of norm 2, and these partition naturally into five sets of 120 under left multiplication by units, and into another five sets of 120 under right multiplication by units. Thus by closing under addition we obtain five left ideals and five right ideals of index 2^4. The intersection of any of these left ideals with any of the right ideals contains exactly 24 of the icosians of norm 2. Since a left ideal is a submodule of the ring considered as a left module over itself, these left ideals have the shape $\mathbb{I}x$ for some x of norm 2. We therefore also say these ideals have *norm* 2. The five left ideals can be generated by $\frac{1}{2}(-\sqrt{5}+i+j+k)$ and its images under changing the sign on an *odd* number of $\{i,j,k\}$. The five right ideals can be generated by the quaternion conjugates of these.

5.6.3 The icosian Leech lattice

Let $\mathbb{I}^3$ denote the left $\mathbb{I}$-module $\{(x,y,z) \mid x,y,z \in \mathbb{I}\}$ with scalar multiplication defined by $\lambda(x,y,z) = (\lambda x, \lambda y, \lambda z)$ for all $\lambda \in \mathbb{I}$. Let Λ denote the submodule of vectors (x,y,z) satisfying

$$x \equiv y \equiv z \mod L_1,$$
$$x + y + z \equiv 0 \mod L_2, \tag{5.60}$$

where L_1 and L_2 are any two distinct ones of the five left ideals just defined. For definiteness, let us take $h = \frac{1}{2}(-\sqrt{5}+i+j+k)$ and $L_1 = \mathbb{I}h$ and $L_2 = \mathbb{I}\overline{h}$.

It is straightforward to check that Λ is invariant under all coordinate permutations, and under right-multiplication by the diagonal matrices $\mathrm{diag}(i,j,k)$ and $\mathrm{diag}(\omega,\omega,\omega)$, where $\omega = \frac{1}{2}(-1+i+j+k)$. Together these generate a monomial group of shape $2^{3+4}{:}(3 \times S_3)$.

The minimal vectors in Λ have norm 4 and fall into just four orbits under the *right* action of the monomial group and *left* scalar multiplication, represented by the following vectors.

Vector	Number	
$(2,0,0)$	$120 \times 3 =$	360
$(0,h,h)$	$120 \times 24 =$	2880
$(\overline{h},1,1)$	$120 \times 192 =$	23040
$(1,\tau\omega,\sigma\overline{\omega})$	$120 \times 96 =$	11520
	Total: $120 \times 315 =$	37800

(5.61)

If r is any of these vectors, define *reflection* in r by analogy with real reflections: it is the map $x \mapsto x - 2(x.r)(r.r)^{-1}r$, where $x.r$ denotes the standard quaternion inner product $x_1\overline{r_1} + x_2\overline{r_2} + x_3\overline{r_3}$, so that in particular $r.r = 4$ and the reflection is

$$x \mapsto x - \tfrac{1}{2}(x.r)r. \tag{5.62}$$

I claim that if $x \in \Lambda$ then $x.r \in 2\mathbb{I}$. For Λ is generated as a left $\mathbb{I}$-module by the three vectors $(2,0,0)$, $(h,h,0)$ and $(\overline{h},1,1)$, and all inner products of these three vectors are obviously in $2\mathbb{I}$ except for $(h,h,0).(\overline{h},1,1) = h^2 + h = -2\sigma\overline{\omega}$. Sesquilinearity of the inner product now proves the claim.

It follows that reflection in r maps x to $x - \lambda r$, where $\lambda \in \mathbb{I}$, and therefore preserves Λ. The group generated by these reflections contains -1, and we can define J_2 as the quotient of this group by $\{\pm 1\}$. See [164] for more on the icosian Leech lattice, and [24] for quaternionic reflection groups in general.

5.6.4 Properties of the Hall–Janko group

To calculate the group order we need an intrinsic definition of a suitable concept, such as a coordinate frame, whose stabiliser we can explicitly determine. It is not as easy as in the commutative cases, as congruence classes invariant under the right action of the group are not invariant under left action of the scalars.

If $R = h\mathbb{I}$ is any of the five right ideals of norm 2 defined in Section 5.6.2, and v_1, $v_2 \in \Lambda$, we say that v_1 and v_2 are *congruent modulo R* (or *modulo h*) if there is a vector $w \in \Lambda$ with $v_1 - v_2 = hw$. Clearly this is independent of the generator h chosen for the ideal R. We wish to determine the congruence classes of vectors of norm 4. Since the reflection group is transitive on these vectors, up to scalar multiplication, we may restrict attention to scalar multiples of $(2, 0, 0)$. With $h = \frac{1}{2}(-\sqrt{5} + i + j + k)$ as above, we see that $(2, 0, 0)$ is congruent modulo h to $(0, \pm h, \pm h)$, and is congruent modulo $\overline{h}$ to $(0, 2, 0)$ and $(0, 0, 2)$. Similarly it is congruent modulo the other ideals to suitable scalar multiples of all the norm 4 vectors it is orthogonal to.

On the other hand, $v_1 = (2, 0, 0)$ cannot be congruent to any vector v_2 of shape $(\overline{h}, 1, 1)$ or $(1, \tau\omega, \sigma\overline{\omega})$, as $v_1 - v_2$ would have a coordinate of unit norm, whereas the coordinates of hw have norm divisible by 2. Similarly $(2, 0, 0)$ cannot be congruent to any vector of shape $(h, h, 0)$ unless it is orthogonal to it. Moreover, the only scalar multiples allowed are an appropriate conjugate of the quaternion group $\{\pm 1, \pm i, \pm j, \pm k\}$.

Therefore, by explicit verification, we see that each congruence class contains exactly 24 vectors of norm 4. Moreover, each congruence class has 15 scalar multiples, three each modulo each of the five ideals. Therefore the total number of *coordinate frames*, consisting of such a set of 15 congruence classes, is $120.315.5/24.15 = 525$.

Next we show that the stabiliser in $2 \cdot J_2$ of a coordinate frame is exactly the monomial group $2^{3+4} {:} (3 \times S_3)$ defined earlier. Equivalently, we show that the stabiliser of a congruence class is $2^{3+4} {:} S_3$. It is enough to determine the diagonal matrices fixing the congruence class, say the class of $(2, 0, 0)$ modulo $\overline{h}$. We have seen that the diagonal entries must all be ± 1, $\pm i$, $\pm j$ or $\pm k$. This amounts to a group $Q_8 \times Q_8 \times Q_8$ of order 2^9. It remains to show that once two diagonal entries are fixed, the last is determined up to sign. Now if $\operatorname{diag}(1, 1, \lambda)$ is a symmetry, then both $(\overline{h}, 1, 1)$ and $(\overline{h}, 1, \lambda)$ belong to the lattice, and therefore so does $(0, 0, 1 - \lambda)$. Hence $\lambda = \pm 1$ as required. We deduce that $2 \cdot J_2$ has order $2^8.3^2.525 = 2^8.3^3.5^2.7$, and therefore J_2 has order $2^7.3^3.5^2.7 = 604800$.

To show it is simple we can use Iwasawa's Lemma in the usual way, for example using the quaternionic reflections as generators. By explicit calculation we show that the action of J_2 on the 315 reflections has rank 6, with suborbit lengths $1 + 10 + 32 + 32 + 80 + 160$. Indeed, each suborbit is defined by the inner product with the fixed vector, being a scalar of norm 1 times 4, 0, 2σ, 2τ, $2h$ and 2 respectively. Hence the action is primitive. It is easy to see that the group is generated by the images of reflections, which are commutators, and hence J_2 is simple by Theorem 3.1.

5.6.5 Identification with the Leech lattice

The icosian ring has its usual norm, with values in $\mathbb{Z}[\sigma]$, and a corresponding inner product, also with values in $\mathbb{Z}[\sigma]$. To identify our lattice with the Leech lattice, however, we need a norm with values in $\mathbb{Z}$. We do this by composing the usual norm, and the inner product, with the $\mathbb{Z}$-linear map on $\mathbb{Z}[\sigma]$ defined by $\sigma \mapsto 0$. The new inner product is easily seen to be $\mathbb{Z}$-valued and $\mathbb{Z}$-bilinear, and positive definite, so gives rise to a $\mathbb{Z}$-norm as required. Indeed, this norm identifies the icosian ring with a scaled copy of the E_8 lattice, in such a way that the 240 roots of E_8 correspond to the 120 icosians of norm 1 and their multiples by σ.

The vectors in Λ of norm 4 under this new norm correspond to those of norm 4, $4\sigma^2$ and $2 + 2\sigma^2$ under the original norm. Thus we have $120 \times 315 = 37800$ of norm 4, and their multiples by σ, together with $120 \times 1008 = 120960$ of norm $2 + 2\sigma^2$ which are the images under the monomial group and scalar multiplication of the following vectors.

Vector	Number		
$(h, \sigma\omega h, 0)$	$120 \times 48 =$	5760	
$(1, 1, \sigma\overline{\omega}\overline{h})$	$120 \times 192 =$	23040	
$(\sigma\overline{\omega}, \sigma\overline{\omega}, \overline{h})$	$120 \times 192 =$	23040	(5.63)
$(1, \sigma\overline{\omega}, \sigma\omega^j + \omega^k)$	$120 \times 576 =$	69120	
Total:	$120 \times 1008 =$	120960	

The norms of all lattice vectors are even, since this is true of the spanning vectors, and all inner products are integers. Now we can count the vectors of small norm and use the characterisation of the Leech lattice in Theorem 5.1 to show that this is just another way of looking at the Leech lattice (see Exercise 5.20). In particular, we have that $2 \cdot A_5 \circ 2 \cdot J_2$ is a subgroup of the automorphism group $2 \cdot Co_1$ of the Leech lattice. Thus $A_5 \times J_2$, and indeed $(A_5 \times J_2){:}2$, is a subgroup of the Conway group.

5.6.6 J_2 as a permutation group

The original construction by Marshall Hall, of J_2 as a permutation group on 100 points, was obtained roughly as follows. Start with the group $PSU_3(3)$, which contains a conjugacy class $\mathcal{L}$ of 36 subgroups isomorphic to $\mathrm{PSL}_3(2)$, and a conjugacy class $\mathcal{J}$ of 63 involutions. Construct a graph Γ on the vertex set $\{*\} \cup \mathcal{L} \cup \mathcal{J}$ by joining $*$ to all vertices of $\mathcal{L}$, joining two involutions if their product has order 4, joining two copies of $\mathrm{PSL}_3(2)$ if their intersection is S_4, and joining a copy of $\mathrm{PSL}_3(2)$ to the 21 involutions it contains. The structure of the graph is summarised in the following picture.

$$
\begin{array}{ccccc}
 & 14 & & 24 & \\
\overset{}{\underset{1}{\bigcirc}} \overset{36}{\rule{1cm}{0.4pt}} \overset{1}{\underset{36}{\bigcirc}} \overset{21}{\rule{1cm}{0.4pt}} \overset{12}{\underset{63}{\bigcirc}} & & &
\end{array}
\tag{5.64}
$$

Fixing a point in the 36-orbit $\mathcal{L}$, this orbit breaks up as $1 + 7 + 7 + 21$ under the action of $\mathrm{PSL}_3(2)$, and the 63-orbit $\mathcal{J}$ breaks up as $21 + 21 + 21$. It is then possible with a moderate amount of computation to show that there is a symmetry which fixes the two orbits of length 7 pointwise, and swaps the other orbits in pairs, commuting with the action of $\mathrm{PSL}_3(2)$. As the vertex stabiliser is $\mathrm{PSU}_3(3){:}2$, of order 12096, we obtain that the order of $\mathrm{Aut}(\Gamma)$ is 1209600. But the involution centralising $\mathrm{PSL}_3(2)$ is an odd permutation, so $\mathrm{Aut}(\Gamma)$ has a subgroup of index 2, which is in fact J_2.

5.6.7 Subgroups of $\mathbf{J_2}$

The maximal subgroups of J_2 are mostly visible in one or other of the representations we have given above. These include the point stabiliser $\mathrm{PSU}_3(3)$ and the edge stabiliser $\mathrm{PSL}_3(2){:}2$ in the graph, and the coordinate frame stabiliser $2^{2+4}{:}(3 \times S_3)$ and the reflection centraliser $2^{1+4}{:}A_5$ in the icosian Leech lattice (modulo scalars). Also we have (modulo scalars in each case) $3{\cdot}\mathrm{PGL}_2(9)$ which is the normaliser of $\mathrm{diag}(\omega, \omega, \omega)$, and $A_5 \times D_{10}$ and A_5 which are the stabilisers of certain 1-dimensional subspaces, $A_4 \times A_5$ which is the normaliser of the quaternion group generated by $\mathrm{diag}(i, j, k)$ and $\mathrm{diag}(j, k, i)$, and finally the Sylow 5-normaliser $5^2{:}D_{12}$. See Table 5.4. More detailed descriptions of the maximal subgroups can be found in [179].

5.6.8 The exceptional double cover of $\mathbf{G_2(4)}$

Let $\mathbb{T}$ denote the 'tetrian' ring of quaternions, also know as the Hurwitz ring of integral quaternions, generated by i, j, k and $\omega = \frac{1}{2}(-1 + i + j + k)$, which was described in some detail in Section 4.4.1. The group of units of $\mathbb{T}$ is a group of order 24 consisting of ± 1, $\pm i$, $\pm j$, $\pm k$ and $\frac{1}{2}(\pm 1 \pm i \pm j \pm k)$, which is isomorphic to $2{\cdot}A_4$. (This group is sometimes called the *binary tetrahedral group*, since it is a double cover of the rotation group of the tetrahedron; hence our designation $\mathbb{T}$.) The (left or right) ideal generated by $1 + i$ is easily seen to be a 2-sided ideal, and is the unique ideal of index 4. The quotient $\mathbb{T}/(1+i)\mathbb{T}$ is isomorphic to $\mathbb{F}_4$ in such a way that $0, 1, \omega, \overline{\omega}$ in $\mathbb{T}$ map to $0, 1, \omega, \overline{\omega}$ in $\mathbb{F}_4$.

In Section 5.2.1 we constructed a *hexacode* of dimension 3 and length 6 over $\mathbb{F}_4$. Here we modify the construction so that the vector $(1, 1, 1, 1, 1, 1)$

Table 5.4. Maximal subgroups of J_2 and Suz

Suz	J_2
$G_2(4)$	$PSU_3(3)$
$J_2{:}2$	$GL_3(2){:}2$
$3{\cdot}PSU_4(3){:}2$	$3{\cdot}PGL_2(9)$
$(A_4 \times PSL_3(4)){:}2$	$A_4 \times A_5$
$(A_5 \times A_6){\cdot}2$	$A_5 \times D_{10}$
$(A_6 \times 3^2{:}4){\cdot}2$	
$2^{1+6}{\cdot}PSU_4(2)$	$2^{1+4}{:}A_5$
$2^{2+8}{\cdot}(A_5 \times S_3)$	$2^{2+4}{:}(3 \times S_3)$
$2^{4+6}.3{\cdot}A_6$	
$PSL_2(25)$	$5^2{:}D_{12}$
A_7	A_5
$PSU_5(2)$	
$3^5{:}M_{11}$	
$M_{12}{:}2$	
$3^{2+4}{:}(2A_4 \times 2^2).2$	
$PSL_3(3){:}2$	
$PSL_3(3){:}2$	

belongs to the code. The simplest way to do this is to multiply the coordinates respectively by $(\overline{\omega}, \omega, \overline{\omega}, \omega, \overline{\omega}, \omega)$, so that the code is spanned over $\mathbb{F}_4$ by $(1, 1, \omega, \overline{\omega}, \omega, \overline{\omega})$, $(\omega, \overline{\omega}, 1, 1, \omega, \overline{\omega})$ and $(\omega, \overline{\omega}, \omega, \overline{\omega}, 1, 1)$. If we then label the coordinates in order $\infty, 0, 1, 2, 4, 3$ we obtain a group $PSL_2(5)$ of coordinate permutations preserving the code. (You can check that the maps $t \mapsto t + 1$ and $t \mapsto -1/t$ do indeed preserve the code.) There are three orbits of this group on the non-zero hexacode words (up to scalar multiplications): one word of shape $(1, 1, 1, 1, 1, 1)$, five of shape $(1, 1, \omega, \overline{\omega}, \omega, \overline{\omega})$, and fifteen of shape $(0, 0, \overline{\omega}, \omega, \overline{\omega}, \omega)$.

Define Λ to be the subset of $\mathbb{T}^6$ consisting of all vectors $(t_\infty, t_0, t_1, t_2, t_4, t_3)$ satisfying the three conditions

$$t_k \equiv m \bmod (1 + i)\mathbb{T} \text{ for some } m,$$
$$\sum_k t_k \equiv 2m\overline{\omega} \bmod 2(1 + i)\mathbb{T}, \text{ and}$$

$$((1 + i)^{-1}t_x \bmod (1 + i)\mathbb{T})_{x \in PL(5)} \text{ is a word in the hexacode.} \quad (5.65)$$

It is straightforward to check that this lattice Λ is invariant under right multiplication by the diagonal matrices $\operatorname{diag}(i^{t_k})$, where (t_k) is a hexacode word and i^ω is interpreted as j, and so on. It is then a straightforward calculation to show that the minimal vectors of Λ have norm 8 and are the images under this diagonal group and $PSL_2(5)$ of the following vectors.

Representative		Number
$(2+2i,0,0,0,0,0)$	$6\times 24=$	144
$(2,2,0,0,0,0)$	$15\times 24\times 8=$	2880
$(0,0,1+k,1+j,1+k,1+j)$	$15\times 24\times 8\times 4\times 4=$	46080
$(i+j+k,1,1,1,1,1)$	$6\times 24\times 2^{10}=$	147456
	Total $=$	196560

(5.66)

If we double these vectors and write them in a MOG array with the entries labelled as follows:

$$
\begin{array}{cccccc}
1 & 1 & 1 & 1 & 1 & 1 \\
k & j & k & j & k & j \\
i & k & i & k & i & k \\
j & i & j & i & j & i
\end{array}
$$

(5.67)

(remember that we have multiplied the hexacode words by $(\overline{\omega},\omega,\overline{\omega},\omega,\overline{\omega},\omega)$), then we obtain the 196560 minimal vectors of the Leech lattice.

Let G be the automorphism group of this quaternionic lattice Λ, that is, the group of right multiplications by 6×6 matrices over the skew-field of fractions of $\mathbb{T}$ which preserve the lattice. We shall show in Section 5.6.9 that G is isomorphic to a double cover $2{\cdot}G_2(4)$ of $G_2(4)$. It follows that $2{\cdot}G_2(4)$ is a subgroup of $2{\cdot}\mathrm{Co}_1$, and $G_2(4)$ is a subgroup of Co_1. Indeed, this group is centralised by the group $2{\cdot}A_4$ of left-multiplications by units of $\mathbb{T}$, and normalised by the map $\omega\mapsto\overline{\omega}$ (suitably interpreted), to give a subgroup $(2{\cdot}A_4\circ 2{\cdot}G_2(4)){:}2$ in $2{\cdot}\mathrm{Co}_1$, and $(A_4\times G_2(4)){:}2$ in Co_1.

Note that the ring $\mathbb{T}$ of quaternions which we have used in the construction of $2{\cdot}G_2(4)$ is a subring of the ring $\mathbb{I}$ of icosians used in the construction of $2{\cdot}\mathrm{J}_2$ in Section 5.6.3. It follows fairly easily from this that J_2 is contained in $G_2(4)$, and its index is 416. It can be shown that this action has rank 3, with suborbit lengths 1, 100, and 315 (i.e. these are the orbit lengths of the point stabiliser J_2). This gives rise to an action of $G_2(4)$ on a graph on 416 vertices, with each vertex joined to exactly 100 others. The full automorphism group of this graph is $G_2(4){:}2$. This was exploited by Suzuki in his construction of the sporadic Suzuki group acting on a graph with 1782 vertices (see Section 5.6.11). The structure of the graph is summarised in the following picture:

(5.68)

5.6.9 The map onto $G_2(4)$

The minimal norm vectors in Λ fall naturally into 4095 congruence classes modulo $1+i$ (where v_1 and v_2 are said to be *congruent modulo* $1+i$ whenever $v_1-v_2\in(1+i)\Lambda$). Under left multiplication by ω these congruence classes fall into 1365 triples, forming 'coordinate frames'.

The stabiliser (in $\mathrm{Aut}(\Lambda)$) of a coordinate frame is a group of shape $2^{5+6}{:}(3 \times \mathrm{PSL}_2(5))$, generated by the diagonal action of the hexacode, conjugation by ω, and the permutation group $\mathrm{PSL}_2(5)$. The four orbits of this group on minimal vectors, displayed in (5.66), give rise to four orbits on coordinate frames, of lengths $1 + 20 + 320 + 1024$. Moreover, writing $\theta = \omega - \overline{\omega} = i + j + k$ for brevity, the element

$$\zeta = \frac{1}{4}(i - j)\begin{pmatrix} \theta & 1 & 1 & -1 & -1 & 1 \\ 1 & 1 & \theta & -1 & 1 & -1 \\ 1 & \theta & 1 & 1 & -1 & -1 \\ -1 & -1 & 1 & -1 & \theta & -1 \\ -1 & 1 & -1 & \theta & -1 & -1 \\ 1 & -1 & -1 & -1 & -1 & \theta \end{pmatrix} \tag{5.69}$$

preserves the quaternionic Leech lattice Λ and fuses the four orbits on coordinate frames. Thus the automorphism group of Λ acts transitively on the 1365 coordinate frames, and we deduce the order of this group. Factoring out by -1 we obtain a group of order $251596800 = 2^{12}.3^3.5^2.7.13$ which we shall show is isomorphic to $G_2(4)$.

To see this, look at the 6-dimensional $\mathbb{F}_4$-space $\Lambda/(1 + i)\Lambda$. The ordinary inner product divided by $2 + 2i$ and reduced modulo $1 + i$ gives a symplectic form over $\mathbb{F}_4$. Take the images of the following vectors as the basis $\{x_1, x_2, x_3, x_6, x_7, x_8\}$ of this 6-dimensional symplectic space:

$$\begin{aligned}
\widehat{x_1} &= (2i - 2j, 0, 0, 0, 0, 0), \\
\widehat{x_2} &= (2, 0, 0, 2, 0, 0), \\
\widehat{x_3} &= (2, 0, 0, 0, 2, 0), \\
\widehat{x_6} &= (i - j)(\omega, 1, 0, \omega, -1, 0) = (k - i, i - j, 0, k - i, j - i, 0), \\
\widehat{x_7} &= (i - j)(\omega, 0, 1, -1, \omega, 0) = (k - i, 0, i - j, j - i, k - i, 0), \\
\widehat{x_8} &= (i + j + k, 1, 1, -1, -1, 1).
\end{aligned} \tag{5.70}$$

It is straightforward to check that in the induced action, the subgroup $2^{5+6}{:}(3 \times \mathrm{PSL}_2(5))$ acts on this basis in the same way that the maximal parabolic subgroup $2^{4+6}{:}(3 \times \mathrm{PSL}_2(5)) \cong 2^{4+6}{:}\mathrm{GL}_2(4)$ of $G_2(4)$ acts on the basis described in Section 4.4.3. Moreover, the element ζ defined in (5.69) maps to the element $x_i \mapsto x_{9-i}$ of $G_2(4)$. It follows that the automorphism group of Λ is a double cover of $G_2(4)$.

5.6.10 The complex Leech lattice

Suzuki's original construction [161] of the sporadic group Suz was in terms of the permutation action on the cosets of the subgroup $G_2(4)$ of index 1782. However, there is an alternative construction in terms of the Leech lattice.

We begin by modifying the ternary Golay code (see Section 5.3.5) so that it contains the word $(1, 1, 1, 1, 1, 1, 1, 1, 1, 1, 1, 1)$. Relative to the MINIMOG there are essentially two different ways of doing this—either change sign on the top row, or change sign on two of the columns. The latter approach is probably easier, as we retain a subgroup $3^2{:}SD_{16}$ of the group $3^2{:}2S_4$ preserving the columns of the MINIMOG. For the sake of argument, let us change sign on the last two columns of the MINIMOG. In any case, the group of coordinate permutations which preserve the code is isomorphic to M_{11} (acting transitively on the 12 coordinates). We construct a monomial group of shape $2 \times 3^6{:}M_{11}$ by letting each Golay code word $(x_1, \ldots, x_{12})$ act on complex 12-space as $\mathrm{diag}(\omega^{x_1}, \ldots, \omega^{x_{12}})$, and adjoining an overall sign-change.

Now the complex Leech lattice Λ may be defined as the sublattice of $\mathbb{Z}[\omega]^{12}$ consisting of vectors $(z_1, \ldots, z_{12})$ satisfying the three conditions

$$z_i \equiv m \bmod \theta,$$

$$\sum_{i=1}^{12} z_i \equiv -3m \bmod 3\theta, \text{ and}$$

$$((z_i - m)/\theta \bmod \theta)_{1 \leqslant i \leqslant 12} \text{ is in the (new) ternary Golay code,} \quad (5.71)$$

where $\theta = \omega - \overline{\omega} = \sqrt{-3}$.

It is easy to check that the complex Leech lattice is spanned by its minimal norm vectors, which are the images under the monomial group of the following vectors. (Note that there are two orbits of M_{11} on the hexads, of lengths 22 and 110.)

Shape		Number
$(3, -3, 0^{10})$	$12.11.3^2 =$	1188
$(\theta^6, 0^6)$	$22.2.3^5 =$	10692
$(\theta^6, 0^6)$	$110.2.3^5 =$	53460
$(-2^2, 1^{10})$	$12.11.3^6 =$	96228
$(-2 - 3\omega, 1^{11})$	$2.12.3^6 =$	17496
$(-2 - 3\overline{\omega}, 1^{11})$	$2.12.3^6 =$	17496
	Total $=$	196560

$$(5.72)$$

Since these are the minimal norm vectors in the lattice, we scale the inner product, by multiplying by $2/9$, so that they have norm 4. This is the smallest scale on which the real parts of the inner products are integers.

Again, we can calculate the numbers of vectors of norms 6 and 8, and verify that they are the same as in the Leech lattice (see Exercise 5.21). Therefore, by Theorem 5.1, as a real lattice this lattice Λ is isomorphic to the Leech lattice.

Now return to considering Λ as a 12-dimensional complex lattice, and consider congruence classes of vectors modulo $\theta\Lambda$, where $\theta = \omega - \overline{\omega} = \sqrt{-3}$. Clearly any vector is congruent to its multiples by ω and $\overline{\omega}$. On the other hand, if u and v are congruent, and not scalar multiples of each other, then by multiplying by ω if necessary we may assume that the real part of their inner product is non-negative, and therefore the norm of $u - v$ is at most the sum of the norms of u and v. But if u and v are congruent, then $u - v$ has norm at least 12. In particular, the norm 4 vectors are congruent only to their multiples by ω and $\bar{\omega}$, while the norm 6 vectors are congruent only if they are scalar multiples of each other or orthogonal. Therefore the vectors of norm 0, 4 and 6 account for at least

$$1 + \frac{196560}{3} + \frac{16773120}{36} = 531441 = 3^{12} \tag{5.73}$$

congruence classes, which is all there are.

Next we need to show that the automorphism group of Λ acts transitively on the congruence classes of vectors of norm 6. This can be done for example by showing that the matrix which acts as $A = \dfrac{\theta}{3} \begin{pmatrix} 1 & 1 & 1 \\ 1 & \omega & \bar{\omega} \\ 1 & \bar{\omega} & \omega \end{pmatrix}$ on the first two columns of the MINIMOG, and as $-A$ on the last two columns, preserves the lattice.

5.6.11 The Suzuki group

The definition of the complex Leech lattice Λ shows that the monomial subgroup of its automorphism group is $2 \times 3^6{:}M_{11}$. This subgroup preserves a pair of congruence classes of norm 6 vectors, which are negatives of each other. Since there are $16773120/72 = 232960$ such pairs of congruence classes, the order of $\mathrm{Aut}(\Lambda)$ is $232960.2.3^6.7920$. On the other hand, $\mathrm{Aut}(\Lambda)$ contains a normal subgroup of scalars of order 6, generated by $-\omega$, and the quotient is a group of order $|\mathrm{Aut}(\Lambda)|/6 = 448\,345\,497\,600$. This is the sporadic Suzuki group, denoted Suz, so we have

$$\begin{aligned} |\mathrm{Suz}| &= 448\,345\,497\,600 \\ &= 2^{13}.3^7.5^2.7.11.13. \end{aligned} \tag{5.74}$$

Simplicity of Suzuki's group may be obtained by an easy application of Iwasawa's Lemma (Theorem 3.1). For example, there is a subgroup $3^5{:}M_{11}$, obtained from the monomial automorphisms of the complex Leech lattice by factoring out the scalars. This subgroup is generated by elements of order 3 which are conjugate in Suz to elements in the normal subgroup of order 3^5. The suborbit lengths for the action of Suz on the coordinate frames can readily be calculated (see Exercise 5.21), and we deduce without difficulty that the action is primitive. Therefore Suz is generated by conjugates of the given 3-element, which is a commutator. Thus simplicity of Suz follows immediately from Iwasawa's Lemma.

The embedding of the ring $\mathbb{Z}[\omega]$ into $\mathbb{T} = \mathbb{Z}[\omega, i]$ gives rise to an embedding of $G_2(4)$ in Suz. This subgroup has index 1782. It can be shown that the corresponding action of Suz on 1782 points has rank 3, with suborbit lengths 1, 416 and 1365. This action was constructed from scratch by Suzuki, along the following lines. Take $G_2(4)$ acting on (i) the 416 conjugates of a subgroup

J_2 and (ii) the 1365 conjugates of a maximal parabolic subgroup of shape $2^{2+8}{:}(3 \times A_5)$. Then restrict to J_2 and show that the orbit of length 416 breaks up as $1 + 100 + 315$, and that the orbit of length 1365 breaks up as $315 + 1050$. Finally extend to $J_2{:}2$ swapping the two fixed points, and the two orbits of length 315.

In order to prove that this group has the desired order (and is not for example A_{1782}) it is necessary to construct a graph on 1782 points and show it is fixed by both $G_2(4)$ and $J_2{:}2$. This graph may be defined in terms of $G_2(4)$ as follows. We join two copies of J_2 if they intersect in $\mathrm{PSU}_3(3)$, and join J_2 to $2^{2+8}{:}(3 \times A_5)$ if they intersect in $2^{1+4}A_5$, and join two copies H_1 and H_2 of $2^{2+8}{:}(3 \times A_5)$ if the normal 2^2 in H_1 is contained in H_2 but not in $O_2(H_2) \cong 2^{2+8}$. The structure of the graph is summarised in the following picture:

$$\tag{5.75}$$

The full proof relies on detailed knowledge of the structure of $G_2(4)$, and is omitted.

The subgroups $G_2(4)$ and $J_2{:}2$ are maximal subgroups of Suz. Some other maximal subgroups may be obtained as the stabiliser $2^{2+8}{:}(S_3 \times A_5)$ of a pair of non-adjacent vertices in the graph, or as the stabiliser $\mathrm{PSU}_5(2)$ of a minimal vector in the complex Leech lattice. The full list of maximal subgroups is given in Table 5.4. More on the complex Leech lattice and the Suzuki group can be found in [172].

5.6.12 An octonion Leech lattice

Recently I have discovered a 3-dimensional *octonionic* version of the Leech lattice [185, 186], which provides a uniform way of constructing all the Suzuki-chain subgroups. First recall the non-associative algebra of (real) octonions from Section 4.3.2, and the Coxeter–Dickson integral octonions from Section 4.4.2. In the latter section, the unit integral octonions were identified with the roots of the E_8 root system. For this section, denote these integral octonions by A_0, and define

$$
\begin{aligned}
L &= (1 + i_0)A_0, \\
R &= A_0(1 + i_0), \\
B &= \tfrac{1}{2}(1 + i_0)A_0(1 + i_0).
\end{aligned}
\tag{5.76}
$$

It follows immediately from the Moufang law $(xy)(zx) = x(yz)x$ that $LR = 2B$, and the other two Moufang laws imply that $BL = L$ and $RB = R$.

It is straightforward to check that the 240 roots of L are the 112 octonions $\pm i_t \pm i_u$ for any distinct $t, u \in \mathrm{PL}(7) = \{\infty\} \cup \mathbb{F}_7$, and the 128 octonions $\tfrac{1}{2}(\pm 1 \pm i_0 \pm \cdots \pm i_6)$ which have an odd number of minus signs. Similarly, the roots of B are $\pm i_t$ for $t \in \mathrm{PL}(7)$ together with $\frac{1}{2}(\pm 1 \pm i_t \pm i_{t+1} \pm i_{t+3})$ and $\frac{1}{2}(\pm i_{t+2} \pm i_{t+4} \pm i_{t+5} \pm i_{t+6})$ for $t \in \mathbb{F}_7$.

Let $s = \frac{1}{2}(-1 + i_0 + \cdots + i_6)$, so that $s \in L$ and $\overline{s} \in R$. Since $\overline{s} \in R$ we have $L\overline{s} \subseteq LR = 2B$, but $2B$ is spanned by its 240 roots, all of which lie in $L\overline{s}$, so $L\overline{s} = 2B$. Indeed, the same argument shows that if ρ is any root in R then $L\rho = 2B$. Hence $2L \subset 2B \subset L$, that is $2L \subset L\overline{s} \subset L$, and therefore $2L \subset Ls \subset L$. Moreover, $L\overline{s} + Ls = L$, so by self-duality of L we have $L\overline{s} \cap Ls = 2L$.

Define the *octonionic Leech lattice* $\Lambda = \Lambda_{\mathbb{O}}$ to be the set of triples (x, y, z) of octonions, with norm $N(x, y, z) = \frac{1}{2}(x\overline{x} + y\overline{y} + z\overline{z})$, such that

$$x, y, z \in L,$$
$$x + y, x + z, y + z \in L\overline{s},$$
$$x + y + z \in Ls. \tag{5.77}$$

It is clear that the definition of Λ is invariant under permutations of the three coordinates. The fact that $2L$ lies in both Ls and $L\overline{s}$ immediately shows that it is invariant under sign-changes as well. We show next that it is invariant under the map $r_t : (x, y, z) \mapsto (x, yi_t, zi_t)$. Certainly $Li_t = L$, so the first condition of the definition is preserved. Then $y(1 - i_t) \in LR = 2B = L\overline{s}$, so the second condition is preserved. Finally $(y + z)(1 - i_t) \in 2BL = 2L \subset Ls$, so the third condition is preserved also.

If λ is any root in L, then it is easy to check that $(\lambda s, \lambda, -\lambda)$ lies in Λ. Therefore Λ contains the vectors $(\lambda s, \lambda, \lambda) + (\lambda, \lambda s, -\lambda) = -(\lambda\overline{s}, \lambda\overline{s}, 0)$, that is, all vectors $(2\beta, 2\beta, 0)$ with β a root in B. Hence Λ also contains

$$(\lambda(1 + i_0), \lambda(1 + i_0), 0) + (\lambda(1 - i_0), -\lambda(1 + i_0), 0) = (2\lambda, 0, 0).$$

Applying the above symmetries it follows at once that Λ contains the following 196560 vectors of norm 4, where λ is an arbitrary root of L and $j, k \in J = \{\pm i_t \mid t \in \mathrm{PL}(7)\}$.

Vectors		Number
$(2\lambda, 0, 0)$	$3 \times 240 =$	720
$(\lambda\overline{s}, (\lambda\overline{s})j, 0)$	$3 \times 240 \times 16 =$	11520
$((\lambda s)j, \lambda k, (\lambda j)k)$	$3 \times 240 \times 16 \times 16 =$	184320

It is easy to show that these 196560 vectors span Λ. For if $(x, y, z) \in \Lambda$, then by adding suitable vectors of the third type, we may reduce z to 0. Then we know that $y \in L\overline{s}$, so by adding suitable vectors of the second type we may reduce y to 0 also. Finally we have that $x \in L\overline{s} \cap Ls = 2L$ so by adding vectors of the first type we can reduce x to 0 also.

To identify Λ with the standard Leech lattice, label the coordinates of each brick of the MOG (see Section 5.2.2) as follows.

$$
\begin{array}{|cc|}
\hline
-1 & i_0 \\
\\
i_4 & i_5 \\
\\
i_2 & i_6 \\
\\
i_1 & i_3 \\
\hline
\end{array}
$$

We have seen that the map $i_t \mapsto i_{t+1}$ is a symmetry of the standard Leech lattice. Now L is spanned by $1 \pm i_t$ and s, and it is trivial to verify that the vectors $(1 \pm i_0)(s, 1, 1)$ and $s(s, 1, 1)$ are in this Leech lattice, and therefore so are $(1 \pm i_t)(s, 1, 1)$. These together with coordinate permutations, sign-changes and addition are enough to give all the minimal vectors, which span the lattice. Therefore the octonionic Leech lattice is indeed a copy of the standard Leech lattice.

Next I shall describe some automorphisms of the octonionic Leech lattice which enable us to see the Suzuki-chain subgroups easily. The reflection in the root r (scaled to norm 1) in E_8 can be expressed by octonion multiplication as

$$ x \mapsto -r\bar{x}r. $$

Since $1 + i_t$ is perpendicular to s we have $s = -\frac{1}{2}(1 + i_t)\bar{s}(1 + i_t)$. Using R_a to denote right-multiplication by a, that is $R_a : x \mapsto xa$, the Moufang law $((xa)b)a = x(aba)$ is just $R_a R_b R_a = R_{aba}$. Therefore $R_s = -\frac{1}{2}R_{1+i_t}R_{\bar{s}}R_{1+i_t}$, which can be re-written in the equivalent forms

$$
\begin{aligned}
R_s R_{-1+i_t} &= R_{1+i_t} R_{\bar{s}}, \\
R_{-1+i_t} R_s &= R_{\bar{s}} R_{1+i_t}.
\end{aligned}
\tag{5.78}
$$

Hence

$$
\begin{aligned}
R_s R_{1-i_0} R_{1+i_t} &= R_{1+i_0} R_{1-i_t} R_s, \\
R_{\bar{s}} R_{1-i_0} R_{1+i_t} &= R_{1+i_0} R_{1-i_t} R_{\bar{s}}.
\end{aligned}
\tag{5.79}
$$

Therefore

$$
\begin{aligned}
(L(1 - i_0))(1 + i_t) &= (LR)L = 2BL = 2L, \\
((L\bar{s})(1 - i_0))(1 + i_t) &= ((L(1 + i_0))(1 - i_t))\bar{s} = 2L\bar{s}, \\
((Ls)(1 - i_0))(1 + i_t) &= ((L(1 + i_0))(1 - i_t))s = 2Ls.
\end{aligned}
\tag{5.80}
$$

In other words the map $\frac{1}{2}R_{1-i_0}R_{1+i_t}$ acting simultaneously on all three coordinates of the octonion Leech lattice preserves the lattice.

Next I shall show that these symmetries generate a double cover $2{\cdot}A_8$ of A_8. Note first that the roots $1 - i_t$ for $0 \leqslant t \leqslant 6$ form a copy of the root system of type A_7, whose Weyl group is the symmetric group S_8. The product of the reflections in $1 - i_0$ and $1 - i_t$ is the map

$$ x \mapsto \tfrac{1}{4}(1 - i_t)((1 + i_0)x(1 + i_0))(1 - i_t) $$

that is the product of two bi-multiplications $\frac{1}{2}B_{1+i_0}\frac{1}{2}B_{1-i_t}$. These elements generate the rotation part A_8 of the Weyl group, and applying the triality automorphism which takes bimultiplications $B_{\bar{u}}$ by units $\bar{u}$ of norm 1 to right-multiplications R_u by the octonion conjugate u, we see that the maps $\frac{1}{2}R_{1-i_0}R_{1+i_t}$ generate $2 \cdot A_8$, as required.

Since this group commutes with coordinate permutations and sign changes we get a group $2 \cdot A_8 \times S_4$. Adjoining the symmetry $r_t : (x, y, z) \mapsto (x, yi_t, zi_t)$ described above, this extends to a group of shape $2^{3+12}(A_8 \times S_3)$. This group is a maximal subgroup of the automorphism group $2 \cdot \mathrm{Co}_1$ of the Leech lattice, so we only need one more symmetry to generate the whole of $2 \cdot \mathrm{Co}_1$.

Regard s as a complex number $\frac{1}{2}(-1+\sqrt{-7})$ and consider the 3-dimensional vector space over $\mathbb{Q}(s) = \mathbb{Q}(\sqrt{-7})$. It is well-known (see for example [28], or Exercise 5.22) that the lattice spanned over $\mathbb{Z}[s]$ by the vectors $(\pm 2, 0, 0)$, $(\pm \bar{s}, \pm \bar{s}, 0)$ and $(\pm s, \pm 1, \pm 1)$ has 42 vectors of minimal norm, and the 21 reflections in these vectors generate the automorphism group of this lattice, which is isomorphic to $2 \times L_3(2)$. Indeed, it is generated by the monomial subgroup $2^3{:}S_3 \cong 2 \times S_4$ together with reflection in $(s, 1, 1)$, which may be represented by the negative of the matrix

$$\frac{1}{2}\begin{pmatrix} 0 & \bar{s} & \bar{s} \\ s & -1 & 1 \\ s & 1 & -1 \end{pmatrix}.$$

We now show that this matrix, interpreted as the map

$$(x, y, z) \mapsto \tfrac{1}{2}((y+z)s, x\bar{s} - y + z, x\bar{s} + y - z),$$

is also a symmetry of the octonion Leech lattice. For convenience, write (x', y', z') for the image of (x, y, z) under this map, and note that $s = \frac{1}{2}(-1+\sqrt{-7})$ satisfies $s^2 + s + 2 = 0$, so that $s + \bar{s} = -1$, $s^2 = -2 - s = \bar{s} - 1$ and $\bar{s}^2 = -2 - \bar{s} = s - 1$.

To prove the claim, observe that we may change signs arbitrarily on any of the coordinates x, y, z, x', y', z' to make the calculations easier. Note also that the third condition in the definition has a number of equivalent formulations

$$\begin{aligned} x + y + z \in Ls &\Leftrightarrow (x + y + z)\bar{s} \in 2L \\ &\Leftrightarrow (x\bar{s} - y - z) - (y + z)s \in 2L \\ &\Leftrightarrow x\bar{s} - y - z \in 2L. \end{aligned}$$

Now

(i) $x' = \frac{1}{2}(y+z)s \in L$,

(ii) $y' = \frac{1}{2}(x\bar{s} - y + z) \in L$ and similarly $z' \in L$,

(iii) $x' + y' = \frac{1}{2}(x\bar{s} + y(1 + s) + z(1 - s)) = \frac{1}{2}(x - y - z\bar{s}) \in L\bar{s}$ and similarly $x' + z' \in L\bar{s}$,

(iv) $y' + z' = x\bar{s} \in L\bar{s}$,

(v) $x'\bar{s} + y' + z' = (y + z) + x\bar{s} \in 2L$,

and the claim is proved.

At this stage we have found enough symmetries to generate the full automorphism group $2\!\cdot\!\mathrm{Co}_1$ of the Leech lattice.

The subgroup $2\!\cdot\!A_8$ generated by the maps $\frac{1}{2}R_{1-i_0}R_{1+i_t}$ contains in particular the elements

$$\frac{1}{8}R_{1-i_0}R_{1+i_1}R_{1-i_0}R_{1+i_t}R_{1-i_0}R_{1+i_t} = \frac{1}{8}R_{(1-i_0)(1+i_1)(1-i_0)}R_{(1+i_t)(1-i_0)(1+i_t)}$$
$$= \frac{1}{2}R_{i_1-i_0}R_{i_t-i_0}$$

These maps (for $t \in \{2,3,4,5,6\}$) generate a subgroup $2\!\cdot\!A_7$ which commutes with the reflection group $2 \times \mathrm{GL}_3(2)$ just described, giving rise to a subgroup $\mathrm{GL}_3(2) \times 2\!\cdot\!A_7$.

This group has index 2 in a maximal subgroup of shape $(\mathrm{GL}_3(2) \times 2\!\cdot\!A_7).2$, which may be obtained by adjoining an element such as $\frac{1}{2}R_{i_0-i_1}R_s^*$, where we define

$$R_s^* = R_s\frac{1}{2}\begin{pmatrix} s & 1 & 1 \\ 1 & s & 1 \\ 1 & 1 & s \end{pmatrix}. \tag{5.81}$$

It is left as an exercise (see Exercise 5.23) to show that this element is a symmetry of the octonionic Leech lattice.

The Suzuki chain of subgroups of $2\!\cdot\!\mathrm{Co}_1$ was given in (5.59). We have already described two groups in this list, namely the groups $2\!\cdot\!A_8 \times S_4$ and $(2\!\cdot\!A_7 \times \mathrm{GL}_3(2)).2$. To obtain $2\!\cdot\!A_9 \times S_3$, we take the S_3 of coordinate permutations, together with the group $2\!\cdot\!A_8$ which is generated by $\frac{1}{2}R_{1-i_0}R_{1+i_t}$, and extend $2\!\cdot\!A_7$ to $2\!\cdot\!S_7$ by adjoining the element $\frac{1}{2}R_{i_0-i_1}R_s^*$ as above. The map onto A_9 permuting the points $\{*, \infty, 0, 1, 2, 3, 4, 5, 6\}$ is given by mapping $\frac{1}{2}R_{1-i_t}R_{1+i_u}$ onto (∞, t, u), as we have already seen, and mapping the extra element to $(*, \infty)(0, 1)$. In terms of the root system of L, the factor R_s^* of the new element corresponds to the root s, and extends the root system of type A_7 spanned by the $1 - i_t$ to one of type A_8.

To obtain the remaining groups in the list, all we have to do is adjoin to the complex reflection group $2 \times \mathrm{GL}_3(2)$ the part of $2\!\cdot\!A_9$ which commutes with the appropriate subgroup $2\!\cdot\!A_n$.

5.7 The Fischer groups

Fischer's discovery of these groups arose out of the observation that the symmetric groups are generated by the conjugacy class of transpositions, which has the property that the product of any two transpositions has order 1, 2 or 3.

One might well have been tempted to ask whether there are any other groups with this property. The answer is yes: for example, the Weyl groups of types D_n and E_n (see Section 2.8). More generally, the orthogonal groups

$\mathrm{GO}_{2n}^{\varepsilon}(2)$ have a conjugacy class of orthogonal transvections such that any two transvections either commute or generate $\mathrm{GO}_2^-(2) \cong S_3$ (see Section 3.8). Similarly, the unitary groups over $\mathbb{F}_4$ have a class of unitary transvections, such that any two such transvections either commute or generate $\mathrm{SU}_2(2) \cong S_3$, and in the same way the symplectic transvections in $\mathrm{Sp}_{2n}(2)$ either commute or generate $\mathrm{Sp}_2(2) \cong S_3$.

Fischer [62, 63] defined a 3-*transposition group* to be a finite group G generated by a conjugacy class D of involutions, such that any two elements of D have product of order at most 3, and, for technical reasons to do with the inductive classification of such groups, such that $G' = G''$ and any normal 2- or 3-subgroup of G is central. The elements of D are known as 3-*transpositions*, or just *transpositions* if there is no risk of confusion. Fischer proved that any 3-transposition group is, modulo its centre, either S_n, $\mathrm{Sp}_{2n}(2)$, $\mathrm{PSU}_n(2)$, $\mathrm{GO}_{2n}^{\varepsilon}(2)$, an orthogonal group $\mathrm{P\Omega}_{2n}^{\varepsilon}(3){:}2$ or $\Omega_{2n+1}(3)$ or $\mathrm{SO}_{2n+1}(3)$ over $\mathbb{F}_3$, or one of three new groups, two of which (Fi_{22} and Fi_{23}) are simple and one of which (Fi_{24}) has a simple subgroup of index 2. In the case of the orthogonal groups in characteristic 3, the 3-transpositions are (modulo sign) any fixed conjugacy class of reflections.

There is no really easy way to construct the sporadic Fischer groups. Fischer's method, on which most later constructions are based, is to build a graph with vertices corresponding to the transpositions, joining two transpositions when they commute. (Some people use the complementary graph, where transpositions are joined when their product has order 3, by analogy with the Coxeter groups—see Section 2.8.) Then by many complicated calculations the automorphism group of the graph is determined. One can use the fact that the transpositions are not only vertices of the graph, but also act on the graph by permuting the vertices. Using a known subgroup to define the graph and the action of the transpositions, the crucial step is then to show that the transpositions act as graph automorphisms, i.e. they map edges to edges. Since the vertex stabiliser can be identified by induction, and the group generated by the transpositions is easily seen to be transitive in each case, the group order and other properties now follow.

5.7.1 A graph on 3510 vertices

We begin by sketching the construction of Fi_{22}, using the group $\Omega_7(3)$ to define everything we need. For technical reasons I choose a basis for the orthogonal 7-space such that the quadratic form is given by

$$Q(x_1, x_2, \ldots, x_7) = \sum_{i=1}^{6} x_i{}^2 - x_7{}^2 \tag{5.82}$$

and write vectors in the format $(x_1, x_2, \ldots, x_6; x_7)$ to emphasise that the last coordinate has norm -1 (or 2, if you prefer). The group $\Omega_7(3)$ has a permutation action on the 351 reflections in vectors of norm 2, and we turn these reflections into the vertices of a graph by joining two reflections when they commute. It is easy to see that each such reflection commutes with exactly 126 others, so the graph is regular with valency 126.

Next we take an embedding of the Weyl group of type E_7 in $GO_7(3)$, obtained by reducing the coordinates of the root vectors modulo 3: this makes sense since $\frac{1}{2} = -1 \bmod 3$. This gives rise to an embedding of $W(E_7)'$ in $SO_7(3)$, and to two conjugacy classes of subgroups $W(E_7)' \cong Sp_6(2)$ in $\Omega_7(3)$, and we choose arbitrarily one of these classes. For the sake of argument, we take our standard copy of E_7 to be defined by the 63 reflections in the vectors $(0,0,0,0,0,0;1)$, $(1,1,0,0,0,0;0)$ and $(-1,1,1,1,1,1;1)$ of norm 2, and their images under the monomial group $2^5{:}S_6$ of permutations and even signchanges on the first 6 coordinates. These correspond, in the notation of Section 3.12.4, to the roots of E_8 perpendicular to $(0,0,0,0,0,0;1,-1)$, which are the images under the monomial group of $(0,0,0,0,0,0;1,1)$, $(1,1,0,0,0,0;0,0)$ and $\frac{1}{2}(-1,1,1,1,1,1;1,1)$.

We next show that the action of $\Omega_7(3)$ on the 3159 conjugates of this subgroup $W(E_7)'$ has rank 4 and suborbit lengths $1 + 288 + 630 + 2240$. First observe that the standard copy of $W(E_7)'$ is transitive on the 288 reflections in vectors of norm 2 not contained in the standard E_7 root system: for these fall into orbits of sizes $32 + 96 + 160$ under the monomial group, represented by $(1,1,1,1,1,1;1)$, $(0,1,1,1,1,1;0)$ and $(0,0,0,1,1,1;1)$; and these orbits are easily seen to be fused by $W(E_7)'$. The 288 images of the standard E_7 under these reflections form the orbit of length 288. Clearly the intersection of one of these copies with the standard E_7 consists of the roots which are orthogonal to the reflecting vector, and it is then easy to see that they form a root system of type A_6.

The orbit of length 630 consists of 315 pairs, each associated with one of the 315 sub-root systems of type $A_1A_1A_1D_4$. To show that there are 315 such subsystems, choose the first A_1 in 63 ways, and the second in 30 ways, since it is perpendicular to the first. For example we may choose the reflecting vectors $(1,1,0,0,0,0;0)$ and $(1,-1,0,0,0,0;0)$, and then we see that the orthogonal complement is of type A_1D_4 with reflections given by $(0,0,0,0,0,0;1)$ and the 12 images of $(0,0,1,1,0,0;0)$ under the remaining monomial group. Thus we have $63.30/3! = 315$ such subsystems. To make the corresponding orbit of E_7s, apply a triality automorphism of the D_4 component such as

$$\begin{pmatrix} 1 & -1 & -1 & -1 \\ 1 & -1 & 1 & 1 \\ 1 & 1 & -1 & 1 \\ 1 & 1 & 1 & -1 \end{pmatrix} \tag{5.83}$$

to the standard E_7. It is easy to verify that the new E_7 intersects the standard one in exactly the subsystem $A_1A_1A_1D_4$ just described.

Finally, the orbit of length 2240 consists of 1120 pairs, each associated with one of the 1120 subsystems of type $A_2A_2A_2$. To count these subsystems, choose the first reflection in 63 ways, and the second one in an A_2 in 32 ways.

Since A_2 contains three reflections, there are $63.32/3! = 336$ ways to choose the first A_2. Its orthogonal complement is easily seen to be an A_5 root system. For example, if the first A_2 is spanned by $(0,0,0,0,0,0;1)$ and $(1,1,1,1,1,1;1)$ then the perpendicular A_5 consists of the roots of shape $(1,-1,0,0,0,0;0)$. So to pick the second A_2, choose the first reflection in 15 ways and the second in 8 ways, making $15.8/3! = 20$ choices for the A_2. Then the third A_2 is determined. Thus we obtain $336.20/3! = 1120$ subsystems of type $A_2A_2A_2$. Modulo 3, these systems do not span a 6-space as you might expect, but only a 5-space. The radical of the quadratic form on this 5-space has dimension 2, and therefore defines a long root subgroup of order 3 (see Section 3.7.3), which fixes the 5-space pointwise. This root subgroup maps the standard E_7 to the pair of E_7s in the orbit of length 2240. For example, if the $A_2A_2A_2$ is generated by the reflections in the six vectors

$$
\begin{aligned}
&(0,0,0,0,0,0;1),\\
&(1,1,1,1,1,1;1),\\
&(1,-1,0,0,0,0;0),\\
&(0,1,-1,0,0,0;0),\\
&(0,0,0,1,-1,0;0),\\
&(0,0,0,0,1,-1;0),
\end{aligned}
\tag{5.84}
$$

then the isotropic 2-space is spanned by $(1,1,1,0,0,0;0)$ and $(0,0,0,1,1,1;0)$ and we can apply the formula (3.33).

We make a graph on these 3159 copies of E_7 by joining each vertex to the vertices in the 630-orbit. In other words, we join them when they intersect in $A_1A_1A_1D_4$. Finally we join each copy of E_7 to the 63 reflections in its Weyl group, and join two of the 351 reflections in vectors of norm 2 in $GO_7(3)$ when they commute. Thus we obtain a regular graph of degree 693 on $351 + 3159 = 3510$ vertices, which may be summarised in the following picture:

$$
\overset{126}{\underset{351}{\bigcirc}} \overset{567}{\rule{3cm}{0.4pt}} \overset{63}{} \overset{630}{\underset{3159}{\bigcirc}}
\tag{5.85}
$$

5.7.2 The group Fi$_{22}$

It is possible to define Fi$_{22}$ as the (unique) subgroup of index 2 in the automorphism group of this graph. However, I prefer a more concrete definition. For the 3510 vertices of the graph are not just vertices, they are also 3-transpositions in the Fischer group. Thus we may, for each vertex of the graph, describe the corresponding 3-transposition as a permutation of the 3510 vertices, and prove that this permutation is actually a graph automorphism. We have two cases, corresponding to the two orbits of $\Omega_7(3)$ on vertices.

A transposition in the 351-orbit is the negative of a reflection in $\Omega_7(3)$ and acts in the natural way by conjugation on the other reflections, and on the 3159 copies of E_7. In particular it fixes the vertices it is joined to, and moves the rest. Indeed, it is immediate from the definition of the graph that these transpositions preserve the edges of the graph. Notice that if t, u are two non-commuting transpositions then the third transposition in the S_3 they generate is $t^u = u^t$.

A transposition in the 3159-orbit acts as follows: it fixes the 63 reflections it is joined to, and takes the other 288 reflections to the 288 copies of $W(\mathrm{E}_7)'$ which are its conjugates by the given reflections. Its action on the other $630 + 2240$ copies of $W(\mathrm{E}_7)'$ can be easily computed inside $\Omega_7(3)$: it fixes the vertices in the 630-orbit and interchanges the other 2240 vertices in the 1120 pairs which have already been described. It is easy to see from this definition that this transposition preserves the set of edges between the 351-orbit and the whole graph. But two vertices in the 3159-orbit are joined if and only if they have 15 common neighbours in the 351-orbit, so these edges are preserved also. Therefore the graph is preserved by all 3510 transpositions.

We can now define Fi_{22} to be the group generated by these 3510 transpositions. Since the $\Omega_7(3)$-orbit of 351 transpositions generates $\Omega_7(3)$, which is already transitive on the two orbits of 351 and 3159 vertices, the group generated by all 3510 transpositions is a transitive group of graph automorphisms.

It is still not entirely clear from this definition how to calculate the order of Fi_{22}. To do so, we shall determine the vertex stabiliser, which turns out to be a double cover of $\mathrm{PSU}_6(2)$. The proof consists of a number of steps. First we interpret the neighbours of a vertex in terms of $\Omega_7(3)$. The transpositions which fix a given vertex are the 693 which are joined to it. Taking our fixed vertex to be in the 351-orbit, say $(0, 0, 0, 0, 0, 0; 1)$, we have a visible group of symmetries isomorphic to $\Omega_6^-(3){:}2 \cong 2{\cdot}\mathrm{PSU}_4(3){:}2$. This group has orbits of lengths 126 and 567 on the 693 neighbours of the fixed vertex, and by restricting the edge-definitions from $\Omega_7(3)$ all the edges between them can be defined in terms of $\Omega_6^-(3)$-concepts.

To be more explicit, the vertices in the 126-orbit are just the reflections in $\Omega_6^-(3){:}2$ defined by vectors of norm 2. The vertices in the 567-orbit are copies of E_7 which contain the fixed reflection (in $(0, 0, 0, 0, 0, 0; 1)$), so can be identified with copies of D_6 in the perpendicular 6-space. The edges of this subgraph on $126 + 567 = 693$ vertices are as follows. Two reflections are joined if the reflecting vectors are perpendicular. A copy of D_6 is joined to the 30 reflections it contains. Now if two copies of E_7 inside $\Omega_7(3)$ intersect in $\mathrm{A}_1\mathrm{A}_1\mathrm{A}_1\mathrm{D}_4$, then we can take the fixed vector of $\Omega_6^-(3)$ either in an A_1 component or in the D_4 component. Thus the two copies of D_6 intersect in either $\mathrm{A}_1\mathrm{A}_1\mathrm{D}_4$ or $\mathrm{A}_1\mathrm{A}_1\mathrm{A}_1\mathrm{A}_1\mathrm{A}_1$, and we obtain two orbits of edges between these D_6s. In summary, the graph on 693 vertices looks like this.

$$
\overset{45}{\underset{126}{\bigcirc}} \overset{135}{\rule{4cm}{0.4pt}} \overset{30}{\rule{4cm}{0.4pt}} \overset{30+120}{\underset{567}{\bigcirc}}
\tag{5.86}
$$

Moreover, the actions of the 693 transpositions on this graph can also be defined in terms of $\Omega_6^-(3)$. The transpositions in the 126-orbit act as the corresponding reflections. Given a standard copy of D_6 in the 567-orbit, there are now four orbits, of lengths $30 + 96 + 120 + 320$, on the other 566. The ones which intersect the standard D_6 in $A_1A_1D_4$ or $A_1A_1A_1A_1A_1$ are the orbits of lengths 30 and 120 respectively. The ones in the 96-orbit intersect it in A_6, while those in the 320-orbit intersect it in A_2A_2. As before, the transposition corresponding to the standard copy of D_6 interchanges the D_6s in the 96-orbit with the corresponding reflections, and interchanges the 160 pairs of D_6s in the 320-orbit, each pair being defined by the property that they both intersect the standard copy in the same A_2A_2.

The next step is to re-interpret these vertices, edges and 3-transpositions, in terms of the complex reflection group $3 \cdot \Omega_6^-(3).2 \cong 6 \cdot \mathrm{PSU}_4(3).2$ described in Section 3.12.2. We use the second basis described there, so that the vectors of norm 8 are the images under the monomial group $3.2^5 S_6$ of $(2,2,0,0,0,0)$ and $(\theta,1,1,1,1,1)$, mapping modulo $\theta = \omega - \overline{\omega} = \sqrt{-3}$ to the vectors $(-1,-1,0,0,0,0)$ and $(0,1,1,1,1,1)$ of norm 2. The $30 + 96 = 126$ 1-spaces spanned by these vectors give rise to 126 complex reflections, which correspond naturally to the 126 reflections in vectors of norm 2 in $\Omega_6^-(3):2$. The 567 copies of D_6 correspond to 'coordinate frames' of norm 16 vectors, defined as congruence classes modulo 2Λ, where Λ is the lattice spanned over $\mathbb{Z}[\omega]$ by the norm 8 vectors. For example, the standard copy of D_6 corresponds to the standard coordinate frame consisting of vectors of shape $(\pm 4, 0, 0, 0, 0, 0)$.

The edges are as follows. Clearly two complex reflections are joined if and only if they commute, that is if and only if the corresponding vectors are perpendicular. A coordinate frame is joined to the reflections in those vectors which when written with respect to that frame have shape $(\pm 2, \pm 2, 0, 0, 0, 0)$. To describe the edges between coordinate frames, we first calculate orbit representatives for the nontrivial orbits (of lengths $30 + 96 + 120 + 320$) under the monomial group. These are as follows.

$(2, 2\theta, 0, 0, 0, 0)$	$(1, \theta, \theta, \theta, \theta, \theta)$
$(2\theta, 2, 0, 0, 0, 0)$	$(\theta, -3, 1, 1, 1, 1)$
$(0, 0, 2, 2, 2, 2)$	$(\theta, 1, -3, 1, 1, 1)$
$(0, 0, 2, 2, -2, -2)$	$(\theta, 1, 1, -3, 1, 1)$
$(0, 0, 2, -2, 2, -2)$	$(\theta, 1, 1, 1, -3, 1)$
$(0, 0, 2, -2, -2, 2)$	$(\theta, 1, 1, 1, 1, -3)$

$(2, 2, 2\omega, 2\omega, 0, 0)$	$(-2 - \theta, \theta, \theta, 1, 1, 1)$
$(2, 2, -2\omega, -2\omega, 0, 0)$	$(\theta, -2 - \theta, \theta, 1, 1, 1)$
$(0, 0, 2\overline{\omega}, -2\overline{\omega}, 2, 2)$	$(\theta, \theta, -2 - \theta, 1, 1, 1)$
$(0, 0, 2\overline{\omega}, -2\overline{\omega}, -2, -2)$	$(1, 1, 1, \theta, \theta, 2 - \theta)$
$(2\overline{\omega}, -2\overline{\omega}, 0, 0, 2\omega, -2\omega)$	$(1, 1, 1, \theta, 2 - \theta, \theta)$
$(2\overline{\omega}, -2\overline{\omega}, 0, 0, -2\omega, 2\omega)$	$(1, 1, 1, 2 - \theta, \theta, \theta)$

We now see that two coordinate frames are joined if and only if the inner products of the vectors in one frame with those in the other are divisible by

8. This completes the interpretation of the graph in terms of the complex reflection group.

The next step is to reduce this complex reflection group modulo 2, so that we can re-interpret the vertices, edges, and 3-transpositions in terms of $\mathrm{PSU}_6(2)$. We saw in Section 3.12.2 that $\Lambda/2\Lambda$ becomes a unitary 6-space over $\mathbb{F}_4$ on dividing the inner products by 4 and reducing them mod 2, to give a non-singular conjugate-symmetric sesquilinear form. Thus the $126 + 567$ vertices of the graph become the 693 isotropic 1-spaces in $\mathbb{F}_4{}^6$, and they are joined precisely when they are perpendicular with respect to this sesquilinear form. It is clear that the 693 corresponding 3-transpositions preserve the norm and fix precisely those isotropic 1-spaces they are perpendicular to. Therefore they are exactly the unitary transvections in $\mathrm{SU}_6(2)$, and it follows immediately that the vertex stabiliser in Fi_{22} contains a subgroup which acts on the 693 neighbours of that vertex as $\mathrm{PSU}_6(2)$ acts on the isotropic 1-spaces.

Since the pointwise stabiliser of this set of 693 transpositions lies inside the setwise stabiliser of the 126-orbit illustrated in (5.86), which in turn lies inside $\Omega_7(3)$, it is easy to see that it is generated by the single transposition corresponding to the fixed vertex, and it follows that the vertex stabiliser contains $2 \cdot \mathrm{PSU}_6(2)$. (It is a non-split extension because, as we have already seen, it contains $\Omega_6^-(3) \cong 2 \cdot \mathrm{PSU}_4(3)$.) Hence the vertex stabiliser is either $2 \cdot \mathrm{PSU}_6(2)$ or $2 \cdot \mathrm{PSU}_6(2){:}2$. It is possible to show that elements in the outer half of the latter group act as odd permutations on the 3510 vertices, whereas the 3-transpositions are even permutations as we have seen. We deduce that

$$\begin{aligned} |\mathrm{Fi}_{22}| &= 3510.|2 \cdot \mathrm{PSU}_6(2)| \\ &= 2^{17}.3^9.5^2.7.11.13 \\ &= 64\,561\,751\,654\,400. \end{aligned} \tag{5.87}$$

Simplicity of Fi_{22} follows easily by Iwasawa's Lemma, since it acts primitively on the vertices of the graph, and is generated by the 3-transpositions, which lie in a normal abelian subgroup of a vertex stabiliser, which is itself perfect.

By this stage we have completed the proof of the existence of a simple group Fi_{22} of the specified order, generated by a conjugacy class of 3510 3-transpositions, each with centraliser $2 \cdot \mathrm{PSU}_6(2)$. Notice that we obtain for free the existence of this exceptional double cover of $\mathrm{PSU}_6(2)$.

We can now re-interpret the graph on 3510 vertices in terms of concepts related to $\mathrm{PSU}_6(2)$. The 693 transvections lift to two classes of 693 involutions in $2 \cdot \mathrm{PSU}_6(2)$. We take one of these classes and let $\mathcal{T}$ be a set of 693 vertices corresponding to the elements in the class: it matters which class we take, as there is no automorphism interchanging them. Now we saw in Section 3.12.2 that $\mathrm{PSU}_6(2)$ contains three conjugacy classes of subgroups $\mathrm{PSU}_4(3).2$, of index 1408. Two of these lift to non-split extensions $2 \cdot \mathrm{PSU}_4(3).2$ in $2 \cdot \mathrm{PSU}_6(2)$, while the other lifts to $2 \times \mathrm{PSU}_4(3).2$. Therefore $2 \cdot \mathrm{PSU}_6(2)$ has a subgroup $\mathrm{PSU}_4(3).2$ of index 2816: we take the 2816 cosets of this subgroup as a set $\mathcal{U}$ of vertices. Again, there are two classes of subgroups of this shape, and it matters which one we take. Our picture of the graph now looks like this.

$$\begin{array}{ccccccc}
& 180 & & & 567 \\
\boxed{1} \; \overset{693}{\rule{1.5cm}{0.4pt}} \; \boxed{693} \; \overset{512}{\rule{1cm}{0.4pt}} \; \overset{126}{\rule{1cm}{0.4pt}} \; \boxed{2816}
\end{array} \qquad (5.88)$$

The full automorphism group of the graph is obtained by extending $2 \cdot \mathrm{PSU}_6(2)$ to $2 \cdot \mathrm{PSU}_6(2){:}2$. This extends Fi_{22} to its automorphism group $\mathrm{Fi}_{22}{:}2$.

Now in $\mathrm{SU}_6(2)$, as in any unitary group, two unitary transvections $T_v(\lambda)$ and $T_w(\mu)$ commute if and only if v and w are perpendicular to each other. The same is true for their images in the quotient $\mathrm{PSU}_6(2)$. Therefore the maximal commuting sets of transvections correspond to maximal isotropic subspaces. Such a space has dimension 3 over $\mathbb{F}_4$, so contains exactly 21 subspaces of dimension 1, each of which gives rise to a single 3-transposition (i.e. unitary transvection) in $\mathrm{PSU}_6(2)$. It follows immediately that a maximal commuting set of 3-transpositions in Fi_{22} has size 22. Moreover, the image in $\mathrm{PSU}_6(2)$ of the stabiliser of a maximal isotropic subspace is $2^9{:}\mathrm{PSL}_3(4)$, and $\mathrm{PSL}_3(4)$ is the point stabiliser in M_{22} (see Sections 5.2.6 and 5.2.9). It follows (though not entirely trivially: see Exercise 5.26) that the stabiliser in Fi_{22} of a maximal commuting set of 3-transpositions is of shape $2^{10}.\mathrm{M}_{22}$. In fact this is a split exension $2^{10}{:}\mathrm{M}_{22}$.

5.7.3 Conway's description of Fi_{22}

Perhaps the simplest construction of Fi_{22} (in the sense of writing down generators) is Conway's construction [27] of the 77-dimensional representation over $\mathbb{F}_3$. This uses the fact that the maximal commuting sets of transpositions have size 22, and each such set generates an elementary abelian subgroup of order 2^{10}, and is fixed by a subgroup M_{22} permuting the 22 transpositions in the natural manner. Fixing such a *base* of 22 commuting transpositions, the 3510 transpositions are of three types:

$$\begin{aligned}
22 \; &\textit{basic,} \\
77.2^5 = 2464 \; &\textit{hexadic,} \\
2^{10} = 1024 \; &\textit{anabasic,}
\end{aligned} \qquad (5.89)$$

where the basic transpositions lie in the base, the hexadic transpositions each commute with exactly a hexad of basic transpositions, and the anabasic transpositions commute with no basic transposition. Note that the product of the six basic transpositions corresponding to the points in a hexad is trivial.

The representation has a basis $\{e_H\}$ where H runs over the 77 hexads of the Steiner system $S(3, 6, 22)$. The 22 basic transpositions i act diagonally, mapping e_H to $-e_H$ if $i \notin H$ and fixing e_H otherwise. A particular anabasic transposition S can be taken to act by

$$S : e_H \mapsto -e_H + \sum_{|L \cap H| = 2} e_L.$$

Together with the symmetries of M_{22} this gives enough information to generate the group and construct the action of all the transpositions. In principle this construction could be used to give an independent proof of existence of Fi_{22}, but the amount of computation required is probably more than could be reasonably done by hand.

Indeed, there is a 78-dimensional representation of Fi_{22} in characteristic 0, which gives rise to the above representation on reduction modulo 3. To describe this representation, take 77 coordinate vectors e_H indexed by the 77 hexads of the Steiner system $S(3,6,22)$ invariant under M_{22}, together with an extra coordinate vector e_*. The 22 basic transpositions i act by negating the 56 coordinates e_H such that $i \notin H$. The anabasic transposition S acts as follows:

$$S : 16e_* \mapsto -5e_* + \sqrt{3}\sum_H e_H,$$

$$16e_H \mapsto \sqrt{3}e_* - 7e_H - 3\sum_{H \cap K = \emptyset} e_K + \sum_{|L \cap H| = 2} e_L. \qquad (5.90)$$

5.7.4 Covering groups of Fi_{22}

There is a double cover $2 \cdot Fi_{22}$ also generated by 3510 transpositions. Just as in the case of $2 \cdot PSU_6(2)$, there are two classes of such elements in $2 \cdot Fi_{22}$, but in contrast to the situation in $2 \cdot PSU_6(2)$, there is an automorphism interchanging them. The double cover $2 \cdot Fi_{22}$ contains the exceptional cover $2^2 \cdot PSU_6(2)$ of $PSU_6(2)$, and is contained in Fi_{23}. Thus it is effectively constructed by the method described in Section 5.7.6.

There is a triple cover $3 \cdot Fi_{22}$ also generated by 3510 tranpsositions. This group has a 27-dimensional representation over $\mathbb{F}_4$, which is unitary in the sense that its dual is equivalent to the representation obtained by applying the field automorphism. In other words, $3 \cdot Fi_{22}$ embeds in $SU_{27}(2)$. This embedding was explicitly constructed by Richard Parker. It turns out that $3 \cdot Fi_{22}$ contains a proper triple cover $3 \cdot \Omega_7(3)$ of the orthogonal group $\Omega_7(3)$, and in turn $3 \cdot \Omega_7(3)$ contains $3 \cdot G_2(3)$. Thus we can obtain two more of the exceptional covers of groups of Lie type from $3 \cdot Fi_{22}$. Although these covers can be constructed directly, these constructions are quite difficult.

In fact, we can say more than this, as $3 \cdot Fi_{22}$ embeds in the exceptional group $3 \cdot {}^2E_6(2)$ of Lie type. The latter has an outer automorphism of order 3 which fuses three classes of subgroups $3 \cdot Fi_{22}$ in $3 \cdot {}^2E_6(2)$. The group ${}^2E_6(2)$ has an exceptional cover of shape $2^{2} \cdot {}^2E_6(2)$, in which there are three classes of subgroups of shape $2 \times 2 \cdot Fi_{22}$.

The adjoint representation (i.e. the representation on the Lie algebra) of the corresponding simple group ${}^2E_6(2)$ has dimension 78, and this remains irreducible on restriction to Fi_{22}. This representation may also be described as the reduction modulo 2 of the characteristic 0 representation described in Section 5.7.3.

5.7.5 Subgroups of Fi$_{22}$

Some of the maximal subgroups of Fi$_{22}$ are the stabilisers of sets of commuting transpositions. The fact that any such set lies in a maximal set, which has M$_{22}$ acting on it, means that any three commuting transpositions determine a hexad. That is, their product abc can be factorised in exactly two ways as the product of three transpositions, $abc = def$. Moreover, any larger set of commuting transpositions lies in a unique base. Thus we obtain the stabilisers of one transposition, a pair, a hexad, and a base, all of which turn out to be maximal subgroups, of shapes $2{\cdot}\mathrm{PSU}_6(2)$, $(2{\times}2^{1+8}).\mathrm{PSU}_4(2).2$, $2^{5+8}{:}(S_3{\times}A_6)$ and $2^{10}{:}\mathrm{M}_{22}$.

Enright [57] calculated all the subgroups H generated by transpositions, subject to Fischer's conditions that $H' = H''$ and every normal 2-subgroup or 3-subgroup is contained in $Z(H)$. (He called these subgroups 'D-subgroups', since the class of 3-transpositions was traditionally called D.) He found a number of maximal subgroups of this form, in particular $\Omega_7(3)$ (two conjugacy classes, interchanged by the outer automorphism of Fi$_{22}$), and two classes of S_{10}. The maximal subgroups $S_3 \times \mathrm{PSU}_4(3){:}2$ and $\Omega_8^+(2){:}S_3$ are also generated by transpositions, although they do not satisfy $H' = H''$. Similarly, the maximal subgroup $2^6{:}\mathrm{Sp}_6(2)$ is generated by transpositions, but in this case there is a non-central normal 2-subgroup. The complete list of maximal subgroups is given in Table 5.5 (see [111] for a proof).

5.7.6 The group Fi$_{23}$

It is possible to construct Fi$_{23}$ in much the same way as Fi$_{22}$, creating the graph on the 31671 transpositions out of the action of $\mathrm{P}\Omega_8^+(3){:}S_3$ on the 3240 conjugates of a maximal subgroup $2 \times \Omega_7(3)$ and the 28431 conjugates of a maximal subgroup $\Omega_8^+(2){:}S_3$. The former correspond to a conjugacy class of 3-transpositions in $\mathrm{P}\Omega_8^+(3){:}S_3$, so it is clear how to join them, and how they act on the 31671 vertices. The latter are joined to the 360 transpositions they contain, and to a certain collection of 3150 of each other. The details can be calculated inside $\mathrm{P}\Omega_8^+(3){:}S_3$, and will be given below. We summarise in the following picture.

$$
\overset{\displaystyle 351}{\underset{}{\boxed{3240}}} \;\overset{3159}{\rule{3cm}{0.4pt}}\; \overset{360}{} \overset{\displaystyle 3150}{\boxed{28431}} \tag{5.91}
$$

To describe the edges in this graph more precisely, we study the embedding of $\Omega_8^+(2){:}S_3$ in $\mathrm{P}\Omega_8^+(3){:}S_3$. In fact we may as well restrict to the subgroups of index 3, and lift to the orthogonal group $\mathrm{GO}_8^+(3)$, so that we are studying the embedding of $W(\mathrm{E}_8) \cong 2{\cdot}\Omega_8^+(2){:}2$ in $\Omega_8^+(3){:}2$ (or, perhaps better, the embedding of $W(\mathrm{E}_8)' \cong 2{\cdot}\Omega_8^+(2)$ in $\Omega_8^+(3)$), in much the same way as we studied the embedding of $W(\mathrm{E}_7)'$ in $\Omega_7(3)$ in order to construct Fi$_{22}$.

Thus the 28431 vertices in the second orbit may be described as the conjugates of a fixed copy of $W(\mathrm{E}_8)'$. The action of $\mathrm{P}\Omega_8^+(3){:}S_3$ on these 28431

Table 5.5. Maximal subgroups of Fischer groups

Fi_{22}	Fi_{23}	Fi_{24}'
		Fi_{23}
	$2\cdot\mathrm{Fi}_{22}$	$2\cdot\mathrm{Fi}_{22}{:}2$
$2\cdot\mathrm{PSU}_6(2)$	$2^2\cdot\mathrm{PSU}_6(2).2$	$2^2\cdot\mathrm{PSU}_6(2){:}S_3$
$(2\times2^{1+8}).\mathrm{PSU}_4(2).2$	$(2^2\times2^{1+8})(3\times\mathrm{PSU}_4(2)).2$	$2^{1+12}\cdot3\cdot\mathrm{PSU}_4(3).2$
$2^{5+8}{:}(A_6\times S_3)$	$2^{6+8}\cdot(A_7\times S_3)$	$2^{6+8}(A_8\times S_3)$
$2^{10}{:}\mathrm{M}_{22}$	$2^{11}\cdot\mathrm{M}_{23}$	$2^{11}\cdot\mathrm{M}_{24}$
$\Omega_7(3)$	$\mathrm{P}\Omega_8^+(3){:}S_3$	$(3\times\mathrm{P}\Omega_8^+(3){:}3){:}2$
$\Omega_7(3)$	$S_3\times\Omega_7(3)$	
$S_3\times\mathrm{PSU}_4(3){:}2$	$S_4\times\mathrm{Sp}_6(2)$	$(A_4\times\Omega_8^+(2){:}3){:}2$
$\Omega_8^+(2){:}S_3$	$\mathrm{Sp}_8(2)$	$\Omega_{10}^-(2)$
S_{10}	S_{12}	$(A_5\times A_9){:}2$
S_{10}		
$2^6{:}\mathrm{Sp}_6(2)$	$\mathrm{Sp}_4(4){:}2$	
M_{12}	$\mathrm{PSL}_2(23)$	$\mathrm{He}{:}2$
$^2F_4(2)'$		$\mathrm{He}{:}2$
$3^{1+6}{:}2^{3+4}{:}3^2{:}2$	$3^{1+8}{:}2^{1+6}{:}3^{1+2}{:}2S_4$	$3^{1+10}{:}\mathrm{PSU}_5(2){:}2$
		$3^2.3^4.3^8.(A_5\times2\cdot A_4).2$
	$3^3.[3^7].\mathrm{GL}_3(3)$	$3^3.[3^{10}].\mathrm{GL}_3(3)$
		$3^7\cdot\Omega_7(3)$
		$(3^2{:}2\times G_2(3))\cdot2$
		$2^{3+12}\cdot(\mathrm{GL}_3(2)\times A_6)$
		$7{:}6\times A_7$
		$29{:}14$
		$\mathrm{PSU}_3(3){:}2$
		$\mathrm{PSU}_3(3){:}2$
		$\mathrm{PGL}_2(13)$
		$\mathrm{PGL}_2(13)$
		$A_6\times\mathrm{PSL}_2(8){:}3$

points has rank 4 and suborbit lengths $1+2880+3150+22400$. The edges join a fixed copy to the conjugates in the 3150-orbit. On restricting to $\Omega_8^+(2){:}2$, the 2880-orbit splits up into orbits of lengths $960+1920$, while the other orbits are unchanged. The 3150-orbit is characterised by the property that the two copies of E_8 intersect in D_4D_4. For a given intersection, the three copies of E_8 are related by a triality automorphism, as in (5.83), applied to one of the copies of D_4.

Moreover, the action of one of the corresponding 28431 transpositions of Fi_{23} is to interchange the 2880-orbit of transpositions of $\mathrm{P}\Omega_8^+(3){:}S_3$ with the 2880-orbit of subgroups $\Omega_8^+(2){:}S_3$, in the analogous way as in Fi_{22}, and to swap the 22400 vertices in pairs similarly. This last orbit is characterised by its intersection with the fixed E_8 being $A_2A_2A_2A_2$, and on reduction modulo

3 this subsystem spans only a 6-space, which is fixed pointwise by a unique (root) subgroup of order 3. This group of order 3 enables us to construct representatives of this orbit. As in the case of Fi_{22}, the 22400 vertices come in 11200 pairs, intersecting the standard E_8 in the same subsystem, and the action of the transposition is to swap all these pairs.

It is straightforward to show that the transpositions thus defined preserve the graph and generate a transitive group on the 31671 vertices. This group is defined to be Fi_{23}. To calculate the order of Fi_{23}, we show that the graph on the 3510 neighbours of a vertex is isomorphic to the graph on the 3-transpositions of Fi_{22}, and deduce that the vertex stabiliser is $2 \cdot \mathrm{Fi}_{22}$. This is quite straightforward, as the neighbours of a fixed reflection in $\mathrm{P}\Omega_8^+(3){:}2$ consist of

(i) the 351 transpositions in the centraliser $\Omega_7(3)$, and
(ii) the 3159 copies of E_7 which are the orthogonal complements of the reflecting vector in the copies of E_8 which contain it.

Thus we easily see that

(i) two reflections in $\Omega_7(3)$ are joined if and only if they commute,
(ii) a reflection in $\Omega_7(3)$ is joined to those copies of E_7 which contain it, and
(iii) two copies of E_7 are joined if and only if they intersect in $\mathrm{A}_1\mathrm{A}_1\mathrm{A}_1\mathrm{D}_4$.

Hence the group Fi_{23} generated by the 31671 transvections, which in this case turns out to be the full automorphism group of the graph, is a group of order $31671.|2 \cdot \mathrm{Fi}_{22}|$. Thus

$$
\begin{aligned}
|\mathrm{Fi}_{23}| &= 4\,089\,470\,473\,293\,004\,800 \\
&= 2^{18}.3^{13}.5^2.7.11.13.17.23.
\end{aligned}
\tag{5.92}
$$

The vertex stabiliser $2 \cdot \mathrm{Fi}_{22}$ has orbits $1 + 3510 + 28160$ on the 31671 vertices, and the 2-point stabilisers are $2^2 \cdot \mathrm{PSU}_6(2)$ and $\Omega_7(3)$, contained in maximal subgroups $2^2 \cdot \mathrm{PSU}_6(2){:}2$ and $S_3 \times \Omega_7(3)$. Simplicity of Fi_{23} follows as in the case of Fi_{22}. The structure of the graph is summarised in this picture.

$$\tag{5.93}$$

It is clear now that a maximal commuting set of transpositions (a *base*) in Fi_{23} has size 23, and normaliser of shape $2^{11}.\mathrm{M}_{23}$. This time, however, this turns out to be a non-split extension $2^{11} \cdot \mathrm{M}_{23}$, which makes any construction along the lines of Conway's construction of Fi_{22} (Section 5.7.3) somewhat more problematic. Nevertheless, analogous to the 77-dimensional representation of Fi_{22}, there is a 253-dimensional representation of Fi_{23} over $\mathbb{F}_3$, in which $2^{11} \cdot \mathrm{M}_{23}$ acts monomially. The coordinates correspond to the 253 heptads of the Steiner system $S(4,7,23)$ (see Section 5.2.8), and a transposition in the 2^{11} negates the 176 heptads it is not in.

5.7.7 Subgroups of Fi$_{23}$

Just as in Fi$_{22}$ we can find a number of maximal subgroups as stabilisers of
sets of commuting transpositions. The fact that M$_{23}$ preserves a Steiner system $S(4, 7, 23)$ implies that any four commuting transpositions determine a
heptad. That is, their product $abcd$ can be factorised uniquely as a product
of three commuting transpositions, $abcd = efg$. Any larger set of commuting
transpositions lies in a unique base. Thus the subsets which give rise to maximal subgroups are singletons, pairs, triads, and the heptads and bases. These
are the first five groups listed under Fi$_{23}$ in Table 5.5.

Enright [57] classified the 'D-subgroups' (see Section 5.7.5), generated by
3-transpositions, and found a number of other maximal subgroups. Of particular note is S_{12}, which he uses to label all 31671 transpositions and to give
a complete table of t^u for all pairs of transpositions t and u. Another interesting maximal D-subgroup is Sp$_8(2)$. The maximal subgroup P$\Omega_8^+(3){:}S_3$ is
generated by 3-transpositions, but fails the condition $H' = H''$. A complete
list of the maximal subgroups is given in Table 5.5 (see also [113]).

5.7.8 The group Fi$_{24}$

This case is rather more difficult and we make no attempt here to prove the
existence of Fi$_{24}$. One approach is to make a graph on $1 + 31671 + 275264 =$
306936 vertices, consisting of $*$, the 31671 transpositions of Fi$_{23}$, and the
275264 cosets of a certain subgroup P$\Omega_8^+(3){:}3$ of Fi$_{23}$. If we define the edges
in terms of the structure of Fi$_{23}$, we can restrict to the subgroup 2·Fi$_{22}$ fixing
two vertices, with orbit lengths $1 + (1 + 3510 + 28160) + (28160 + 247104)$,
and show that there is an automorphism of the graph swapping the two fixed
points and extending 2·Fi$_{22}$ to 2·Fi$_{22}$:2. Once this has been proved, we see that
the automorphism group of the graph is a group Fi$_{24}$ of order 306936.2.|Fi$_{23}$|.
It turns out that this group has a subgroup Fi$_{24}'$ of index 2, which is simple,
and has order

$$|\text{Fi}_{24}'| = 1\,255\,205\,709\,190\,661\,721\,292\,800$$
$$= 2^{21}.3^{16}.5^2.7^3.11.13.17.23.29. \qquad (5.94)$$

The parameters of the graph are summarised in the following picture.

$$(5.95)$$

The 31671 transpositions of Fi$_{23}$ are joined just when they commute. The
275264 other transpositions come in 137632 pairs, each pair corresponding to
a subgroup P$\Omega_8^+(3){:}S_3$ of Fi$_{23}$, and joined to the 3240 transpositions in this
subgroup. When two of these subgroups intersect in P$\Omega_8^+(2){:}S_3$ we draw two of
the four possible edges between the two pairs. The precise joining rule is given
by taking the edges to correspond to cosets of P$\Omega_8^+(2)$, and the endpoints of
the edge to correspond to the two cosets of P$\Omega_8^+(3)$ which contain this coset
of P$\Omega_8^+(2)$.

5.7.9 Parker's loop

The maximal sets (*bases*) of commuting transpositions have size 24, and generate an elementary abelian group 2^{12}. This group supports a Golay code structure, in the sense that the product of a set of transpositions in the base is the identity if and only if the set is (the support of) a word in the Golay code. Modulo the base, there is a group M_{24} permuting the 24 transpositions. However, the resulting group $2^{12}M_{24}$ is non-split. It has a subgroup $2^{11\cdot}M_{24}$ of index 2, which is also a non-split extension.

These subgroups can be described in terms of a certain loop $\mathbb{P}$, discovered by Richard Parker, which is a kind of 'double cover' of the Golay code. This cover is not a group, but behaves rather like the units $\{\pm 1, \pm i_0, \ldots, \pm i_6\}$ of the octonions (see Section 4.3.2), which can be thought of as a non-associative double cover of an elementary abelian group of order 8. Thus the elements of Parker's loop $\mathbb{P}$ come in 2^{12} pairs $\pm d$, corresponding to words $\widetilde{d}$ in the Golay code (identified as usual with their supports). In particular $\widetilde{1}$ is the zero word in the Golay code. An element d squares to ± 1 according as $|\widetilde{d}|$ is 0 or 4 modulo 8. It follows from the fact that $(de)^2 = d^2[d,e]e^2$ that two elements of the loop commute if and only if the corresponding Golay code words intersect in a multiple of 4 points. Similarly, three elements of the loop associate (in the sense that $(de)f = d(ef)$) if and only if the three Golay code words intersect in an even number of points. Thus we may write

$$d^2 = (-1)^{|\widetilde{d}|/4},$$
$$de = (-1)^{|\widetilde{d} \cap \widetilde{e}|/2}ed,$$
$$(de)f = (-1)^{|\widetilde{d} \cap \widetilde{e} \cap \widetilde{f}|}d(ef). \tag{5.96}$$

It is not completely obvious, but it can be checked, that these definitions are consistent, so that Parker's loop is a non-associative inverse loop. Here an *inverse loop* is a set with a binary operation, an identity element and an inverse map, satisfying $x1 = x = 1x$ and $(x^{-1})^{-1} = x$ and $x^{-1}(xy) = y = (yx)x^{-1}$. Indeed, $\mathbb{P}$ is a Moufang loop (Exercise 5.28).

The subgroup $2^{12\cdot}M_{24}$ of Fi_{24} acts as automorphisms of Parker's loop. (Conway calls these 'standard' automorphisms, as they preserve the Golay code structure of $\mathbb{P}/\{\pm 1\}$.) A transposition δ in the normal subgroup 2^{12} acts by negating d whenever $\delta \in \widetilde{d}$. More generally, an element $\widetilde{\pi}$ in M_{24} lifts to 2^{12} automorphisms π of the loop. These may be defined by picking a generating set $d_1, \ldots, d_{12}$ for the loop, and defining d_i^π (arbitrarily) to be either of the two loop elements with $\widetilde{d_i^\pi} = \widetilde{d_i}^{\widetilde{\pi}}$.

As an abstract loop $\mathbb{P}$ has more automorphisms, as there is no abstract distinction between the elements mapping to octads and the elements mapping to 16-ads. The subgroup of standard automorphisms has index 2^{11} in the full automorphism group, which may be obtained by adjoining automorphisms corresponding to even subsets α of $\widetilde{\Omega}$, mapping d to Ωd whenever $|\alpha \cap \widetilde{d}|$ is odd.

5.7.10 The triple cover of Fi'_{24}

There is a triple cover $3 \cdot \mathrm{Fi}'_{24}$ of Fi'_{24} and a corresponding group $3 \cdot \mathrm{Fi}_{24}$. The former group has two 783-dimensional unitary representations in characteristic 0, which can be constructed using Parker's loop. I shall describe this construction, which is given in the Atlas [28], without proof. First we take the group $2^{11} \cdot \mathrm{M}_{24}$, constructed as in Section 5.7.9 using the loop, and let it act on a space of dimension $24 + 759$ as follows. The action on the 24-space is just the permutation action of M_{24}. The 759-space has a double basis consisting of the 759×2 octad elements of $\mathbb{P}$, and the action of $2^{11} \cdot \mathrm{M}_{24}$ is the same as the standard automorphisms of $\mathbb{P}$. That is, the space is spanned by vectors e_d such that $d \in \mathbb{P}$ and $\tilde{d}$ is an octad of the Golay code, with the understanding that $e_{-d} = -e_d$. A standard automorphism π of $\mathbb{P}$ acts as $\pi : e_d \mapsto e_{d^\pi}$.

The best way to describe how this extends to $3 \cdot \mathrm{Fi}'_{24}$ and to $3 \cdot \mathrm{Fi}_{24}$ is to use the transpositions, which correspond to certain vectors. Note however that the transpositions invert the central element of order 3, which acts as the scalar ω, say. Therefore they act semilinearly, in the sense that if t maps v to v^t, then it maps $\lambda v + w$ to $\bar{\lambda} v^t + w^t$, where $\bar{\lambda}$ is the complex conjugate of λ.

Now, in fact, the base stabiliser $2^{12} \cdot \mathrm{M}_{24}$ in Fi'_{24} has three orbits on transpositions, of lengths 24 and $759.2^5 = 24288$ and $276.2^{10} = 282624$. In $3 \cdot \mathrm{Fi}_{24}$ these numbers are multiplied by 3. The transpositions outside the base either commute with just eight of the basic transpositions, forming one of the 759 octads, or with just two of the basic transpositions. For a given octad, there are exactly 2^5 such 'octadic' transpositions, and for a given pair there are exactly 2^{10} such 'duadic' transpositions.

There is a semilinear, commutative, non-associative product $*$ invariant under the action of $3 \cdot \mathrm{Fi}_{24}$, where in this context semilinear means

$$(\lambda x + y) * z = \bar{\lambda}(x * z) + (y * z). \tag{5.97}$$

In fact, the inner product $x.(y * z)$ is a trilinear form symmetric in all three variables, which is perhaps a better way to describe the algebra. If r_t denotes the vector corresponding to the transposition t, then the algebra product is related to the group action by the equation

$$r_t * r_u = r_{t^u} + (r_u.r_t)r_u, \tag{5.98}$$

or more generally

$$x * r_u = x^u + (r_u.x)r_u, \tag{5.99}$$

where $r.x$ is the inner product of r with x. The trilinear form is then given by the formula

$$T(r_t, r_u, r_v) = r_{t^u}.r_v + (r_u.r_t)(r_u.r_v). \tag{5.100}$$

For simplicity, we scale the coordinates so that the first 24 coordinates have norm $\frac{1}{8}$ (as in the Leech lattice), while the rest have norm 1. Thus

$$r.x = \frac{1}{8} \sum_{i=1}^{24} r_i \overline{x}_i + \sum_{\text{octads } \tilde{d}} r_d \overline{x}_d. \tag{5.101}$$

Symmetry of the product $*$ implies that if t and $u \neq t$ commute, then $r_u.r_t = 1$ and $r_t * r_u = r_t + r_u$. Similarly, if t and u do not commute, so that $t^u = u^t$, then $r_u.r_t = 0$ and $r_t * r_u = r_{t^u}$. In fact we have $r_t.r_t = 9$, so $r_t * r_t = 10 r_t$. In particular, it is now easy to see that the basic transpositions correspond to vectors of shape

$$(-7, 1^{23} \mid 0^{759}). \tag{5.102}$$

Moreover, since we know how these transpositions act on the space, (5.99) gives the algebra product of these vectors with any vector in the space. Writing $E_i = 8e_i$ for the 24 vectors of shape $(8, 0^{23} \mid 0^{759})$ and e_d for the 759 vectors of shape $(0^{24} \mid 1, 0^{758})$, we have

$$2E_i * E_i = -81 e_i + 15 \sum_{j \neq i} e_j,$$

$$2E_i * E_j = 15 e_i + 15 e_j - \sum_{k \notin \{i,j\}} e_k,$$

$$2E_i * e_d = 3 e_d \text{ if } i \in \tilde{d},$$
$$2E_i * e_d = -e_d \text{ if } i \notin \tilde{d}. \tag{5.103}$$

In fact, the rest of the algebra is defined by

$$2e_d * e_d = 3 \sum_{i \in \tilde{d}} e_i - \sum_{j \notin \tilde{d}} e_j,$$

$$2e_d * e_f = e_{df} \text{ if } \widetilde{df} \text{ is an octad,}$$
$$2e_d * e_f = e_{df\Omega} \text{ if } \widetilde{df\Omega} \text{ is an octad,}$$
$$2e_d * e_f = 0 \text{ otherwise.} \tag{5.104}$$

The information about inner products given above is almost enough to determine the vectors corresponding to octadic transpositions. They have shape

$$\tfrac{1}{2}(-1^8, 1^{16} \mid \pm\theta, \pm 1^{30}, 0^{728}), \tag{5.105}$$

where $\theta = \omega - \overline{\omega} = \sqrt{-3}$. Here θ is on the coordinate corresponding to the fixed octad, and the $\pm 1^{30}$ are on the 30 coordinates corresponding to octads disjoint from the fixed octad. The signs are determined by the loop $\mathbb{P}$, as follows: we pick a subgroup of $\mathbb{P}$ of order 2^5, containing preimages $O_1, \ldots, O_{30}$ of the 30 octads, and containing ΩO_0, where $\widetilde{O_0}$ is the fixed octad. Then the signs are $+$ in the direction of O_i. (Notice that the signs depend on our choice of Ω: but the outer automorphism of Fi'_{24} swaps Ω with $-\Omega$, so this does not matter.)

Now (5.99) tells us how to calculate the action of any transposition u on the space, in terms of the algebra product of the corresponding vector r_u. In particular, we can calculate the action of an octadic transposition. To obtain a duadic vector, pick two octads intersecting in just two points, and act on an octadic vector for the first octad by a transposition corresponding to the second octad. We find the duadic vectors have shape

$$\tfrac{1}{8}(-7^2, 1^{22} \mid \pm\theta^{77}, \pm1^{330}, 0^{352}), \tag{5.106}$$

where the θ are on the coordinates corresponding to octads containing the two fixed points, and the ±1 correspond to octads containing neither of the fixed points. The signs here are more complicated to describe, but can in principle be calculated from the information given in this section.

5.7.11 Subgroups of Fi$_{24}$

Again we find that most (though not all) of the maximal subgroups of Fi$'_{24}$ (and also of Fi_{24}) can be described nicely in terms of the transpositions. Thus the stabilisers of one, two or three commuting transpositions are maximal subgroups Fi$_{23}$, $2{\cdot}$Fi$_{22}{:}2$ and $2^2{\cdot}$PSU$_6(2){:}S_3$ of Fi$'_{24}$. The centraliser of a product of four mutually commuting transpositions is a maximal subgroup $2^{1+12}{\cdot}3{\cdot}$PSU$_4(3){.}2$, permuting transitively the 126 different ways it can be factorised into four transpositions. Since the product of five commuting transpositions can be written as a product of just three other commuting transpositions, the stabiliser of any set of five or more commuting transpositions is contained in either the octad stabiliser $2^{6+8}(A_8 \times S_3)$ or the base stabiliser $2^{11}{\cdot}$M$_{24}$. There is a series of subgroups S_n generated by transpositions, with normalisers in Fi$_{24}$ as follows (all of which are maximal subgroups):

$$
\begin{aligned}
&S_3 \times \mathrm{P\Omega}_8^+(3){:}S_3,\\
&S_4 \times \mathrm{P\Omega}_8^+(2){:}S_3,\\
&S_5 \times S_9,\\
&S_6 \times \mathrm{PSL}_2(8){:}3,\\
&S_7 \times 7{:}6. \tag{5.107}
\end{aligned}
$$

Other interesting subgroups are two conjugacy classes of He:2 in Fi$'_{24}$, where He is the Held sporadic simple group (see Section 5.8.9 for more details). These two classes are interchanged by the outer automorphism of Fi$'_{24}$. The complete list of maximal subgroups is given in Table 5.5 (see also [121]).

5.8 The Monster and subgroups of the Monster

The embeddings of the 3-transposition groups $2^2{\cdot}$PSU$_6(2) < 2{\cdot}$Fi$_{22} <$ Fi$_{23}$ and the embedding of $2{\cdot}$Fi$_{22}$ in $2^{2{\cdot}2}E_6(2)$ suggested to Fischer that perhaps there were bigger simple groups $\mathbb{B}$ and $\mathbb{M}$ with $2^{2{\cdot}2}E_6(2) < 2{\cdot}\mathbb{B} < \mathbb{M}$. They would not be 3-transposition groups, but $^2E_6(2)$ contains a class of involutions whose products have order at most 4, so this was considered a suitable generalisation. Indeed, $\mathbb{M}$ contains a class of 6-transpositions, i.e. involutions whose products have order at most 6.

These groups, now known as the Baby Monster and the Monster, were eventually proved to exist, by Leon and Sims [154] and Griess [70] respectively. The Monster turned out to contain a number of other interesting subgroups, such as $3{\cdot}\mathrm{Fi}_{24}$, and three more sporadic groups, namely the Held group He (previously known, and contained in Fi_{24}—though this fact was not known until the discovery of the Monster), the Harada–Norton group HN, and the Thompson group Th, centralising elements of orders 7, 5 and 3 respectively.

5.8.1 The Monster

The Monster is so called largely because of its enormous size. Its order is

$$|\mathbb{M}| = 808\,017\,424\,794\,512\,875\,886\,459\,904\,961\,710\,757\,005\,754\,368\,000\,000\,000$$
$$= 2^{46}.3^{20}.5^9.7^6.11^2.13^3.17.19.23.29.31.41.47.59.71. \qquad (5.108)$$

Its smallest real representation is in 196883 dimensions, and this representation has a kind of non-associative algebra structure on it. By adjoining an identity element we obtain an algebra in 196884 dimensions, now usually known as the Griess algebra after Griess used it to produce the first construction of the Monster group in 1981 [70].

I shall sketch Conway's version [29] of this construction, in which he uses Parker's loop and a kind of triality to simplify some of the details. The construction is reminiscent of the use of triality in the construction of the exceptional Jordan algebras (Albert algebras) from the octonions (see Section 4.8). [Really it is an instance of a much more general construction which can be used for almost all simple groups: if, as usually happens, there is an involution with three conjugates whose product is the identity element, then, in most instances, there is a 'triality' automorphism cycling these three conjugates, and usually the involution centraliser and the triality element will together generate the simple group.]

First we construct an analogue of the group of isotopies of the octonions (see Section 4.7.1), but with the Moufang loop of octonions of norm 1 replaced by Parker's loop $\mathbb{P}$ (see Section 5.7.9). Thus we consider the set of triples (a, b, c) of elements of $\mathbb{P}$ which satisfy $abc = 1$, and let the following maps act on them (notice that $d^2 = \pm 1$ so that $d^{-1} = \pm d$, and therefore $d^{-1}ad^{-1}$ can be simplified to dad, and so on):

$$x_d : (a, b, c) \mapsto (dad, db, cd),$$
$$y_d : (a, b, c) \mapsto (ad, dbd, dc),$$
$$z_d : (a, b, c) \mapsto (da, bd, dcd). \qquad (5.109)$$

Notice that $x_d y_d z_d$ is the identity map. It turns out that as d ranges over all elements of $\mathbb{P}$, the maps x_d, y_d and z_d generate a group of order 2^{37}. In particular the latter contains a subgroup of order 2^{11} consisting of maps

$$x_\delta : (a,b,c) \mapsto ((-1)^{|\tilde{a}\cap\delta|}a, (-1)^{|\tilde{b}\cap\delta|}b, (-1)^{|\tilde{c}\cap\delta|}c) \qquad (5.110)$$

where δ is any even subset of the underlying set Ω of 24 elements. These maps x_δ may be thought of as 'inner' automorphisms of $\mathbb{P}$. The group may be extended by adjoining all the 'even' automorphisms π in $2^{11}{\cdot}M_{24}$, thus

$$x_\pi : (a,b,c) \mapsto (a^\pi, b^\pi, c^\pi). \qquad (5.111)$$

Finally, we can adjoin the 'odd' automorphisms provided we invert a, b and c and reverse their cyclic ordering. These are the automorphisms

$$x_\delta : (a,b,c) \mapsto ((-1)^{|\tilde{a}\cap\delta|}a^{-1}, (-1)^{|\tilde{c}\cap\delta|}c^{-1}, (-1)^{|\tilde{b}\cap\delta|}b^{-1}),$$
$$y_\delta : (a,b,c) \mapsto ((-1)^{|\tilde{c}\cap\delta|}c^{-1}, (-1)^{|\tilde{b}\cap\delta|}b^{-1}, (-1)^{|\tilde{a}\cap\delta|}a^{-1}),$$
$$z_\delta : (a,b,c) \mapsto ((-1)^{|\tilde{b}\cap\delta|}b^{-1}, (-1)^{|\tilde{a}\cap\delta|}a^{-1}, (-1)^{|\tilde{c}\cap\delta|}c^{-1}), \qquad (5.112)$$

where δ is an odd subset of Ω. Notice that if δ is odd, then $x_\delta y_\delta = y_\delta z_\delta = z_\delta x_\delta$ and $x_\delta y_\delta : (a,b,c) \mapsto (b,c,a)$ is a 'triality' automorphism, while if δ is even, $x_\delta = y_\delta = z_\delta$.

The group generated by all x_d, y_d, z_d, x_δ, y_δ, z_δ and x_π, y_π, z_π will be denoted N, and has shape $2^2.2^2.2^{11}.2^{22}.M_{24}.S_3$. It has a normal subgroup K of order 4 containing the three non-trivial elements

$$k_1 = y_\Omega z_{-\Omega} = x_\Omega z_{-1} = x_{-\Omega} y_{-1},$$
$$k_2 = z_\Omega x_{-\Omega} = y_\Omega x_{-1} = y_{-\Omega} z_{-1},$$
$$k_3 = x_\Omega y_{-\Omega} = z_\Omega y_{-1} = z_{-\Omega} x_{-1}. \qquad (5.113)$$

The quotient by K turns out to be isomorphic to a maximal subgroup of the Monster. Modulo K, the normal 2^2-subgroup consists of the three involutions x_{-1}, y_{-1}, and z_{-1}.

The next step in the construction is to build the centraliser of an involution, which, modulo K, has shape $2^{1+24}{\cdot}Co_1$. To do this, we recast the centraliser of x_{-1} in N in the shape $2.2^{1+24}.2^{12}M_{24}$, and observe that there is a normal subgroup 2.2^{1+24} possessing the symmetries of the Leech lattice Λ (in a sense which I shall make precise in a moment). This subgroup 2.2^{1+24} is generated by the x_d and x_i, together with k_1, and the four-to-one map onto $\Lambda/2\Lambda$ with kernel $\langle x_{-1}, k_1 \rangle$ is as follows. Write w_d for the reduction modulo 2 of the Leech lattice vector with 2 on $\tilde{d}$ and 0 elsewhere, and write w_i for the reduction modulo 2 of the Leech lattice vector with with -3 on i and 1 elsewhere. Then the map is given by $x_d \mapsto w_d$ and $x_i \mapsto w_i$. Observe that all the vectors w_d and w_i have inner products 0 mod 2, except that $w_d.w_i = 1$ mod 2 if $i \in d$. Correspondingly, all x_d and x_i commute modulo K, except that $x_d x_i = x_{-1} x_i x_d$ if $i \in d$. Similarly, all w_d and w_i have even type (i.e. norm divisible by 4) except that w_d has odd type if $\tilde{d}$ is a dodecad; and correspondingly x_d and x_i square to 1 unless $\tilde{d}$ is a dodecad, in which case $x_d{}^2 = x_{-1}$. Thus the group $\langle x_d, x_i \rangle$ modulo K, which is an extraspecial group 2^{1+24}_+, has its square and commutator maps defined by the Leech lattice.

We extend our notation to label all the elements of the group generated by the x_d and x_i as follows. Each element is a product of the x_d and x_i and we write

$$x_{d.e.....i.j....} = x_d x_e.....x_i x_j.... \tag{5.114}$$

The corresponding vector in $\Lambda/2\Lambda$ is obtained as the sum of the vectors w_d, $w_e, \ldots, w_i, w_j, \ldots$, corresponding to $x_d, x_e, \ldots, x_i, x_j, \ldots$, and the preimages in 2.2^{1+24} of the type 2 vectors are as follows:

$$
\begin{aligned}
x_{\pm i.j} &\mapsto (4, -4, 0^{22}), \\
x_{\pm \Omega.i.j} &\mapsto (4, 4, 0^{22}), \\
x_{\pm \Omega.d.i} &\mapsto (-3, 1^{23}), \text{ sign changed on } \widetilde{d}, i \in \widetilde{d}, \\
x_{\pm \Omega.d.i.j} &\mapsto (2^6, -2^2, 0^{16}), i, j \in \widetilde{d}, \text{ an octad}, \\
x_{\pm d.i.j.k.l} &\mapsto (2^4, -2^4, 0^{16}), i, j, k, l \in \widetilde{d}, \text{ an octad}, \tag{5.115}
\end{aligned}
$$

and similarly for the multiples by k_1.

To make the Monster explicitly, it is necessary to make explicit matrices generating the various subgroups. We can start with the group $2^2.2^{1+24}.\text{Co}_1$, which has a centre of order 2^3 generated by K and x_{-1}, and has various interesting quotients, as follows.

(i) The quotient by K is the involution centralizer $2^{1+24}.\text{Co}_1$ in M.
(ii) The quotient by $\langle y_\Omega, z_{-\Omega} \rangle$ is another group of shape $2^{1+24}.\text{Co}_1$ which is *not* isomorphic to the first one. This group has a representation of degree $2^{12} = 4096$ which extends the natural representation of 2^{1+24} (see Section 3.10.2 for a description of this representation).
(iii) The quotient by $\langle x_{-1}, y_\Omega, z_{-\Omega} \rangle$ has shape $2^{24}.\text{Co}_1$.
(iv) The quotient by $\langle x_d, x_i, k_1 \rangle$ is $2\cdot\text{Co}_1$, which has a natural 24-dimensional representation coming from the Leech lattice (see Section 5.4.1).
(v) The quotient by $\langle x_d, x_i, y_\Omega, z_{-\Omega} \rangle$ is Co_1.

Using these various quotients, we make a 196884-dimensional representation in three pieces, as follows.

98280 = the monomial representation of the second quotient (ii) on the 2×98280 conjugates of x_d, with the convention that $x_{-d} = -x_d$. Since x_{-1} acts trivially, this is actually a representation of (iii).

98304 = $4096 \otimes 24$, where 24 denotes the representation of $2\cdot\text{Co}_1$ on the Leech lattice, and 4096 denotes the natural representation of 2^{1+24} extended to $2^{1+24}\cdot\text{Co}_1$ of type (ii).

299 + 1 = symmetric square of 24, representing the quotient (v). (5.116)

Conway constructs most of these representations, with a careful choice and labelling of basis vectors. See also the Atlas [28, pp. 228–30]. However, neither Conway nor Griess, nor the Atlas, provides an explicit construction of the

4096-representation. They do however construct the action of the subgroup $2^{1+24}.2^{11}M_{24}$ and show that there is a unique way to extend the action to $2^{1+24}Co_1$.

There is a double basis for 98280 consisting of the elements listed in (5.115), with the convention that $x_{-r} = -x_r$. The action of $2^{1+24}Co_1$ on this space is precisely defined above, so can be explicitly computed.

We write the 24-space in such a way that the Leech lattice takes its usual form, i.e. the basis vectors i $(i \in \Omega)$ are mutually orthogonal and have norm $\frac{1}{8}$. The 300-space then has basis consisting of

$$(ii) = i \otimes i,$$
$$(ij) = i \otimes j + j \otimes i, \tag{5.117}$$

and the symmetric square action of Co_1. The 4096-space is spanned by 4×4096 vectors d^+ and d^- (for $d \in \mathbb{P}$) of norm 1 subject to the relations

$$(-d)^+ = -d^+,$$
$$(-d)^- = -d^-,$$
$$(\Omega d)^+ = d^+,$$
$$(\Omega d)^- = -d^-. \tag{5.118}$$

The action of $2^{1+24}2^{12}M_{24}$ is given by the second and third coordinates of (5.109–5.112), that is

$$x_d : f^+ \mapsto (df)^+,$$
$$f^- \mapsto (fd)^-,$$
$$y_d : f^+ \mapsto (dfd)^+,$$
$$f^- \mapsto (df)^-,$$
$$z_d : f^+ \mapsto (fd)^+,$$
$$f^- \mapsto (dfd)^-,$$
$$x_i : f^+ \mapsto \pm(f^{-1})^-, \text{ minus sign just when } i \in f,$$
$$f^- \mapsto \pm(f^{-1})^+, \text{ minus sign just when } i \in f,$$
$$x_\pi : f^+ \mapsto (f^\pi)^+,$$
$$f^- \mapsto (f^\pi)^- \text{ for } \pi \text{ even.} \tag{5.119}$$

As we have just noted, this is sufficient to determine the action of $2^{1+24}\cdot Co_1$ on the 4096-space, although it is cumbersome to compute this and write it down explicitly. Now on restriction from $2^{1+24}\cdot Co_1$ to $2^{1+24}2^{11}M_{24}$ these pieces break up into irreducibles as follows:

$$98280 = 552 + 48576 + 49152,$$
$$98304 = 98304,$$
$$299 + 1 = 276 + 23 + 1, \tag{5.120}$$

and on further restriction to the subgroup of index 2 generated by the x_d and even x_π, the irreducibles of degrees 552 and 98304 break up into two pieces of equal size, and we have

$$98280 = 276 + 276 + 48576 + 49152,$$
$$98304 = 49152 + 49152,$$
$$299 + 1 = 276 + 23 + 1. \tag{5.121}$$

Finally to make the Monster, it is necessary to label the bases in such a way that the triality symmetry is visible, fusing the three constituents of degree 276 and the three of degree 49152. Conway does this by providing a 'dictionary' between three different labellings of the basis. Equally it could be done by writing out explicitly the action of a triality symmetry, as follows.

$$(ii) \mapsto (ii),$$
$$(ij) \mapsto x_{ij} + x_{\Omega.ij}$$
$$\mapsto x_{ij} - x_{\Omega.ij},$$
$$\text{for } i \notin \tilde{d}, \ x_{d.i} \mapsto d^+ \otimes i$$
$$\mapsto d^- \otimes i,$$
$$x_{\Omega d.\delta} \mapsto \tfrac{1}{8} \sum_{\varepsilon \subseteq d} (-1)^{\frac{1}{2}|\tilde{\varepsilon}|} (-1)^{\frac{1}{2}|\tilde{\delta}\varepsilon|} x_{\Omega d.\varepsilon}$$
$$\mapsto \tfrac{1}{8} \sum_{\varepsilon \subseteq d} (-1)^{\frac{1}{2}|\tilde{\delta}|} (-1)^{\frac{1}{2}|\tilde{\delta}\varepsilon|} x_{\Omega d.\varepsilon}, \tag{5.122}$$

where in the last two lines the sum is over all cosets of the Golay code which have representatives ε which are *even* subsets of d.

To prove that the resulting group really is the Monster it is necessary to show that it preserves some algebraic or combinatorial structure on the 196884-space. Indeed, there is a commutative non-associative 'algebra' structure invariant under the Monster.

5.8.2 The Griess algebra

The definition of the algebra is simplified if we take into account the fact that it is really a symmetric trilinear form: that is the inner product of $x*y$ with z equals the inner product of x with $y*z$. First we may consider the 300-space as being the space of symmetric 24×24 matrices. Given two such matrices A and B, we define their Griess product to be 4 times the Jordan product: $A * B = 2(AB + BA)$.

Next pick a basis vector x_r in the 98280-space, corresponding to a type 2 vector w_r in the Leech lattice. Then $w_r \otimes w_r$ is a symmetric 24×24 matrix, so let $\lambda = \frac{1}{64} \text{Tr}(A.(w_r \otimes w_r))$ be the natural inner product of A with $w_r \otimes w_r$. Then $x_r * A = \lambda x_r$. The symmetry of the trilinear form now gives $x_r * x_r = w_r \otimes w_r$. The other products among the x_r are given by $x_r * x_s = x_{rs}$ if rs corresponds to a type 2 vector in the Leech lattice: here rs denotes the product inside 2^{1+24}. If rs is not of this type, then $x_r * x_s = 0$.

Now we can choose basis vectors in the 98304-space of the form $W \otimes v$, where W is a vector in 4096-space, and v is in 24-space. In particular, since A is a 24×24 matrix, it can act on v, and we set

$$A * (W \otimes v) = W \otimes vA + (\tfrac{1}{8}\mathrm{Tr}A)(W \otimes v).$$

At this stage we have defined the products of A with all vectors in 196884-space.

Next consider the product of x_r with $W \otimes v$. Let $v' = v - 2(v, w_r)w_r$ and let W' denote the image of W under the action of r, considered as an element of 2^{1+24} acting on the 4096-space. Then $(W \otimes v) * x_r = W' \otimes v'$.

Finally, the product of two elements of the form $W \otimes v$ has no component in the 98304-space, so this product is determined by the above formulae and the symmetry of the trilinear form.

The Monster can be defined as the automorphism group of this algebra.

5.8.3 6-transpositions

The largest subgroup of the Monster turns out to be a double cover $2\text{·}\mathbb{B}$ of the Baby Monster, of index $97\,239\,461\,142\,009\,186\,000$, i.e. just under 10^{20}. This subgroup fixes a unique non-zero vector in the 196883-dimensional representation. Of course it is out of the question to compute directly with such permutations. However, this permutation representation can be thought of in a number of different ways, either permuting this orbit of vectors, or permuting the involutions in the smallest conjugacy class.

These involutions have the property that the product of any two of them has order at most 6, so they are called 6-*transpositions*. Indeed, their product lies in one of just nine conjugacy classes in $\mathbb{M}$, which in Atlas notation [28] are $1A$, $2A$, $2B$, $3A$, $3B$, $4A$, $4B$, $5A$ and $6A$. In each case, the pair (a, b) of 6-transpositions is unique up to conjugacy, and we tabulate here, for fixed a, the number of b in the given orbit, together with the normaliser of $\langle ab \rangle$ and the normaliser of $\langle a, b \rangle$.

Class	Suborbit length	$N(\langle ab \rangle)$	$N(\langle a, b \rangle)$
$1A$	1	$\mathbb{M}$	$2\text{·}\mathbb{B}$
$2A$	$27\,143\,910\,000$	$2\text{·}\mathbb{B}$	$2^{2\cdot 2}E_6(2){:}S_3$
$2B$	$11\,707\,448\,673\,375$	$2^{1+24}\text{·}\mathrm{Co}_1$	$2^{1+24}\text{·}\mathrm{Co}_2$
$3A$	$2\,031\,941\,058\,560\,000$	$3\text{·}\mathrm{Fi}_{24}$	$S_3 \times \mathrm{Fi}_{23}$
$3C$	$91\,569\,524\,834\,304\,000$	$S_3 \times \mathrm{Th}$	$S_3 \times \mathrm{Th}$
$4A$	$1\,102\,935\,324\,621\,312\,000$	$2^{1+24}\text{·}\mathrm{Co}_3$	$2^{1+24}\text{·}\mathrm{McL.2}$
$4B$	$1\,254\,793\,905\,192\,960\,000$	$(D_8 \times F_4(2)).2$	$(D_8 \times F_4(2)).2$
$5A$	$30\,434\,513\,446\,055\,706\,624$	$(D_{10} \times \mathrm{HN})\text{·}2$	$(D_{10} \times \mathrm{HN})\text{·}2$
$6A$	$64\,353\,605\,265\,653\,760\,000$	$(S_3 \times 2\text{·}\mathrm{Fi}_{22}).2$	$(S_3 \times 2\text{·}\mathrm{Fi}_{22}).2$

5.8.4 Monstralisers and other subgroups

As might be expected of such a large group, the subgroup structure of the Monster is very rich. For example there are interesting subgroups obtained as normalisers of various cyclic subgroups:

$$2 \cdot \mathbb{B},$$
$$3 \cdot \mathrm{Fi}_{24}':2,$$
$$S_3 \times \mathrm{Th},$$
$$(D_{10} \times \mathrm{HN}) \cdot 2,$$
$$(7{:}3 \times \mathrm{He}){:}2. \tag{5.123}$$

These all turn out to be maximal subgroups, where $\mathbb{B}$, Th, HN and He are sporadic simple groups described respectively in Sections 5.8.6, 5.8.7, 5.8.8, 5.8.9 below. In these cases there are two subgroups which are the centralisers of each other. For example, $D_{10} = C_{\mathbb{M}}(\mathrm{HN})$ and $\mathrm{HN} = C_{\mathbb{M}}(D_{10})$, or $3 \cdot \mathrm{Fi}_{24}' = C_{\mathbb{M}}(3)$ and $3 = C_{\mathbb{M}}(3 \cdot \mathrm{Fi}_{24}')$. Norton considered much more general cases of two subgroup H_1 and H_2 of $\mathbb{M}$ such that $C_{\mathbb{M}}(H_1) = H_2$ and $C_{\mathbb{M}}(H_2) = H_1$. He called such pairs (H_1, H_2) *Monstraliser pairs*, where 'Monstraliser' was meant as a contraction of 'Monster centraliser'.

In many cases it turns out that $N_{\mathbb{M}}(H_1)$ $(= N_{\mathbb{M}}(H_2))$ is a maximal subgroup of $\mathbb{M}$. For example there are maximal subgroups

$$(A_5 \times A_{12}){:}2,$$
$$(A_5 \times \mathrm{PSU}_3(8){:}3){:}2,$$
$$(\mathrm{PSL}_2(11) \times \mathrm{M}_{12}){:}2. \tag{5.124}$$

In three cases there are extra symmetries: $S_5 \wr S_3$, $(A_6 \times A_6 \times A_6) \cdot (2 \times S_4)$ and $(\mathrm{PSL}_2(11) \times \mathrm{PSL}_2(11)).4$. In some cases, H_1 is abelian, so $H_1 < H_2$, for example $2^{2 \cdot 2} E_6(2){:}S_3$.

There is another series of cyclic subgroup of prime order, with normalisers

$$2^{1+24} \cdot \mathrm{Co}_1,$$
$$3^{1+12}_+ \cdot 2 \cdot \mathrm{Suz}{:}2,$$
$$5^{1+6}_+ {:} 2 \cdot \mathrm{J}_2{:}4,$$
$$7^{1+4}_+ {:} (3 \times 2 \cdot A_7{:}2),$$
$$13^{1+2} {:} (3 \times 2 \cdot A_4{:}4). \tag{5.125}$$

These are also Monstralisers, though in a rather trivial sense. Most of the rest of the maximal subgroups are p-local subgroups for some p. (A *p-local subgroup* is the normaliser of a non-trivial p-subgroup.) The (very difficult) classification of the non-local subgroups, apart from a few cases still open, has been obtained by Beth Holmes [82, 83, 84, 81] by extensive calculations in a computer version of the Griess construction [80]. The known maximal subgroups are listed in Table 5.6.

5.8.5 The Y-group presentations

The subgroup $(A_5 \times A_{12}){:}2$ leads to an interesting presentation of the Monster. It has a subgroup $(A_5 \times (A_5 \times A_7){:}2){:}2$ in which the two A_5 factors can be interchanged, yielding a subgroup of $\mathbb{M}$ of shape $((A_5 \times A_5){:}2^2 \times A_7){:}2$. If we take Coxeter generators (see Section 2.8.2) for a group $S_5 \times S_{12}$, with the

Table 5.6. Known maximal subgroups of the Monster

$2 \cdot \mathbb{B}$	$2^{1+24} \cdot \mathrm{Co}_1$
$2^{2 \cdot 2} E_6(2){:}S_3$	$2^2.2^{11}.2^{22} \cdot (S_3 \times \mathrm{M}_{24})$
	$2^3.2^6.2^{12}.2^{18}.(\mathrm{GL}_3(2) \times 3S_6)$
$2^{10}.2^{16}.\Omega_{10}^+(2)$	$2^5.2^{10}.2^{20}.(\mathrm{GL}_5(2) \times S_3)$
$3 \cdot \mathrm{Fi}_{24}$	$3^{1+12} \cdot 2 \cdot \mathrm{Suz}{:}2$
$(3^2{:}2 \times \mathrm{P}\Omega_8^+(3)) \cdot S_4$	$3^2.3^5.3^{10} \cdot (2 \cdot S_4 \times \mathrm{M}_{11})$
$3^8 \cdot \Omega_8^-(3) \cdot 2$	$3^3.3^2.3^6.3^6{:}(\mathrm{SL}_3(3) \times SD_{16})$
$S_3 \times \mathrm{Th}$	
$(D_{10} \times \mathrm{HN}) \cdot 2$	$5^{1+6}{:}2J_2.4$
$(5^2{:}4 \cdot 2^2 \times \mathrm{PSU}_3(5)){:}S_3$	$5^2.5^2.5^4{:}(S_3 \times \mathrm{GL}_2(5))$
$5^4{:}(3 \times \mathrm{SL}_2(25)){:}2$	$5^3 \cdot 5^3 \cdot (2 \times \mathrm{SL}_3(5))$
$(7{:}3 \times \mathrm{He}){:}2$	$7^{1+4}{:}(3 \times 2 \cdot S_7)$
$(7^2{:}(3 \times 2A_4) \times \mathrm{PSL}_2(7)).2$	$7^2.7.7^2{:}\mathrm{GL}_2(7)$
	$7^2{:}\mathrm{SL}_2(7)$
$11^2{:}(5 \times 2 \cdot A_5)$	
$(13{:}6 \times \mathrm{PSL}_3(3)) \cdot 2$	$13^{1+2}{:}(3 \times 4S_4)$
	$13^2{:}\mathrm{SL}_2(13).4$
$41{:}40$	
$(A_5 \times A_{12}){:}2$	$(A_7 \times (A_5 \times A_5){:}2^2){:}2$
$(\mathrm{PSL}_2(11) \times \mathrm{M}_{12}){:}2$	$\mathrm{PSL}_2(11)^2{:}4$
$\mathrm{M}_{11} \times A_6 \cdot 2^2$	$A_6{}^3 \cdot (2 \times S_4)$
$(A_5 \times \mathrm{PSU}_3(8){:}3){:}2$	$S_5{}^3{:}S_3$
$(\mathrm{GL}_3(2) \times \mathrm{Sp}_4(4){:}2) \cdot 2$	
$\mathrm{PSL}_2(71)$	$\mathrm{PSL}_2(59)$
$\mathrm{PGL}_2(29)$	$\mathrm{PGL}_2(19)$

Note: according to [82, 83, 84, 81] any other maximal subgroup G of $\mathbb{M}$ satisfies $S \leqslant G \leqslant \mathrm{Aut}(S)$ where S is one of the following simple groups: $\mathrm{PSL}_2(13)$, $\mathrm{PSU}_3(4)$, $\mathrm{PSU}_3(8)$, $\mathrm{Sz}(8)$, $\mathrm{PSL}_2(8)$, $\mathrm{PSL}_2(16)$, $\mathrm{PSL}_2(27)$. The cases $\mathrm{PSL}_2(8)$, $\mathrm{PSL}_2(16)$ and $\mathrm{PSL}_2(27)$ have reportedly been eliminated in unpublished work of Holmes.

understanding that only products of an even number of generators are in $\mathbb{M}$, then swapping the two copies of A_5 yields the following diagram

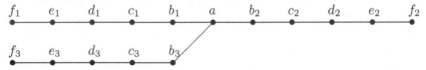

where only the order of the product of b_1 and b_3 remains to be determined. But in fact the subgroup $(A_5 \times (A_6 \times A_6){:}2^2){:}2$ of $(A_5 \times A_{12}){:}2$ extends to $(A_6 \times (A_6 \times A_6){:}2^2){:}2$ so $b_1 b_3$ has order 2 in $\mathbb{M}$. This diagram is commonly known as Y_{555} on account of its shape and the fact that there are five nodes on each arm.

Now the nodes of the diagram represent involutions which are not in the Monster, and which must therefore interchange one copy of the Monster with another. The nodes can therefore be thought of as generators for a wreath product $M \wr 2$.

It can be shown that $f_1 = (ab_1b_2b_3c_1c_2d_1)^9$, and similarly for any permutation of the subscripts, so the generators f_1, f_2 and f_3 are redundant. Moreover, the subgroup of M generated by even products of $a, b_1, b_2, b_3, c_1, c_2, c_3$ turns out to have shape $3^5{:}SO_5(3)$. This implies that $(ab_1c_1ab_2c_2ab_3c_3)^{10} = 1$. It is a remarkable fact that these relations are sufficient to define a presentation for $M \wr 2$.

5.8.6 The Baby Monster

We have seen that the Baby Monster $\mathbb{B}$ has a double cover $2{\cdot}\mathbb{B}$ which is a subgroup of the Monster. The original construction of $\mathbb{B}$ by Leon and Sims [118] (see also [154]) was a tour-de-force of computational group theory, effectively constructing the group as a permutation group on $13\,571\,955\,000$ points. However, these permutations were much too big to be written down explicitly, and various clever techniques were used to manipulate them.

The simple group $\mathbb{B}$ has a rather easier description as a group of 4371×4371 matrices (in characteristic 2 the dimension can be reduced to 4370). In every characteristic except 2, the maximal subgroup $2^{1+22}{\cdot}Co_2$ acts as a direct sum of three irreducible representations, $2048 \oplus 2300 \oplus 23$, where 2048 denotes the unique extension of the faithful irreducible representation of 2^{1+22} to $2^{1+22}{\cdot}Co_2$, and 2300 denotes a monomial representation of the quotient $2^{22}{\cdot}Co_2$, and 23 is the irreducible representation of Co_2 derived from the Leech lattice.

Now in the group $2^{22}{\cdot}Co_2$, the action of Co_2 on the 2^{22} is obtained by taking the Leech lattice modulo 2, that is $\Lambda/2\Lambda$ in the notation of Section 5.4.1, fixing the image x of a Leech lattice vector of norm 4, and taking the 22-space $V = x^\perp/\langle x \rangle$. In particular, V contains an orbit of 46575 vectors coming from Leech lattice vectors of norm 8, such that the stabiliser of one such vector in Co_2 is $2^{10}M_{22}{:}2$ (see Table 5.2). This gives a subgroup

$$2^{1+22}.2^{10}M_{22}.2 \cong 2^2.2^{10}.2^{20}(2 \times M_{22}.2)$$

of $2^{1+22}Co_2$, and its derived group is a subgroup $2^2.2^{10}.2^{20}.M_{22}$ of index 4. Restricting to this latter subgroup, the representation breaks up as

$$(1024 \oplus 1024) \oplus (1024 \oplus 1232 \oplus 22 \oplus 22) \oplus (22 \oplus 1).$$

It is now possible to fuse the three constituents of dimension 1024, and the three of dimension 24, by adjoining a 'triality' automorphism of the subgroup $2^{2+10+20}M_{22}$, in order to generate the Baby Monster. This construction is described in more detail in [144].

It is not easy to deduce the order of the group from this construction, even with computational assistance, as the smallest permutation representation of $\mathbb{B}$ is on $13\,571\,955\,000$ points. Indeed, there is an orbit of $13\,571\,955\,000$

vectors in the 4371-dimensionsal representation, permuted transitively by $\mathbb{B}$. The stabiliser of one of these vectors is a subgroup $2 \cdot {}^2 E_6(2){:}2$. Hence we can calculate the order of the Baby Monster as

$$|\mathbb{B}| = 4\,154\,781\,481\,226\,426\,191\,177\,580\,544\,000\,000$$
$$= 2^{41}.3^{13}.5^6.7^2.11.13.17.19.23.31.47. \tag{5.126}$$

Of course, these vectors are in one-to-one correspondence with the involutions in the centre of their stabilisers, that is the involutions in the conjugacy class labelled $2A$ in the Atlas [28]. These involutions are 4-*transpositions*, in the sense that the product of any two of them has order at most 4. Indeed, the action of the Baby Monster on these involutions has rank 5, and the 5 suborbits can be distinguished by the conjugacy class of the product of the two involutions, which may be either $1A$, $2B$, $2C$, $3A$ or $4B$. The corresponding suborbits lengths are

$$1 + 3968055 + 23113728 + 2370830336 + 11174042880.$$

The stabilisers of the unordered pairs of transpositions are respectively $2^{1+22}\mathrm{PSU}_6(2).2$, $(2^2 \times F_4(2)).2$, $S_3 \times \mathrm{Fi}_{22}{:}2$ and $2^{1+20}\mathrm{PSU}_4(3).D_8$. The normaliser of the cyclic group generated by the product of the two transpositions is larger than this in the first and last cases, namely $2^{1+22}\mathrm{Co}_2$ and $2^{1+22}\mathrm{PSU}_4(3).D_8$ respectively.

Many of the maximal subgroups of $\mathbb{B}$ can be read off from Norton's list of Monstralisers, by centralising a suitable involution in one of the factors. However, some of the smaller non-local subgroups cannot be obtained in this way, and have only been constructed by an exhaustive search using a computer. This includes the maximal subgroups $\mathrm{PSL}_2(49) \cdot 2$, $\mathrm{PSL}_2(31)$ and M_{11}. For a complete list of maximal subgroups, see Table 5.7 and [181].

5.8.7 The Thompson group

This group was first discovered as a subgroup of the Monster, by taking an element x of order 3 from $2^{1+24} \cdot \mathrm{Co}_1$, such that x maps to an element of Co_1 with centraliser $3 \times A_9$, and asking what is the centraliser of x in the Monster. Modulo $\langle x \rangle$, this group would have an involution centraliser of shape $2^{1+8} \cdot A_9$, and at that time no suitable group was known. Eventually it turned out that $N_{\mathrm{M}}(\langle x \rangle) \cong S_3 \times \mathrm{Th}$, where Th is the Thompson sporadic simple group of order

$$|\mathrm{Th}| = 90\,745\,943\,887\,872\,000$$
$$= 2^{15}.3^{10}.5^3.7^2.13.19.31. \tag{5.127}$$

This simple group was first constructed by Smith and Thompson (see [163], [155], [156]) as a subgroup of $E_8(3)$, several years before Griess's construction of the Monster. The fact that Th centralises an involution in the Monster implies that Th is a subgroup of the Baby Monster.

Table 5.7. Maximal subgroups of the Baby Monster

$2 \cdot (^2E_6(2)){:}2$	$2^{1+22} \cdot \mathrm{Co}_2$
$(2^2 \times F_4(2)){:}2$	$2^2.2^{10}.2^{20}.(\mathrm{M}_{22}{:}2 \times S_3)$
$S_4 \times {}^2F_4(2)$	$2^3.[2^{32}] \cdot (S_5 \times \mathrm{GL}_3(2))$
$2^9.2^{16}\mathrm{Sp}_8(2)$	$2^5.2^5.2^{10}.2^{10} \cdot \mathrm{GL}_5(2)$
$S_3 \times \mathrm{Fi}_{22}{:}2$	$3^{1+8}{:}2^{1+6} \cdot \mathrm{PSU}_4(2).2$
$(3^2{:}D_8 \times \mathrm{PSU}_4(3).2^2).2$	$3^2.3^3.3^6.(S_4 \times 2S_4)$
$5{:}4 \times \mathrm{HS}{:}2$	$5^{1+4}{:}2^{1+4}A_5.4$
$5^2{:}4S_4 \times S_5$	$5^3 \cdot \mathrm{SL}_3(5)$
$47{:}23$	
$S_5 \times \mathrm{M}_{22}{:}2$	
$(S_6 \times \mathrm{PSL}_3(4){:}2).2$	
$(S_6 \times S_6).4$	
Fi_{23}	$\mathrm{PSL}_2(49) \cdot 2$
Th	$\mathrm{PSL}_2(31)$
$\mathrm{HN}{:}2$	$\mathrm{PGL}_2(17)$
M_{11}	$\mathrm{PGL}_2(11)$
	$\mathrm{L}_3(3)$

In fact, Th has a 248-dimensional representation over $\mathbb{Q}$, and when this representation is reduced modulo 3, it acquires an invariant Lie algebra structure. The basic idea of the Smith–Thompson construction is to make the subgroups $2^{1+8} \cdot A_9$ and $2^5 \cdot \mathrm{GL}_5(2)$ intersecting in $2^{1+8}A_8$. Now the so-called 'Dempwolff group' $2^5 \cdot \mathrm{GL}_5(2)$ is a subgroup of $E_8(\mathbb{C})$ and it is possible (though not easy) to obtain generators for it in this way. Reducing modulo 3 so that we are working in $E_8(3)$, we find the involution centraliser $2^{1+8}A_8$ in $2^5\mathrm{GL}_5(2)$, and the full centraliser $2 \cdot \mathrm{P\Omega}_{16}^+(3)$ in $E_8(3)$ of the same involution.

Thus the problem now is to extend 2^8A_8 to 2^8A_9 inside $\mathrm{P\Omega}_{16}^+(3)$. Note that the Schur multiplier of $\mathrm{P\Omega}_{16}^+(3)$ is 2^2, so there are three double covers: one is $\Omega_{16}^+(3)$, and the other two are both isomorphic to the involution centraliser in $E_8(3)$. Working inside $\Omega_{16}^+(3)$ for simplicity, our group 2^8A_8 lifts to $2^{1+8}A_8$, and the normaliser of this extraspecial 2^{1+8} in $\Omega_{16}^+(3)$ is $2_+^{1+8} \cdot \Omega_8^+(2)$. Moreover, there is a unique A_9 containing the given subgroup A_8 inside $\Omega_8^+(2)$. Mapping back to the simple group $\mathrm{P\Omega}_{16}^+(3)$ we see that there is a unique 2^8A_9 containing the given 2^8A_8 in $\mathrm{P\Omega}_{16}^+(3)$.

The corresponding subgroup $2^{1+8}A_9$ in $2 \cdot \mathrm{P\Omega}_{16}^+(3)$ acts as $120 \oplus 128$ on the Lie algebra for $E_8(3)$. So far, what we have done does not depend on the characteristic being 3. However, in all other characteristics $p > 3$ there are two 128-dimensional irreducible representations of $2^{1+8}A_9$, and it turns out that one of these lies in the Lie algebra for $E_8(p)$ and the other one is in the 248-dimensional representation of the Thompson group. In characteristic 3, exceptionally, there is only one such 128-dimensional representation, and therefore Th is a subgroup of $E_8(3)$.

The maximal subgroups of Th were determined by Linton [122] and are listed in Table 5.8. The fact that they are all relatively small makes the group hard to study. The largest maximal subgroup is $^3D_4(2){:}3$, which has index 143127000.

Table 5.8. Maximal subgroups of the 'small monsters'

Th	HN	He
$^3D_4(2){:}3$	A_{12}	$\mathrm{Sp}_4(4){:}2$
$2^5{\cdot}\mathrm{GL}_5(2)$	$2{\cdot}\mathrm{HS}{:}2$	$2^2{\cdot}\mathrm{PSL}_3(4){:}S_3$
$2^{1+8}{\cdot}A_9$	$\mathrm{PSU}_3(8){:}3$	$2^6{:}3{\cdot}S_6$
$\mathrm{PSU}_3(8){:}6$	$2^{1+8}_+{\cdot}(A_5 \times A_5).2$	$2^6{:}3{\cdot}S_6$
$(3 \times G_2(3)){:}2$	$(D_{10} \times \mathrm{PSU}_3(5)){\cdot}2$	$2^{1+6}{:}\mathrm{GL}_3(2)$
$3.[3^8].2S_4$	$5^{1+4}{:}2^{1+4}.5.4$	$7^2{:}\mathrm{SL}_2(7)$
$3^2.[3^7].2S_4$	$2^6{\cdot}\mathrm{PSU}_4(2)$	$3{\cdot}S_7$
$3^5{:}2{\cdot}S_6$	$(A_6 \times A_6){\cdot}D_8$	$7^{1+2}_+{:}(S_3 \times 3)$
$5^{1+2}_+{:}4S_4$	$2^3.2^2.2^6.(3 \times \mathrm{GL}_3(2))$	$S_4 \times \mathrm{GL}_3(2)$
$5^2{:}\mathrm{GL}_2(5)$	$5^2.5^{1+2}.4{\cdot}A_5$	$7{:}3 \times \mathrm{GL}_3(2)$
$7^2{:}(3 \times 2S_4)$	$\mathrm{M}_{12}{:}2$	$5^2{:}4A_4$
$\mathrm{PGL}_2(19)$	$\mathrm{M}_{12}{:}2$	
$\mathrm{PSL}_3(3)$	$3^4{:}2{\cdot}(A_4 \times A_4).4$	
$A_6{\cdot}2$	$3^{1+4}_+{:}4{\cdot}A_5$	
$31{:}15$		
S_5		

Some of these subgroups can be obtained by centralising in $\mathbb{M}$ certain subgroups containing the S_3 which centralises Th. Here again Norton's Monstraliser list (see Section 5.8.4 and [142]) is very useful. For example, the maximal subgroup $\mathrm{PSU}_3(8){:}6$ of Th comes from

$$S_3 \times \mathrm{PSU}_3(8){:}6 < (A_5 \times \mathrm{PSU}_3(8){:}3){:}2 < \mathbb{M}.$$

5.8.8 The Harada–Norton group

If x is an element of order 5 in $2^{1+24}{\cdot}\mathrm{Co}_1$, mapping to an element of class $5B$ in the quotient group Co_1, then its centraliser in $2^{1+24}{\cdot}\mathrm{Co}_1$ has shape $5 \times 2^{1+8}{\cdot}(A_5 \times A_5).2$. Just as in the case of the Thompson group, we expect to find that $C_{\mathbb{M}}(x)/\langle x \rangle$ is a new simple group HN, with involution centraliser of shape $2^{1+8}{\cdot}(A_5 \times A_5).2$. This indeed turns out to be the case, and HN is a simple group of order

$$|\mathrm{HN}| = 2^{14}.3^6.5^6.7.11.19$$
$$= 273\,030\,912\,000\,000. \qquad (5.128)$$

The normaliser of x in the Monster has shape $(D_{10} \times \mathrm{HN}){\cdot}2$. The fact that HN centralises D_{10} in the Monster implies that HN is also a subgroup of the

Baby Monster. Moreover, this D_{10} extends to an A_5 which has normaliser $(A_5 \times A_{12})$:2 in the Monster (see Sections 5.8.4 and 5.8.5), and therefore HN has a subgroup A_{12}. In fact this is the largest subgroup, and has index 1140000.

The corresponding permutation representation of HN on 1140000 points was constructed (implicitly) by Norton in his Ph. D. thesis [140]. The points are identified with certain vectors in a 133-dimensional space over $\mathbb{Q}[\sqrt{5}]$, on which the group acts irreducibly. One can define a graph of degree 462 on the 1140000 points by joining two conjugates of A_{12} just when they intersect in $(A_6 \times A_6).2.2$. The automorphism group of this graph is HN:2, which has rank 10 in its action on the vertices. The suborbit lengths are

$$1 + 462 + 5040 + 10395 + 16632 + 30800 + 69300 + 311850 + 332640 + 362880.$$

On restriction to HN the suborbits of length 5040 and 332640 each split into two orbits of half the size, and the rank of the action increases to 12.

Again, as in the case of the Thompson group, many of the maximal subgroups may be obtained by centralising (in the Monster) subgroups containing the D_{10} which centralises HN. For example, besides the A_{12} already mentioned, there is a maximal subgroup $\mathrm{PSU}_3(8)$:3 obtained by centralising another A_5 in the Monster, and $(A_6 \times A_6).D_8$ obtained by centralising A_6, and M_{12}:2 obtained by centralising $\mathrm{PSL}_2(11)$. The full list of maximal subgroups is given in Table 5.8.

The fact that the smallest representation of HN has degree 133, which is the dimension of the adjoint representation of the groups of type E_7, has suggested to many people that HN should be a subgroup of such a group, most probably $E_7(5)$. However, it is not hard to show that this is not the case.

5.8.9 The Held group

The Held group He not only centralises an element of order 7 in the Monster, but also an element of order 3 which normalises the given element of order 7. Thus we see He as a subgroup of $3 \cdot \mathrm{Fi}_{24}$ and therefore as a subgroup of the simple Fischer group Fi_{24}'. Indeed, the full normaliser of the group of order 7 in the Monster is $(7{:}3 \times \mathrm{He})$:2, containing $\mathrm{Aut}(\mathrm{He}) = \mathrm{He}$:2, so there are two conjugacy classes of He:2 in Fi_{24}', interchanged by the outer automorphism. It turns out that both He and He:2 have orbits of lengths $2058 + 24990 + 281946$ on the 306936 transpositions of Fi_{24}.

It is possible, though not easy, to build the Held group from scratch by constructing a graph on 2058 vertices. The point stabiliser in this permutation representation is $\mathrm{Sp}_4(4)$:2, the split extension of $\mathrm{Sp}_4(4)$ by its field automorphism, and the suborbit lengths are

$$1 + 136 + 136 + 425 + 1360.$$

The two suborbits of length 136 are interchanged by the outer automorphism of He, which extends $Sp_4(4){:}2$ to $Sp_4(4){:}4$. The stabiliser in $Sp_4(4)$ of a point in one 136-orbit is $GO_4^+(4)$, and in the other is $Sp_2(4) \wr 2$, and these subgroups are interchanged by the graph automorphism (see Section 3.5.5). Thus there is a directed graph of in-degree and out-degree 136 whose automorphism group turns out to be He. Ignoring the directions gives a graph of degree 272 whose automorphism group is He:2. Once all this has been proved, it follows that the order of the Held group is

$$\begin{aligned}
|\text{He}| &= 2058.|Sp_4(4){:}2| \\
&= 2^{10}.3^3.5^2.7^3.17 \\
&= 4\,030\,387\,200.
\end{aligned} \tag{5.129}$$

The Held group was originally discovered in an attempt to characterise M_{24} as the unique simple group with involution centraliser $2^{1+6}{:}GL_3(2)$. However, it was found that there are exactly three simple groups with this involution centraliser, namely M_{24}, $GL_5(2)$, and a previously unknown simple group, now known as the Held group. Moreover, this new group also contains two conjugacy classes of subgroups $2^6{:}3 \cdot S_6$, isomorphic to the sextet stabiliser in M_{24} (see Section 5.2.1). These two classes are interchanged by an outer automorphism, which may be chosen to centralise a complement $3 \cdot S_6$. The maximal subgroups of the Held group, determined by Greg Butler [18] are listed in Table 5.8.

Now the permutation character on 2058 points contains two (complex conjugate) irreducible characters of degree 51, so each of these two 51-dimensional representations contains an orbit of 2058 vectors under the action of the Held group. Moreover, a basis can be chosen so that a subgroup $2^6{:}3 \cdot S_6$ acts monomially. Explicit matrices for these representations have been constructed by Stephen Rogers [146].

5.8.10 Ryba's algebra

Each of these two 51-dimensional representations of the Held group, reduced modulo 7, contains a 50-dimensional irreducible constituent. The latter extends to He:2, and Ryba [149] gives a concise description of the action of some generators. For simplicity he takes a 51-dimensional space with basis vectors v_i indexed by 6 points i and v_x indexed by the 45 hexacode words x of weight 4 (see Section 5.2.1). The 50-space may then be taken as the subspace consisting of vectors such that the coefficients of $v_1, \ldots, v_6$ add to 0 (modulo 7). Then $3 \cdot S_6$ permutes the v_i naturally, and permutes the v_x in the same way that it permutes the vectors x in the hexacode. The 2^6 is identified with the dual of the hexacode, regarded as a 6-dimensional space over $\mathbb{F}_2$, and may be generated by six elements t_i, defined to negate v_x if and only if the hexacode word x has ω or $\overline{\omega}$ in the ith coordinate.

It remains to define the action of the extra generator. Since it commutes with $3 \cdot S_6$, it is enough to specify the images of v_1 and v_{001111}. It turns out that

$$v_1 \mapsto 5v_1 + 2\sum_{j \neq i} v_j + \sum_{x_i = 0} v_x + 3\sum_{x_i \neq 0} v_x,$$

and if $x = 001111$ then

$$v_x \mapsto v_1 + v_2 + 3(v_3 + v_4 + v_5 + v_6) + v_{\bar{x}} + 4v_{\omega x} + 4v_{\bar{\omega}x}$$
$$+ 3\sum_{y \in Y} v_y + \sum_{z \in Z} (5v_z + 6v_{\omega z} + 6v_{\bar{\omega}z}), \qquad (5.130)$$

where Y is the set of 24 hexacode words of shape $\lambda(01 \mid 01\omega\bar{\omega})$ (so that if $y \in Y$ then $x + \lambda y$ is a word of weight 4 for every λ) and Z is the set of 6 hexacode words of shape $(11 \mid 1100)$ or $(\omega\bar{\omega} \mid 0101)$ (so that if $z \in Z$ then $x + z$ has weight 4 but $x + \omega z$ does not).

Of course, we cannot use this definition effectively without some geometric structure of which He:2 is the automorphism group. Ryba constructs an algebra structure and checks it is invariant under the group. It is easy to check that the inner product defined by saying that the given basis of the 51-space is orthonormal, is invariant. Now the algebra product $*$ is symmetric, and satisfies the rule $a * b.c = a.b * c$. In other words the form $t(a, b, c) = (a * b).c$ is a symmetric trilinear form. It may be defined by

$$\begin{aligned}
v_i * v_i &= -v_i, \\
v_i * v_j &= 0, \\
v_i * v_x &= 0 \text{ if } x_i = 0, \\
v_i * v_x &= 5v_x \text{ if } x_i \neq 0, \\
v_x * v_{\omega x} &= 3v_{\bar{\omega}x}, \\
v_x * v_y &= v_{x+y} \text{ if } x + \lambda y \text{ has weight 4 for every } \lambda, \\
v_x * v_y &= 2v_{x+y} \text{ if } x + y \text{ has weight 4 but } x + \omega y \text{ does not}, \\
v_x * v_y &= 0 \text{ otherwise.} \qquad (5.131)
\end{aligned}$$

For details of the proof that this defines the Held group, see [149].

5.9 Pariahs

There are just six sporadic simple groups which are not involved in the Monster. The first of these, J_1, was found by Janko [99] in 1965, and caused a storm of excitement at the time, as it was the first sporadic simple group to be discovered for nearly 100 years. The last of the sporadic simple groups, J_4, was also discovered by Janko, ten years later.

The basic properties of these groups are listed in Table 5.9 and 5.10. Their maximal subgroups are listed in Table 5.11. We shall consider them in order, from the smallest to the largest.

Table 5.9. Basic information about the pariahs

G	M	A	$\|G\|$
J_1	1	1	$175\,560 = 2^3.3.5.7.11.19$
J_3	3	2	$50\,232\,960 = 2^7.3^5.5.17.19$
Ru	2	1	$145\,926\,144\,000 = 2^{14}.3^3.5^3.7.13.29$
O'N	3	2	$460\,815\,505\,920 = 2^9.3^4.5.7^3.11.19.31$
Ly	1	1	$51\,765\,179\,004\,000\,000 = 2^8.3^7.5^6.7.11.31.37.67$
J_4	1	1	$86\,775\,571\,046\,077\,562\,880 = 2^{21}.3^3.5.7.11^3.23.29.31.37.43$

Note: M is the Schur multiplier (see Section 2.7.1) and A the outer automorphism group.

Table 5.10. Minimal representations of the pariahs

G	d	q	Notes	n	H
J_1	7	11	$J_1 < G_2(11)$	266	$PSL_2(11)$
J_3	9	4	$3{\cdot}J_3 < SU_9(2)$	6\,156	$PSL_2(16){:}2$
Ru	28	2	$Ru < \Omega_{28}^-(2)$	4\,060	$^2F_4(2)$
O'N	45	7	$3{\cdot}O'N < SL_{45}(7)$	122\,760	$PSL_3(7){:}2$
Ly	111	5	$Ly < \Omega_{111}(5)$	8\,835\,156	$G_2(5)$
J_4	112	2	$J_4 < \Omega_{112}^+(2)$	173\,067\,389	$2^{11}{:}M_{24}$

Note: d is the minimum degree of a matrix representation of a covering group of G, and q is the order of the field; n is the minimum permutation degree, and H is the point stabiliser.

Table 5.11. Maximal subgroups of the Pariahs

J_1	J_3	Ru	O'N	Ly	J_4
$PSL_2(11)$	$PSL_2(16){:}2$	$^2F_4(2)$	$PSL_3(7){:}2$	$G_2(5)$	$2^{11}{:}M_{24}$
$2^3{:}7{:}3$	$PSL_2(19)$	$(2^6{:}PSU_3(3)){:}2$	$PSL_3(7){:}2$	$3{\cdot}McL{:}2$	$2^{1+12}{\cdot}3{\cdot}M_{22}{:}2$
$2 \times A_5$	$PSL_2(19)$	$(2^2 \times Sz(8)){:}3$	J_1	$5^3{\cdot}SL_3(5)$	$2^{10}{:}GL_5(2)$
19:6	$2^4{:}(3 \times A_5)$	$2^{3+8}{:}GL_3(2)$	$4{\cdot}PSL_3(4){:}2$	$2{\cdot}A_{11}$	$2^{3+12}{\cdot}(S_5 \times GL_3(2))$
11:10	$PSL_2(17)$	$PSU_3(5){:}2$	$(3^2{:}4 \times A_6){\cdot}2$	$5^{1+4}{:}4S_6$	$PSU_3(11){:}2$
$D_6 \times D_{10}$	$(3 \times A_6){:}2$	$2{\cdot}2^{4+6}{:}S_5$	$3^4{:}2_-^{1+4}D_{10}$	$3^5{:}(2 \times M_{11})$	$M_{22}{:}2$
7:6	$3^2.3^{1+2}{:}8$	$PSL_2(25){\cdot}2^2$	$PSL_2(31)$	$3^{2+4}{:}2A_5.D_8$	$11_+^{1+2}{:}(5 \times 2S_4)$
	$2^{1+4}A_5$	A_8	$PSL_2(31)$	67:22	$PSL_2(32){:}5$
	$2^{2+4}{:}(3 \times S_3)$	$PSL_2(29)$	$4^3{\cdot}GL_3(2)$	37:18	$PSL_2(23){:}2$
		$5^2{:}GL_2(5)$	M_{11}		$PSU_3(3)$
		$3{\cdot}A_6{\cdot}2^2$	M_{11}		29:28
		$5^{1+2}{:}[2^5]$	A_7		43:14
		$PGL_2(13)$	A_7		37:12
		$A_6{\cdot}2^2$			
		$5{:}4 \times A_5$			

5.9.1 The first Janko group J_1

The easiest construction of J_1 is as a subgroup of $G_2(11)$, that is as a group of automorphisms of the octonions over $\mathbb{F}_{11}$ (see Section 4.3.2). Take first the monomial group $2^3{:}7{:}3$ generated by the sign change a on i_0, i_3, i_5 and i_6, and the coordinate permutations $b = (0,1,2,3,4,5,6)$ and $c = (1,2,4)(3,6,5)$. Next we adjoin an involution d which inverts b and commutes with c. Now the normaliser of $\langle b \rangle$ in $G_2(11)$ is contained in the maximal subgroup $\mathrm{SL}_3(11){:}2$ (see Section 4.3.6) and has shape $(7 \times 19){:}6$. Therefore, inside $G_2(11)$ there is a unique such involution, which happens to be the map

$$d : i_t \mapsto 9i_{-t} + (i_{1-t} + i_{2-t} + i_{4-t}) + 3(i_{3-t} + i_{6-t} + i_{5-t}). \quad (5.132)$$

It is worth remarking that the same construction inside $G_2(3^{2n+1})$ yields the Ree group ${}^2G_2(3^{2n+1})$ (see Section 4.5): in this case the 7-normaliser in $G_2(3^{2n+1})$ lies inside $\mathrm{SU}_3(3^{2n+1}){:}2$ and has shape $(q^2 - q + 1){:}6$ so again the extension is unique. There are other similarities between J_1 and the small Ree groups, which led Janko to conjecture that J_1 might be the first in a series of simple groups of order $q(q^3 - 1)(q + 1)$ for q a power of 11.

Of course, it is not easy to determine many properties of the group J_1 by hand from the above construction, but with a computer it is very easy to calculate almost everything one might want. The order of the group is

$$|J_1| = 175560 = 2^3.3.5.7.11.19, \quad (5.133)$$

and the monomial subgroup $2^3{:}7{:}3$ used in the construction is maximal, as is the subgroup $7{:}6$ generated by b, c, d. The other maximal subgroups are $\mathrm{PSL}_2(11)$, $2 \times A_5$, $S_3 \times D_{10}$, $11{:}10$ and $19{:}6$.

A useful technique for finding an involution centraliser is the so-called *Bray trick* [15]: if a is an involution and g is any element, then the group generated by a and a^g is dihedral, of order $2k$, say. If k is odd, say $k = 2m + 1$, then $g(aa^g)^m$ centralises a, while if k is even, say $k = 2m$, then both $(aa^g)^m$ and $(aa^{g^{-1}})^m$ centralise a. In our case, we find that aa^{db} has order 5, so that $db(aa^{db})^2$ is in the centraliser of a, and in fact

$$\langle a, a^b, c, db(aa^{db})^2 \rangle \cong 2 \times A_5 \quad (5.134)$$

which is the full involution centraliser.

Inside this subgroup there is an element e acting on coordinates 0, 3, 5,

$$6 \text{ as } -\begin{pmatrix} 3 & -1 & -1 & -1 \\ -1 & 3 & 2 & 3 \\ -1 & 2 & 3 & 3 \\ -1 & 3 & 3 & 2 \end{pmatrix}, \text{ and as } \begin{pmatrix} 9 & 7 & 5 \\ 7 & 5 & 9 \\ 5 & 9 & 7 \end{pmatrix} \text{ on coordinates } 1, 2, 4.$$

Further calculation reveals that

$$\langle a, c, d, e \rangle \cong S_3 \times D_{10},$$
$$\langle a^{b^3}, a^d, c, e \rangle \cong \mathrm{PSL}_2(11). \quad (5.135)$$

The latter group acts irreducibly on the 7-space, and its normaliser in $G_2(11)$ is a maximal subgroup $\mathrm{PGL}_2(11)$.

It turns out that $\mathrm{PSL}_2(11)$ is the largest subgroup of J_1 and it has index 266. Indeed, Livingstone [123] constructed J_1 as the automorphism group of a graph of degree 11 on 266 vertices, thus providing an independent proof that J_1 contains $\mathrm{PSL}_2(11)$. (See the Atlas [28] for a concise description of this graph and the action of the group on it.) The structure of the graph is summarised in the following picture.

$$
\underset{}{\boxed{1}} \;\overset{11}{\text{———}}\; \underset{}{\boxed{11}} \;\overset{10}{\text{———}}\; \underset{}{\boxed{110}} \;\overset{6}{\text{———}}\; \underset{}{\boxed{132}} \;\overset{1}{\text{———}}\; \underset{}{\boxed{12}} \tag{5.136}
$$

There are several nice presentations of J_1 in terms of generators and relations. For example, we may take the Coxeter-type presentation (as defined in Section 2.8.2)

$$
\underset{p}{\bullet} \;\text{———}\; \underset{q}{\bullet} \;\overset{5}{\text{———}}\; \underset{r}{\bullet} \;\text{———}\; \underset{s}{\bullet} \;\overset{5}{\text{———}}\; \underset{t}{\bullet} \tag{5.137}
$$

and factor by the relations $(pqr)^5 = 1$, and $(rst)^5 = p$. This exhibits the subgroups $\mathrm{PSL}_2(11)$ generated by p, q, r, s and $2 \times A_5$ generated by p, q, r, t, as well as $S_3 \times D_{10}$ generated by p, q, s, t. (See [45] for more details.)

It is not hard to show that J_1 has trivial outer automorphism group (see Exercise 5.29) and trivial Schur multiplier (see [99]). It *is* hard to show that J_1 is not contained in the Monster (see [177]). Another interesting but still rather mysterious fact about J_1 is that it is contained in the O'Nan group (see Section 5.9.4), as the centraliser of a non-inner automorphism of order 2 (and is in fact a maximal subgroup).

5.9.2 The third Janko group J_3

This group was first discovered by Janko [100], defined as a simple group with a single conjugacy class of involutions, each involution having centraliser of shape $2^{1+4}A_5$. This contrasts with the second Janko group J_2 (also known as the Hall–Janko group), which also has an involution centraliser of shape $2^{1+4}A_5$, but has two classes of involutions. The group J_3 has order $50232960 = 2^7.3^5.5.17.19$, and was first constructed by Higman and McKay by coset enumeration (see [103]), giving a permutation representation on 6156 points (see [77]). This is the smallest permutation representation, and the point stabiliser is $\mathrm{PSL}_2(16){:}2$, of order 8160. The rank is 8, and since the order of the point stabiliser is only slightly bigger than the number of points, it would be almost impossible to construct the group by hand in this way.

A later construction by Conway and Wales [33] was more geometrical: they constructed an 18-dimensional complex representation of the triple cover $3{\cdot}J_3$, using the subgroup $2^4{:}(3 \times A_5)$ as a 'nearly monomial' subgroup to get started. Indeed, the representation restricts to this subgroup as the direct sum of a

15-dimensional monomial representation and a 3-dimensional representation coming from the action of A_5 as symmetries of a regular dodecahedron.

The nearest we can come to a convenient or elegant construction of the group, is to construct its triple cover inside $SU_9(2)$. This representation was found entirely by accident, when Richard Parker was demonstrating an early version of his Meataxe programs [143] to me, and this previously unsuspected representation emerged. (Actually, he tells me that he found it the previous night, as an 18-dimensional representation over $\mathbb{F}_2$, but did not notice at the time that he had found something new.) Soon afterwards, Benson and Conway produced the version I shall now describe. We have to construct an action of $3 \cdot J_3$ on 9-dimensional unitary space over $\mathbb{F}_4 = \{0, 1, \omega, \overline{\omega}\}$. It is convenient to label the 9 coordinate vectors e_z of an orthonormal basis by elements $z = x + iy$ of the field $\mathbb{F}_9 = \{0, \pm 1, \pm i, \pm 1 \pm i\}$ of order 9. We write these coordinates in the order $0, 1, i - 1, (i - 1)^2, \ldots, (i - 1)^7$.

There is a diagonal subgroup of order 3^4 generated by scalars and the elements $A_1 = \operatorname{diag}(1, \omega, \omega, 1, \omega, \overline{\omega}, \overline{\omega}, 1, \overline{\omega})$, $A_i = \operatorname{diag}(1, 1, \overline{\omega}, \omega, \omega, 1, \omega, \overline{\omega}, \overline{\omega})$ and $E_0 = \operatorname{diag}(1, \omega, \overline{\omega}, \omega, \overline{\omega}, \omega, \overline{\omega}, \omega, \overline{\omega})$, which can be expressed by the formulae

$$A_Z : e_z \mapsto \omega^{Xx+Yy} e_z,$$
$$E_Z : e_z \mapsto \omega^{(X-x)^2+(Y-y)^2} e_z, \tag{5.138}$$

where $Z = X + iY \in \mathbb{F}_9$. Then E_Z is a scalar multiple of $E_0 A_Z$.

Modulo this diagonal subgroup, there is a 2-transitive action on the nine coordinates, forming a quotient group of shape $3^2{:}8$. This may be generated by the 'translations' $D_Z : e_z \mapsto \lambda e_{z+Z}$, where $\lambda = \omega$ if $Z/z = i - 1$, $\lambda = \overline{\omega}$ if $Z/z = -i - 1$, and $\lambda = 1$ otherwise, together with the cyclic permutations of the last 8 coordinates (i.e. the maps $B_Z : e_z \mapsto e_{Zz}$ for $Z \neq 0$).

Thus we obtain a monomial group of order $3^6.8$ (mapping modulo scalars to a group of order $3^5.8$). This is a maximal subgroup of $3 \cdot J_3$, and we can extend to $3 \cdot J_3$ by adjoining some other elements such as

$$C_Z : e_0 \mapsto e_0,$$
$$e_z \mapsto e_{-Z/z} + e_{iZ/z} + e_{-iZ/z} \text{ for } z \neq 0. \tag{5.139}$$

The stabiliser of the 1-space $\langle e_0 \rangle$ is generated modulo scalars by the maps A_Z, B_Z, C_Z and E_0. This group has the shape $3 \times (3 \times A_6){:}2$, and is the normaliser of the cyclic group $\langle E_0 \rangle$ of order 3. It turns out that there are 23256 images of this 1-space, and thus the order of J_3 is 50 232 960.

The stabiliser of the 1-space $\langle e_0 + e_{-1} + e_1 \rangle$ is a group $3 \times PSL_2(17)$ generated modulo scalars by D_1 and C_{-1}. This group $PSL_2(17)$ is the centraliser of the outer automorphism $\lambda e_z \mapsto \lambda^2 e_{\overline{z}}$. There are 20520 images of this 1-space, thereby accounting for all the $20520 + 23256 = 43776 = 2^8(2^9 - 1)/3$ non-isotropic 1-spaces.

The $43605 = (2^8 - 1)(2^9 + 1)/3$ isotropic 1-spaces are all in the same orbit under $3 \cdot J_3$, and the stabiliser of one of them (such as $\langle e_1 + e_{-1} \rangle$) is a group

$3 \times 2^{2+4}{:}(3 \times S_3)$. We can see the elements A_i, E_0, B_{-1}, and $C_{\pm 1}$ in this stabiliser, but these elements do not generate it.

Besides these groups, the other maximal subgroups of J_3 are the involution centraliser $2^{1+4}A_5$, the largest subgroup $\mathrm{PSL}_2(16){:}2$ of index 6156, two conjugacy classes of $\mathrm{PSL}_2(19)$, and a subgroup $2^4{:}(3 \times A_5)$. This was first proved by Finkelstein and Rudvalis [61].

The interesting fact that $3{\cdot}J_3 < 3{\cdot}E_6(4)$ was first shown by Peter Kleidman [114].

5.9.3 The Rudvalis group

The Rudvalis group has order $145\ 926\ 144\ 000 = 2^{14}.3^3.5^3.7.13.29$, and its smallest representations have degree 28. These are actually representations of the double cover $2{\cdot}\mathrm{Ru}$ over the complex numbers, or over any field of odd characteristic containing a square root of -1, but they also give rise to representations of the simple group Ru over the field $\mathbb{F}_2$ of order 2. The original construction by Conway and Wales [34] used the subgroup $\mathrm{PSL}_2(25)$ to help make a 28-dimensional complex representation of $2{\cdot}\mathrm{Ru}$, and simultaneously construct a permutation representation on 16240 points, which gives rise to a permutation representation of the simple group Ru on 4060 points.

This permutation representation has rank 3, and the suborbit lengths are $1 + 1755 + 2304$. Thus the Rudvalis group can be described as the automorphism group of a regular graph of degree 1755 or 2304 on 4060 vertices. The structure of the graph is summarised in the following picture.

$$
\begin{array}{cccccc}
 & & 730 & & 975 & \\
\boxed{1} & \xrightarrow{\ 1755\ } & \underset{1755}{\overset{1}{\boxed{}}} & \xrightarrow{\ 1024\ \ 780\ } & \boxed{2304} & \quad (5.140)
\end{array}
$$

The vertex stabiliser is $^2F_4(2)$, and in the valency 1755 case the stabiliser of a pair of non-adjacent vertices is $\mathrm{PSL}_2(25).2^2$. The stabiliser of an edge is then a non-maximal subgroup $2^{1+4+6}.5.4$, contained in the involution centraliser of shape $2^{1+4+6}S_5$. The orbit of length 1755 corresponds to the points of the generalised octagon for $^2F_4(2)$ (see Section 4.9.4). Two points are joined in the graph just when they are perpendicular in the 26-dimensional natural representation of $^2F_4(2)$.

Lifting to the double cover $2{\cdot}\mathrm{Ru}$, and looking at the 28-dimensional complex representation, we find that the subgroup $2{\cdot}^2F_4(2)$ has a normal simple subgroup $^2F_4(2)'$ of index 4, which fixes a vector. Modulo this subgroup is a group C_4 which multiplies the vector by scalars $\pm 1, \pm i$.

Here I shall instead describe the 28-dimensional complex representations with respect to a basis in which a group $2^6{\cdot}G_2(2)$ appears as the monomial subgroup. The group $G_2(2)$ acts on 2^6 as described in Section 4.4.3 but the extension is non-split. Recall from Section 4.4.4 that $G_2(2) \cong \mathrm{SU}_3(3){:}2$. The required group $2^6{\cdot}G_2(2)$ contains a subgroup $2^6{:}\mathrm{SU}_3(3)$ of index 2 which is a split extension, so there is a subgroup $\mathrm{SU}_3(3)$.

The 28 coordinates are labelled by the isotropic 1-spaces in the natural representation of $SU_3(3)$, subject to the rule $e_{\lambda v} = \lambda^* e_v$, where for $\pm\lambda = 1$, i, $1 + i$, $1 - i$ respectively in $\mathbb{F}_9$, $\lambda^* = 1, -1, -i, i$ in $\mathbb{C}$. (In the language of representation theory, $\lambda^* = \chi(\lambda)$, where χ is a suitable character of the cyclic group of order 8.) There is then a subgroup $SU_3(3)$ acting monomially on the coordinates as $g : e_v \mapsto e_{vg}$ in the natural way.

We can extend $SU_3(3)$ to the full group $(2^6{:}SU_3(3)){:}2$ by adjoining an involution defined as follows. We first choose a basis $\{e_v\}$ by picking a particular vector v in each isotropic 1-space: we pick the first one arbitrarily, and then take the 27 isotropic vectors which have inner product 1 with the first one. For example, writing v with respect to an orthonormal basis of the unitary 3-space, we may pick the 28 vectors $v = (v_1, v_2, v_3)$ which are either $(1, 1, 1)$, or satisfy $v_1 + v_2 + v_3 = 1$. More, explicitly, v is obtained from one of the vectors

$$(\pm 1, 1, 1), \ (1, i, -i), \ (-1 + i, 1 + i, 1 + i), \ (1 + i, -i, 0)$$

by applying coordinate permutations and/or the Frobenius automorphism $(x, y, z) \mapsto (x^3, y^3, z^3)$ which maps i to $-i$. Now define the extending involution on the complex 28-space by $e_v \mapsto e_{\overline{v}}$ for these basis vectors e_v. Note that this map conjugates the original $SU_3(3)$ to a different (indeed, non-conjugate) $SU_3(3)$ in $2^6{:}SU_3(3)$. For this other copy of $SU_3(3)$, the scalars λ^* are replaced by their complex conjugates.

For more details of this description of the Rudvalis group, including an extra element which extends this maximal subgroup to $2{\cdot}Ru$, see [175].

The 4060 distinguished 1-spaces which are permuted by the Rudvalis group fall into three orbits, of lengths $252 + 2016 + 1792$, under the action of $2^6{\cdot}G_2(2)$. Representatives of these orbits can be described as follows:

(i) Fix a non-isotropic vector u, and let v, w, $v + w$, $v - w$ be the four isotropic vectors orthogonal to u (up to scalars). Then take

$$X(u) = \langle e_v + e_w + e_{v+w} + e_{v-w}\rangle.$$

For example, we may take $u = (1, 0, 0)$ and choose $v = (0, 1, 1 + i)$ and $w = (0, i, 1 - i)$. There are (up to scalars) four sign combinations for each of the 63 choices for u, making 252 such 1-spaces which are images under the monomial group of $X(u)$.

(ii) Fix an orthonormal basis $\{u, v, w\}$, ordered so that the matrix with rows u, v, w in that order has determinant 1. Take the 16 isotropic vectors x (up to scalars) which are orthogonal to none of u, v, w. Choose appropriate scalar multiples so that the coordinates of x with respect to the ordered basis $\{u, v, w\}$ are cyclic rotations of $(\pm 1, 1, 1)$, $(1, \pm i, -i)$, $(i, \pm 1, -1)$. Then take

$$Y(u, v, w) = \tfrac{1}{2}\sum_x e_x.$$

There are (modulo scalars) 32 sign combinations for each of the 63 orthonormal bases, making 2016 such 1-spaces, i.e. the images under the group of $Y(u, v, w)$.

(iii) Fix an isotropic vector v, and take

$$Z(v) = \frac{1-2i}{2+2i}e_v + \frac{1}{2+2i}\sum_{v.w=1} e_w.$$

(This is the same choice of basis vectors that we used in constructing the outer half of the group $(2^6{:}SU_3(3)){:}2$.) There are 64 sign combinations for each of the 28 isotropic 1-spaces, making a total of 1792 such 1-spaces, images of $Z(v)$.

Reducing this complex representation of $2{\cdot}Ru$ modulo 2 gives a 28-dimensional orthogonal representation of the simple Rudvalis group. Norton calculated that there are just 10 orbits of the group on non-zero vectors in this space, as listed in Table 5.12. In fact, seven of these vector stabilisers

Table 5.12. Vector stabilisers in the Rudvalis group

Stabiliser	Orbit length	Norm
$^2F_4(2)$	4060	0
$(2^6{:}SU_3(3)){:}2$	188500	0
$(2^2 \times Sz(8)){:}3$	417600	0
$2.2^4.2^6.S_5$	593775	0
$PSL_2(25){\cdot}2^2$	4677120	1
S_7	28953600	1
$3{\cdot}A_6{\cdot}2^2$	33779200	1
$[2^8.3.5]$	38001600	0
$PSL_2(13){:}2$	66816000	1
$[2^9.3]$	95004000	0

are maximal subgroups: the exceptions are S_7 and the groups of orders $2^8.3.5$ and $2^9.3$. The other maximal subgroups are $2^3.2^8{:}GL_3(2)$, $5^{1+2}{:}[2^5]$, $5{:}4 \times A_5$, $5^2{:}GL_2(5)$, $A_6{\cdot}2^2$, A_8, $PSL_2(29)$, and $PSU_3(5){:}2$ (see Table 5.11 and [175]).

The Rudvalis group, like the Higman–Sims group, is a subgroup of $E_7(5)$ (see [109]). Indeed, one is struck by many other similarities between Ru and HS, for example in several of their maximal subgroups (see Section 5.5.1).

5.9.4 The O'Nan group

By the time we reach the O'Nan group, all pretence that we could do the calculations by hand has to be abandoned. The group has order

$$|O'N| = 460\,815\,505\,920 = 2^9.3^4.5.7^3.11.19.31 \tag{5.141}$$

and the smallest faithful permutation representation is on 122760 points. It was this permutation representation which was the basis of the original (unpublished) construction of the O'Nan group by Sims. The point stabiliser in this action is $\mathrm{PSL}_3(7){:}2$, and the action has rank 5, with suborbit lengths $1 + 5586 + 6384 + 52136 + 58653$. The outer automorphism of O'N interchanges two such permutation representations. We can therefore express the group O'N:2 as a group of automorphisms of a bipartite graph on 2×122760 points. In fact we may choose this graph to be regular of degree 456, and O'N:2 is the full automorphism group of this graph. The outer half of the group interchanges the two parts of the bipartite graph, and the stabiliser of an edge is a group of shape $7^{1+2}{:}(3 \times D_{16})$.

A more recent construction of this permutation representation by Soicher [157] uses coset enumeration to enumerate the cosets of a subgroup $\mathrm{PSL}_3(7){:}2$. He starts from a presentation for $\mathrm{PSL}_3(7){:}2$ obtained by adjoining the relation $af = (cd)^4$ to the Coxeter relations given by the following diagram.

$$
\begin{array}{cccccc}
\bullet & \bullet & \bullet & \overset{8}{\bullet} & \bullet & \bullet \\
a & b & c & d & e & f
\end{array}
\tag{5.142}
$$

He then adjoins a further generator g, satisfying the extra relations

$$
1 = [(cg)^2, d] = [c, (dg)^2] = afg^2 = (g^b g^c)^2 b = (g^e g^d)^2 e = (bcdg)^5.
\tag{5.143}
$$

(This is the version given in Appendix 2 of [102], and differs slightly from the version in [157].)

The smallest matrix representation is a representation of the triple cover $3{\cdot}\mathrm{O'N}$ in 45 dimensions over $\mathbb{F}_7$. Generating matrices for this representation were computed by Richard Parker, and proved to generate the required group by Alex Ryba (see [148]). This existence proof relies on the fact that the representation supports an invariant skew-symmetric trilinear form. There is an outer automorphism of $3{\cdot}\mathrm{O'N}$ which interchanges this representation with its dual, and extends the group to $3{\cdot}\mathrm{O'N}{:}2$. The trilinear form gives rise to an anti-commutative, non-associative algebra structure on the underlying 90-dimensional space.

Following Ryba [148] (see also the Atlas [28]), we build $3{\cdot}\mathrm{O'N}{:}2$ from a subgroup M_{11}, and adjoin an involution which commutes with a subgroup $\mathrm{PSL}_2(11)$ of M_{11}. We start by building the 45-dimensional ordinary irreducible representation of M_{11}. This is obtained as the exterior square of the 10-dimensional deleted permutation representation on 11 letters. Taking the 11-set $\mathbb{F}_{11} = \{0, 1, 2, 3, 4, 5, 6, 7, 8, 9, X\}$ as in Section 3.3.5 we take the copy of $\mathrm{PSL}_2(11)$ generated by the permutations $t \mapsto t + 1$, $t \mapsto 3t$ and $(3, 4)(5, 9)(2, X)(6, 7)$, and extend to M_{11} by adjoining the permutation $(3, 9, 4, 5)(2, 6, X, 7)$, which squares to $(3, 4)(5, 9)(2, X)(6, 7)$. The 45-dimensional representation may then be spanned by 55 vectors $v_{st} = -v_{ts}$ for distinct $s, t \in \mathbb{F}_{11}$, subject to the relations $\sum_{s \neq t} v_{st} = 0$ for each $t \in \mathbb{F}_{11}$. Now take another copy spanned by vectors w_{st} subject to the corresponding relations $w_{st} = -w_{ts}$ and $\sum_{s \neq t} w_{st} = 0$ for every t, and let M_{11} act on the subscripts as described above. Then reduce modulo 7, so that the coefficients of these vectors lie in $\mathbb{F}_7$.

Now we must adjoin the extra involution commuting with $\mathrm{PSL}_2(11)$. Since $\mathrm{PSL}_2(11)$ is 2-transitive on the 11 points, it is sufficient to specify the images of v_{01} and w_{01}. These turn out to be as follows:

$$
\begin{aligned}
-2w_{01} \quad &-(w_{08} + w_{05} + w_{09}) + 2(w_{02} + w_{0X} + w_{04} + w_{06} + w_{07} + w_{03}) \\
&+(w_{18} + w_{15} + w_{19}) - 2(w_{12} + w_{1X} + w_{14} + w_{16} + w_{17} + w_{13}) \\
&-(w_{24} + w_{X3} + w_{47}) + 2(w_{82} + w_{8X} + w_{57} + w_{53} + w_{94} + w_{96}) \\
&+(w_{27} + w_{X6} + w_{63}) - 2(w_{86} + w_{87} + w_{5X} + w_{54} + w_{92} + w_{93}), \\
3v_{01} \quad &+3(v_{08} + v_{05} + v_{09}) - 2(v_{02} + v_{0X} + v_{04} + v_{06} + v_{07} + v_{03}) \\
&-3(v_{18} + v_{15} + v_{19}) + 2(v_{12} + v_{1X} + v_{14} + v_{16} + v_{17} + v_{13}) \\
&-3(v_{24} + v_{X3} + v_{47}) + 3(v_{82} + v_{8X} + v_{57} + v_{53} + v_{94} + v_{96}) \\
&+3(v_{27} + v_{X6} + v_{63}) - 3(v_{86} + v_{87} + v_{5X} + v_{54} + v_{92} + v_{93}).
\end{aligned}
$$
$$(5.144)$$

The other ingredient in Ryba's construction is a skew-symmetric trilinear form. This may be defined by the images under M_{11} of the following values, where all other values of the form at the spanning vectors v_{st}, w_{st} are 0:

$$
\begin{aligned}
t(v_{13}, v_{39}, v_{19}) &= 1, \qquad t(w_{13}, w_{39}, w_{19}) = 6, \\
t(v_{13}, v_{39}, v_{95}) &= 1, \qquad t(w_{13}, w_{39}, w_{95}) = 6, \\
t(v_{13}, v_{39}, v_{54}) &= 1, \\
t(v_{13}, v_{19}, v_{54}) &= 2, \qquad t(w_{13}, w_{19}, w_{54}) = 3, \\
t(v_{01}, v_{34}, v_{95}) &= 4, \qquad t(w_{01}, w_{34}, w_{95}) = 5, \\
t(v_{02}, v_{68}, v_{7X}) &= 3.
\end{aligned}
$$
$$(5.145)$$

Ryba then went on to prove (with some computer assistance) that the group generated by the above matrices preserves this trilinear form, and used this to deduce that it is indeed a group of shape $3{\cdot}\mathrm{O'N}{:}2$.

The full centraliser in O'N of a non-inner automorphism of order 2 turns out to be the Janko group J_1, containing the $\mathrm{PSL}_2(11)$ described above as a maximal subgroup. The full list of maximal subgroups of O'N, calculated three times (independently) in [89, 188, 176] is given in Table 5.11. There are a number of other interesting representations of the O'Nan group and its triple cover, for example a 154-dimensional representation of O'N over $\mathbb{F}_3$, and a 153-dimensional representation of $3{\cdot}\mathrm{O'N}$ over $\mathbb{F}_4$ (see [101]). The smallest representations in characteristic 0 however have dimension 342 (for the triple cover $3{\cdot}\mathrm{O'N}$) or 10944 (for the simple group O'N).

5.9.5 The Lyons group

The Lyons group was discovered as a result of classifying simple groups with an involution centraliser $2{\cdot}A_n$ (see Section 2.7.2). The smallest value of n for which $2{\cdot}A_n$ has non-central involutions (a necessary requirement for being an involution centraliser in a finite simple group) is $n = 8$, and indeed the McLaughlin group has an involution centraliser $2{\cdot}A_8$. The only other case which arises is $n = 11$, in the Lyons group Ly. Moreover, a 3-cycle in $2{\cdot}A_{11}$

centralises $2 \cdot A_8$, and it turns out that the full centraliser of this 3-cycle in the Lyons group is the triple cover $3 \cdot McL$ of the McLaughlin group. Hence the normaliser of this group of order 3 is $3 \cdot McL{:}2$.

Lyons's work [126] established the order of his group, and much of the subgroup structure. In particular,

$$
\begin{aligned}
|\mathrm{Ly}| &= 51\,765\,179\,004\,000\,000 \\
&= 2^8.3^7.5^6.11.31.37.67
\end{aligned}
\tag{5.146}
$$

and there are subgroups $5^3 \cdot SL_3(5)$ and $5^{1+4}{:}4S_6$, which contain subgroups isomorphic to the maximal parabolic subgroups of $G_2(5)$ (see Section 4.3.5). Moreover, $3 \cdot McL{:}2$ contains $3 \cdot PSU_3(5){:}2$, another subgroup of $G_2(5)$. Therefore Lyons conjectured that his group contains $G_2(5)$, to index 8835156, and this was the starting point of Sims's construction.

Sims used these subgroups to create a set of generators and relations which he believed were sufficient to define the group. In essence his proof of this conjecture was a coset enumeration (see [103]), though it was far from straightforward in those days (or even today, with vastly superior computer power available). The action of Ly on 8835156 points turns out to have rank 5, and the suborbit lengths are

$$1 + 19530 + 968750 + 2034375 + 5812500.$$

Thus Ly can be described as the automorphism group of a certain regular graph of valency 19530 on 8835156 vertices.

Now the subgroup $3 \cdot McL{:}2$ contains $S_3 \times M_{11}$. This suggested to Meyer and Neutsch [136] that Ly can be generated by subgroups $2 \cdot A_{11}$ and $S_3 \times M_{11}$ intersecting in $2 \times M_{11}$, and indeed this turns out to be the case. Richard Parker constructed the Lyons group in this way as a group of 111×111 matrices over $\mathbb{F}_5$, and Meyer and Neutsch (see [137]) proved that Parker's group contained elements satisfying Sims's presentation, and deduced that it too was isomorphic to the Lyons group. The Atlas [28] contains a description of the 111-dimensional representation of the Lyons group with respect to a basis in which $3^5{:}(2 \times M_{11})$ acts by permuting 55 subspaces of dimension 2. However, it has been alleged that this contains mistakes, as it seems no-one has succeeded in reconstructing the Lyons group from this information.

Eventually the maximal subgroups were completely determined by explicit calculations with these matrices: there were no surprises, only $G_2(5)$ and the local subgroups which were already in Lyons's paper. (See Table 5.11 and [180].)

There is another relatively small permutation representation, on the 9606125 conjugates of the subgroup $3 \cdot McL{:}2$. This also has rank 5, and the suborbit lengths are

$$1 + 15400 + 534600 + 1871100 + 7185024.$$

There is also one more 'interesting' matrix representation (in the sense that it has smaller degree than any characteristic 0 representation), namely a 651-dimensional representation over $\mathbb{F}_3$.

5.9.6 The largest Janko group J_4

Like the other Janko groups, J_4 was found during the proof of the Classification Theorem for Finite Simple Groups (CFSG), while looking for simple groups with given involution centralisers. By this stage, much of the proof of the CFSG had been completed, but some possibilities for involution centralisers remained. One of these was a group of shape $2^{1+12}_+\cdot3\cdot M_{22}{:}2$. The techniques of local analysis showed that if a simple group existed with such an involution centraliser, then it would contain subgroups of shape $2^{11}{:}M_{24}$ and $2^{10}{:}GL_5(2)$ intersecting in $2^{10}{:}2^4{:}A_8$. Moreover, the character table was calculated, and in particular the existence of a 1333-dimensional complex irreducible character proved.

It is interesting to note that three of the sporadic groups, namely Co_1, Fi'_{24} and J_4, contain maximal subgroups of shape $2^{11}.M_{24}$, but no two of the latter are isomorphic. The structure of the subgroup $2^{11}{:}M_{24}$ in J_4 is defined by saying that M_{24} acts on 2^{11} as on the even subsets of 24 points, modulo the Golay code. Thus the non-trivial elements of the 2^{11} are labelled by the 276 pairs of the 24 points, and the 1771 sextets (see Section 5.2.3). Similarly, the characters of the 2^{11} correspond to elements of the Golay code modulo complementation, so are labelled by the 759 octads and the 1288 pairs of complementary dodecads.

Therefore if we restrict a character of J_4 of degree 1333 to $2^{11}{:}M_{24}$, we obtain $45 + 1288$, in which the 45 is an irreducible character for M_{24}, and the 1288 is the character of a monomial representation obtained by inducing up a suitable linear character of $2^{11}{:}M_{12}{:}2$. Similarly, restricting to $2^{10}{:}GL_5(2)$ gives the sum of two monomial representations, $465 + 868$.

Plans were made to try to construct this representation and deduce the existence of J_4. However, the size of the representation was daunting, and this plan was never fulfilled. Much later, Lempken [117] constructed explicitly the 11-modular reduction of this representation. Eventually, Ivanov and Meierfrankenfeld [94] gave a recipe for constructing such a representation over any field of characteristic not 2 containing a square root of -7, and used this to give a new existence proof for J_4, nearly twenty years after the original proof.

In the meantime, however, the existence of a 112-dimensional representation over $\mathbb{F}_2$ had been conjectured by Thompson, which, if true, provided a much better chance of finding a construction and proving the existence of the group, now known as J_4. This was completed in 1980, by building the group out of a (conjectured) subgroup $P\Sigma U_3(11) \cong PSU_3(11){:}2$. This subgroup acts uniserially on the module, with shape 1.110.1. (In other words, as a module for J_4, the 112-space has a unique composition series, with factors of dimensions

1, 110 and 1 in that order.) Norton, Parker and Thackray constructed this representation explicitly, and restricted to a subgroup $11^{1+2}{:}(5 \times SD_{16})$, namely the Borel subgroup, which acts as $1.1 \oplus 110$ (i.e. as the direct sum of the 110-dimensional irreducible with a uniserial 2-dimensional module, on which the group acts as a transvection). This subgroup was extended to $11^{1+2}{:}(5 \times 2S_4)$ in order to generate J_4.

In fact, making the matrices which generate J_4 was the easy part. Proving that they generate J_4 was much harder. No complete account of this proof was ever published, but a sketch is given by Norton [141], and a more detailed version in Benson's Ph. D. thesis [14]. As usual, the proof is accomplished by finding a suitable geometrical/combinatorial structure whose automorphism group is J_4. It turns out that there is an orbit of $173\,067\,389$ vectors on which the group acts transitively, with point stabiliser $2^{11}{:}M_{24}$. Therefore the order of J_4 is

$$|J_4| = 86\,775\,571\,046\,077\,562\,880$$
$$= 2^{21}.3^3.5.7.11^3.23.29.31.37.43. \tag{5.147}$$

We now give a little more detail of this construction. First consider the permutation representation of $PSU_3(11){:}2$ on the $1 + 11^3 = 1332$ isotropic 1-spaces, and reduce this modulo 2. Thus we have an $\mathbb{F}_2$-vector space with basis $\{e_W\}$, as W runs through the isotropic 1-spaces. In fact this is a uniserial module with structure $1.110.1110.110.1$ (here again we give the dimensions of the composition factors, in order). Any two distinct isotropic 1-spaces, W_1 and W_2, say, span a non-singular 2-space, which contains exactly twelve isotropic 1-spaces, say $W_1, W_2, \ldots, W_{12}$. To make the representation of shape 1.110, take the submodule consisting of the vectors $\sum_W \lambda_W e_W$ which satisfy $\sum_{i=1}^{12} \lambda_{W_i} = 0$ for all 13431 such non-singular 2-spaces. Similarly, to make the representation of shape 110.1, take the quotient of the permutation module obtained by imposing the relations $\sum_{i=1}^{12} e_{W_i} = 0$. [In other words, factor out by the submodule spanned by the vectors $\sum_{i=1}^{12} e_{W_i}$.]

Next we must find equivalent bases for the two copies of the 110-dimensional irreducible. These can be chosen so that the subgroup $11^{1+2}{:}(5 \times SD_{16})$ acts by permuting a set of eleven 10-spaces, each of which can be identified with the field of order 2^{10}, on which the unitary transvections (i.e. the elements in the centre of the extraspecial group 11^{1+2}) act by multiplication by scalars. As this is the 'natural' representation of 11^{1+2}, it is straightforward to write down generators for the extension to $11^{1+2}{:}GL_2(11)$, and find the unique copy of $5 \times 2S_4$ inside $GL_2(11)$ which contains the group $11^{1+2}{:}(5 \times SD_{16})$. Similarly on the 2-dimensional summand, we extend C_2 to S_3. At this stage, generators for J_4 have been obtained.

The maximal subgroups of J_4 are listed in Table 5.11, as computed in [112]. The largest four maximal subgroups are all 2-local subgroups, which is reminiscent of the classical or exceptional groups in characteristic 2, where the maximal 2-local subgroups are just the maximal parabolic subgroups.

Indeed, one can construct a 'geometry' for J_4, analogous to the vector spaces (or more accurately, projective spaces), generalised polygons, etc., for the classical and exceptional groups. This geometry can be visualised inside the 112-dimensional representation, in which the maximal subgroup $2^{11}{:}M_{24}$ fixes a unique non-zero vector, which we shall call a *point*. There are 17367389 points. (In the Atlas [28] they are called *special* vectors, and in Benson's thesis [14] they are called *sacred* vectors.) There are sets of five points whose sum is 0, called *pentads*, whose stabiliser is a group of shape $2^{3+12}{\cdot}(S_5 \times GL_3(2))$. If two points are joined when they lie in a pentad, then we obtain a graph of degree 15180 on the points, whose automorphism group is J_4.

Further reading

For further information about the sporadic simple groups, the 'Atlas of Finite Groups' [28] is an indispensible reference. Griess's book 'Twelve sporadic groups' [69] is the only book on this subject which is written at an introductory level, and covers those sporadic groups which are visible inside Conway's group. His book (like mine) becomes more and more sketchy towards the end, and much is left as (very instructive) exercises for the reader. Conway and Sloane's 'Sphere packings, lattices and groups' [31] contains several chapters of interest, particularly covering M_{24}, the Leech lattice, and the Monster.

Aschbacher's book 'Sporadic groups' [3] is written from the point of view of collecting information needed for the proof of CFSG. It contains a great deal of carefully selected information about some of the sporadic groups, but is not easy to read. His other book '3-transposition groups' [4] is in the same vein, treating the three Fischer groups in pursuit of the definitive proof of existence and uniqueness. Neither of these books is comprehensive, as all aspects of the groups which are not explicitly required are ignored.

The two volumes on 'Geometry of sporadic groups' by Ivanov, and by Ivanov and Shpectorov [90, 91] are more specialised, though easier to read. Ivanov's recent book 'J_4' is slower-paced, as it is designed to be introductory, and includes some useful background material, but, disappointingly, makes no attempt to be comprehensive, and leaves out much that is known about the group. His even more recent book 'The Monster group and Majorana involutions' [93] joins a large literature on the Monster, which also includes Gannon's 'Moonshine beyond the Monster' [64]. Curtis's book 'Symmetric generation of groups' [45] describes his combinatorial approach to many of the sporadic groups by finding presentations which are 'symmetric' in a technical sense, and using coset enumeration of some kind to produce a permutation representation.

For background on coding theory, see for example Ray Hill's book 'A first course in coding theory' [78], which contains among other things an alternative construction of the binary Golay codes. There are numerous other books on coding theory, many of which mention the Golay codes. Many books mention the Mathieu groups in greater or lesser detail. I particularly recommend Dixon and Mortimer's 'Permutation groups' [53] which provides a different perspective from mine. Passman's 'Permutation groups' [145] includes a construction of the Mathieu groups by Witt's method.

Exercises

5.1. Show that the permutation α defined in (5.9) in Section 5.2.3 preserves the Golay code, and fuses the four orbits of the sextet group on sextets into a single orbit.

5.2. Prove that the stabiliser of a sextet in the automorphism group of the extended binary Golay code is exactly $2^6{:}3{\cdot}S_6$ (and no larger).

5.3. Show that M_{24} acts transitively on the set of 2576 dodecads in the extended binary Golay code.

5.4. Construct a 'Leech triangle' for the number of dodecads meeting the set $\{1,\ldots,j-1\}$ in $\{1,\ldots,i-1\}$, where $\{1,\ldots,8\}$ is an octad of the Steiner system $S(5,8,24)$.

5.5. Prove simplicity of M_{24} using Iwasawa's Lemma applied to the permutation action on the 759 octads in the extended binary Golay code.

5.6. Prove that $M_{21} \cong PSL_3(4)$.

5.7. Prove that the binary code of length 24 (with coordinates labelled by $\mathbb{F}_{23} \cup \{\infty\}$) spanned by the vector with 1s in the 12 coordinates labelled by x^2 for some $x \in \mathbb{F}_{23}$, and its images under $t \mapsto t+1$, is equivalent to the extended binary Golay code.

5.8. Using the numbering of the MOG from (5.11), show that for each k, the set $\{x^2 + k \mid x \in \mathbb{F}_{23}\}$ is a dodecad. Deduce that $PSL_2(23)$ is a subgroup of M_{24}.

5.9. Prove that M_{22} is simple by applying Iwasawa's Lemma to the action on the 77 hexads.

5.10. Complete the construction of $2{\cdot}M_{22}$ in Section 5.2.10 by proving the isomorphism of the two graphs on 77 vertices.

5.11. Prove that M_{23} is simple by applying Iwasawa's Lemma to the action on the 253 heptads.

5.12. The (unextended) ternary Golay code is defined by deleting one (fixed) coordinate from all the codewords in the extended code (see Section 5.3.5). Determine the weight distribution of this code, and show that it is a perfect 2-error-correcting code (see Section 5.2.8).

5.13. Show that the vectors given in (5.31) satisfy the congruence conditions in (5.32).

5.14. Classify the orbits of 2^{12}:M_{24} on the vectors of norm 8 in the Leech lattice, and on the crosses. Deduce that Co_1 acts primitively on the set of 8292375 crosses.

5.15. Prove that Co_2 fixing the type 2 vector v acts transitively on the 4600 type 2 vectors w which have inner product -2 with v. Deduce the order of the stabiliser in $2 \cdot Co_1$ of the 2-dimensional sublattice spanned by v and w.

5.16. Prove that Co_2 fixing the type 2 vector v acts transitively on the type 3 vectors w which are perpendicular to v. How many such vectors are there? Deduce the order of the stabiliser in $2 \cdot Co_1$ of the 2-dimensional sublattice spanned by v and w.

5.17. (i) Prove that $2 \cdot Co_1$ is transitive on pairs of type 3 vectors with inner product 3.
(ii) Do the same for inner product 1.

5.18. Apply Iwasawa's Lemma to the subgroup 2^{10}:M_{22}:2 of Co_2 to prove that Co_2 is simple.

5.19. Let Γ be the graph on the 100 Leech lattice vectors of norm 4 which have inner product 3 with each of the vectors $(1, 5, 1^{22})$ and $(5, 1, 1^{22})$, defined by joining two vectors if and only if their inner product is 1. Show that Γ is isomorphic to the Higman–Sims graph Δ with vertex set $\{*\} \cup S \cup H$, where S is a 22-element set, H is the set of 77 hexads of a Steiner system $S(3, 6, 22)$ on S, and $s \in S$ is joined to $*$ and the hexads not containing s, and two hexads are joined in Δ if and only if they are disjoint as subsets of S.

5.20. Enumerate the vectors of integral norm 6 and 8 in the icosian Leech lattice (see Section 5.6.3) and hence complete the proof that as a real lattice it is isomorphic to the ordinary Leech lattice.

5.21. (i) Enumerate the vectors of norm 6 and 8 in the complex Leech lattice (see Section 5.6.10) and hence complete the proof that as a real lattice it is isomorphic to the ordinary Leech lattice.
(ii) Deduce the orbits of the monomial group on coordinate frames in the complex Leech lattice, and show that the action of $6 \cdot Suz$ on them is primitive.

5.22. Show that, if s is the complex number $\frac{1}{2}(-1 + \sqrt{-7})$, then the set of 42 vectors obtained from $(2, 0, 0)$, $(\overline{s}, \overline{s}, 0)$ and $(s, 1, 1)$ by coordinate permutations and sign-changes is invariant under reflection in any one of these vectors. Show that they fall into seven congruence classes modulo s, and that the stabiliser of a congruence class is a monomial group 2^3:S_3. Deduce the order of the group generated by the reflections. By numbering the congruence classes suitably, find a homomorphism from this group onto the group $PSL_3(2)$ as defined in Section 3.3.5.

5.23. Prove that the element $\frac{1}{2}R_{i_0-i_1}R_s^*$ defined in (5.81) is a symmetry of the octonion Leech lattice.

5.24. Let G be a group generated by three conjugate involutions a, b, c, such that any two distinct conjugates of a, b, c have product of order 3.

(i) Show that $abcabc = bcabca$.
(ii) Deduce that every element of G can be expressed as a word of length at most 6 in a, b, c.
(iii) By enumerating the words of length at most 6, or otherwise, show that G has order at most 54.

5.25. Let H be the group defined by the presentation

$$\langle x, y, z, t \mid x^3 = y^3 = z^3 = t^2 = 1, [x, y] = z, [x, z] = [y, z] = [t, z] = 1,$$
$$y^t = y^{-1}, x^t = x^{-1}\rangle. \tag{5.148}$$

Show that H has order 54 and is generated by the nine conjugates of t.

5.26. Prove the uniqueness of the Steiner system $S(3, 6, 22)$ and deduce that the normaliser (i.e. stabiliser) of a base in Fi_{22} has shape $2^{10}M_{22}$.
    [Hint: the Leech triangle (Fig. 5.1) may help.]

5.27. Prove that any two symplectic transvections $T_v(\lambda) : x \mapsto x + \lambda f(x, v)v$ in $\mathrm{Sp}_{2n}(2)$ (see Section 3.5.1) either commute (if $f(u, v) = 0$) or generate S_3 (otherwise).
    Deduce that the same holds for orthogonal transvections (Section 3.8.1) in $\mathrm{GO}_{2n}^{\pm}(2)$.

5.28. Prove that Parker's loop (Section 5.7.9) is a Moufang loop.

5.29. Prove that J_1 has trivial outer automorphism group.
    [Hint: Show first that $2^3{:}7{:}3$ has trivial outer automorphism group, and deduce that if there is any non-inner automorphism, then there is one which centralises $2^3{:}7{:}3$ and also $2 \times A_5$.]

References

1. J. F. Adams: Lectures on Lie groups. Univ. of Chicago Press (1996)
2. J. L. Alperin and R. B. Bell: Groups and representations. Springer (1995)
3. M. Aschbacher: Sporadic groups. Cambridge Univ. Press (1994)
4. M. Aschbacher: 3-transposition groups. Cambridge Univ. Press (1997)
5. M. Aschbacher: On the maximal subgroups of the finite classical groups. Invent. Math., **76**, 469–514 (1984)
6. M. Aschbacher: The 27-dimensional module for E_6. I. Invent. Math., **89**, 159–195 (1987)
7. M. Aschbacher: The 27-dimensional module for E_6. II. J. London Math. Soc., **37**, 275–293 (1988)
8. M. Aschbacher: The 27-dimensional module for E_6. III. Trans. Amer. Math. Soc., **321**, 45–84 (1990)
9. M. Aschbacher: The 27-dimensional module for E_6. IV. J. Algebra, **131**, 23–39 (1990)
10. M. Aschbacher: The 27-dimensional module for E_6. V. Unpublished
11. M. Aschbacher: Some multilinear forms with large isometry groups. Geom. Dedicata, **25**, 417–465 (1988)
12. M. Aschbacher and S. D. Smith: The classification of quasithin groups. I. Structure of strongly quasithin K-groups. American Mathematical Society (2004)
13. M. Aschbacher and S. D. Smith: The classification of quasithin groups. II. Main theorems: the classification of simple QTKE-groups. American Mathematical Society (2004)
14. D. J. Benson: The sporadic simple group J_4. Ph. D. thesis, Cambridge (1980)
15. J. N. Bray: A new method for finding involution centralisers. Arch. Math. (Basel), **74**, 241–245 (2000)
16. K. S. Brown: Buildings. Springer (1989)
17. R. B. Brown: Groups of type E_7. J. Reine Angew. Math., **236**, 79–102 (1969)
18. G. Butler: The maximal subgroups of the sporadic simple group of Held. J. Algebra, **69**, 67–81 (1981)
19. P. J. Cameron: Permutation groups. LMS Student Texts 45. Cambridge Univ. Press (1999)
20. P. J. Cameron: Classical groups. http://www.maths.qmul.ac.uk/~pjc/
21. R. W. Carter: Simple groups of Lie type. Wiley (1972; reprinted 1989)

R.A. Wilson, *The Finite Simple Groups*,
Graduate Texts in Mathematics 251,
© Springer-Verlag London Limited 2009

22. A. Cayley: On Jacobi's elliptic functions, in reply to Rev. B. Bronwin; and on quaternions. Philosophical Magazine, **26**, 208–211 (1845)
23. C. Chevalley: Sur certains groupes simples. Tôhoku Math. J., **7**, 14–66 (1955)
24. A. M. Cohen: Finite quaternionic reflection groups. J. Algebra, **64**, 293–324 (1980)
25. J. H. Conway: A group of order 8,315,553,613,086,720,000. Bull. London Math. Soc., **1**, 79–88 (1969)
26. J. H. Conway: Three lectures on exceptional groups. In: M. J. Collins (ed.) Finite simple groups. Academic Press (1971)
27. J. H. Conway: A construction for the smallest Fischer group F_{22}. In: T. Gagen, M. P. Hale and E. E. Shult (eds.) Finite groups '72. North-Holland (1973)
28. J. H. Conway, R. T. Curtis, R. A. Parker, S. P. Norton and R. A. Wilson: Atlas of finite groups. Clarendon Press, Oxford (1985; reprinted with corrections, 2004)
29. J. H. Conway: A simple construction for the Fischer–Griess Monster group. Invent. Math., **79**, 513–540 (1985)
30. J. H. Conway, S. P. Norton and L. H. Soicher: The Bimonster, the group Y_{555}, and the projective plane of order 3. In: M. C. Tangora (ed.) Computers in algebra. Marcel Dekker (1988)
31. J. H. Conway and N. J. A. Sloane: Sphere-packings, lattices and groups. 3rd ed., Springer (1999)
32. J. H. Conway and D. A. Smith: On quaternions and octonions: their geometry, arithmetic, and symmetry. AKPeters (2003)
33. J. H. Conway and D. B. Wales: Matrix generators for J_3. J. Algebra, **29**, 474–476 (1974)
34. J. H. Conway and D. B. Wales: The construction of the Rudvalis simple group of order 145,926,144,000. J. Algebra, **27**, 538–548 (1973)
35. K. Coolsaet: Algebraic structure of the perfect Ree-Tits generalized octagons. Innov. Incidence Geom., **1**, 67–131 (2005)
36. K. Coolsaet: On a 25-dimensional embedding of the Ree-Tits generalized octagon. Adv. Geom., **7**, 423–452 (2007)
37. K. Coolsaet: A 51-dimensional embedding of the Ree-Tits generalized octagon. Des. Codes Cryptogr., **47**, 75–97 (2008)
38. B. N. Cooperstein: Maximal subgroups of $G_2(2^n)$. J. Algebra, **70**, 23–36 (1981)
39. B. N. Cooperstein: The fifty-six dimensional module for E_7. I. A four-form for E_7. J. Algebra, **173**, 361–389 (1995)
40. H. S. M. Coxeter and W. O. J. Moser: Generators and relations for discrete groups. 4th ed., Springer (1980)
41. R. T. Curtis: A new combinatorial approach to M_{24}. Math. Proc. Cambridge Philos. Soc., **79**, 25–42 (1976)
42. R. T. Curtis: The maximal subgroups of M_{24}. Math. Proc. Cambridge Philos. Soc., **81**, 185–192 (1977)
43. R. T. Curtis: On subgroups of ·0. I. Lattice stabilizers. J. Algebra, **27**, 549–573 (1973)
44. R. T. Curtis: On subgroups of ·0. II. Local structure. J. Algebra, **63**, 413–434 (1980)
45. R. T. Curtis: Symmetric generation of groups. With applications to many of the sporadic finite simple groups. Cambridge Univ. Press (2007)
46. C. W. Curtis and I. Reiner: Representation theory of finite groups and associative algebras. Wiley (1962)

47. C. W. Curtis and I. Reiner: Methods of representation theory. Vols. I and II. With applications to finite groups and orders. Wiley (1981, 1987)

48. L. E. Dickson: Linear groups, with an exposition of the Galois field theory. Teubner (1901); reprinted, Dover (1958)

49. L. E. Dickson: A new system of simple groups. Math. Ann., **60**, 137–150 (1905)

50. L. E. Dickson: A class of groups in an arbitrary realm connected with the configuration of the 27 lines on a cubic surface. Quart. J. Pure Appl. Math., **33**, 145–173 (1901)

51. L. E. Dickson: A class of groups in an arbitrary realm connected with the configuration of the 27 lines on a cubic surface (second paper). Quart. J. Pure Appl. Math., **39**, 205–209 (1908)

52. J. Dieudonné: La géométrie des groupes classiques. Springer (1955)

53. J. D. Dixon and B. Mortimer: Permutation groups. Springer GTM 163 (1996)

54. L. Dornhoff: Group representation theory. Part A: Ordinary representation theory; Part B: Modular representation theory. Marcel Dekker (1971, 1972)

55. E. B. Dynkin: Maximal subgroups of the classical groups. (In Russian.) Trudy Moskov. Mat. Obšč., **1**, 39–166 (1952)

56. W. L. Edge: $PGL(2, 11)$ and $PSL(2, 11)$. J. Algebra, **97**, 492–504 (1985)

57. G. M. Enright: Subgroups generated by transpositions in F_{22} and F_{23}. Comm. Algebra, **6**, 823–837 (1978)

58. W. Feit: The representation theory of finite groups. North-Holland (1982)

59. W. Feit and J. G. Thompson: Solvability of groups of odd order. Pacific J. Math., **13**, 775–1029 (1963)

60. L. Finkelstein: The maximal subgroups of Conway's group C_3 and McLaughlin's group. J. Algebra, **25**, 58–89 (1973)

61. L. Finkelstein and A. Rudvalis: The maximal subgroups of Janko's simple group of order $50, 232, 960$. J. Algebra, **30**, 122–143 (1974)

62. B. Fischer: Finite groups generated by 3-transpositions. Invent. Math., **13**, 232–246 (1971)

63. B. Fischer: Finite groups generated by 3-transpositions. Warwick University Notes (1971)

64. T. Gannon: Moonshine beyond the Monster. The bridge connecting algebra, modular forms and physics. Cambridge Univ. Press (2006)

65. M. Geck: An introduction to algebraic geometry and algebraic groups. Oxford Univ. Press (2003)

66. D. Gorenstein, R. Lyons, and R. Solomon: The classification of the finite simple groups. Numbers 1 to 6. American Math. Soc. (1994, 1996, 1998, 1999, 2002, 2005)

67. J. T. Graves: Letters to W. R. Hamilton. (1843/4)

68. J. A. Green: Polynomial representations of GL_n. Springer (1980)

69. R. L. Griess, Jr.: Twelve sporadic groups. Springer (1998)

70. R. L. Griess, Jr.: The friendly giant. Invent. Math., **69**, 1–102 (1982)

71. R. L. Griess, Jr.: A Moufang loop, the exceptional Jordan algebra, and a cubic form in 27 variables. J. Algebra, **131**, 281–293 (1990)

72. L. C. Grove: Classical groups and geometric algebra. Amer. Math. Soc. (2002)

73. L. C. Grove and C. T. Benson: Finite reflection groups. Springer (1971; 2nd ed. 1985)

74. D. Hertzig: Forms of algebraic groups. Proc. Amer. Math. Soc., **12**, 657–660 (1961)

75. D. G. Higman and C. C. Sims: A simple group of order $44,352,000$. Math. Z., **105**, 110–113 (1968)

76. G. Higman: On the simple group of D. G. Higman and C. C. Sims. Illinois J. Math., **13**, 74–80 (1969)

77. G. Higman and J. McKay: On Janko's simple group of order 50,232,960. Bull. London Math. Soc., **1**, 89–94 (1969)

78. R. Hill: A first course in coding theory. Oxford Univ. Press (1986)

79. G. Hiss and G. Malle: Low degree representations of quasi-simple groups. LMS J. Comput. Math., **4**, 22–63 (2001) and **5**, 95–126 (2002)

80. P. E. Holmes and R. A. Wilson: A computer construction of the Monster using 2-local subgroups. J. London Math. Soc., **67**, 349–364 (2003)

81. P. E. Holmes: A classification of subgroups of the Monster isomorphic to S_4 and an application. J. Algebra, **319**, 3089–3099 (2008)

82. P. E. Holmes and R. A. Wilson: A new maximal subgroup of the Monster. J. Algebra, **251**, 435–447 (2002)

83. P. E. Holmes and R. A. Wilson: $PSL_2(59)$ is a subgroup of the Monster. J. London Math. Soc., **69**, 141–152 (2004)

84. P. E. Holmes and R. A. Wilson: On subgroups of the Monster containing A_5's. J. Algebra, **319**, 2653–2667 (2008)

85. J. E. Humphreys: Reflection groups and Coxeter groups. Cambridge Univ. Press (1990)

86. J. E. Humphreys: Linear algebraic groups. Springer (1975)

87. B. Huppert and N. Blackburn: Finite groups. III. Springer (1982)

88. I. M. Isaacs: Character theory of finite groups. Academic Press (1976)

89. A. A. Ivanov, S. V. Tsaranov and S. V. Shpectorov: The maximal subgroups of the O'Nan–Sims sporadic simple group and its automorphism group. (In Russian.) Dokl. Akad. Nauk SSSR, **291**, 777–780 (1986)

90. A. A. Ivanov: Geometry of sporadic groups. I. Petersen and tilde geometries. Cambridge Univ. Press (1999)

91. A. A. Ivanov and S. V. Shpectorov: Geometry of sporadic groups. II. Representations and amalgams. Cambridge Univ. Press (2002)

92. A. A. Ivanov: The fourth Janko group. Oxford Univ. Press (2004)

93. A. A. Ivanov: The monster group and Majorana involutions. Cambridge Univ. Press (2009)

94. A. A. Ivanov and U. Meierfrankenfeld: A computer-free construction of J_4. J. Algebra, **219**, 113–172 (1999)

95. K. Iwasawa: Über die Einfachheit der speziellen projektiven Gruppen. Proc. Imp. Acad. Tokyo, **17**, 57–59 (1941)

96. G. D. James: Representations of general linear groups. LMS Lecture Notes 94, Cambridge Univ. Press (1984)

97. G. James and A. Kerber: The representation theory of the symmetric group. Addison-Wesley (1981)

98. G. D. James and M. W. Liebeck: Representations and characters of groups. Cambridge Univ. Press (1993)

99. Z. Janko: A new finite simple group with Abelian Sylow 2-subgroups, and its characterization. J. Algebra, **3**, 147–186 (1966)

100. Z. Janko: Some new simple groups of finite order. Symp. Math., **1**, 25–64 (1968)

101. C. Jansen and R. A. Wilson: Two new constructions of the O'Nan group. J. London Math. Soc., **56**, 579–583 (1997)

102. C. Jansen, K. Lux, R. Parker and R. Wilson: An Atlas of Brauer Characters. Clarendon Press (1995)
103. D. L. Johnson: Presentations of groups. 2nd edition. Cambridge Univ. Press (1997)
104. C. Jordan: Traité des substitutions et des équations algébriques. Gauthier-Villars (1870)
105. R. W. Kaye and R. A. Wilson: Linear algebra. Oxford Univ. Press (1998)
106. P. B. Kleidman: The maximal subgroups of the Chevalley groups $G_2(q)$ with q odd, the Ree groups $^2G_2(q)$, and their automorphism groups. J. Algebra, **117**, 30–71 (1988)
107. P. B. Kleidman: The maximal subgroups of the Steinberg triality groups $^3D_4(q)$ and of their automorphism groups. J. Algebra, **115**, 182–199 (1988)
108. P. B. Kleidman and M. W. Liebeck: The subgroup structure of the finite classical groups. Cambridge Univ. Press (1990)
109. P. B. Kleidman, U. Meierfrankenfeld and A. J. E. Ryba: Ru < $E_7(5)$. Comm. Algebra, **28**, 3555–3583 (2000)
110. P. B. Kleidman, U. Meierfrankenfeld and A. J. E. Ryba: HS < $E_7(5)$. J. London Math. Soc., **60**, 95–107 (1999)
111. P. B. Kleidman and R. A. Wilson: The maximal subgroups of Fi_{22}. Math. Proc. Cambridge Philos. Soc., **102**, 17–23 (1987), and **103**, 383 (1988)
112. P. B. Kleidman and R. A. Wilson: The maximal subgroups of J_4. Proc. London Math. Soc., **56**, 484–510 (1988)
113. P. B. Kleidman, R. A. Parker and R. A. Wilson: The maximal subgroups of the Fischer group Fi_{23}. J. London Math. Soc., **39**, 89–101 (1989)
114. P. B. Kleidman and R. A. Wilson: $J_3 < E_6(4)$ and $M_{12} < E_6(5)$. J. London Math. Soc., **42**, 555–561 (1990)
115. A. I. Kostrikin and P. H. Tiep: Orthogonal decompositions and integral lattices. de Gruyter (1994)
116. J. Leech: Notes on sphere packings. Canad. J. Math., **19**, 251–267 (1967)
117. W. Lempken: Constructing J_4 in $GL_{1333}(11)$. Comm. Algebra, **21**, 4311–4351 (1993)
118. J. S. Leon and C. C. Sims: The existence and uniqueness of a simple group generated by $\{3,4\}$-transpositions. Bull. Amer. Math. Soc., **83**, 1039–1040 (1977)
119. V. M. Levchuk and Y. N. Nuzhin: Structure of Ree groups. Alg. i Log., **24**, 26–41 (1985)
120. M. W. Liebeck, C. E. Praeger and J. Saxl: A classification of the maximal subgroups of the finite alternating and symmetric groups. J. Algebra, **111**, 365–383 (1987)
121. S. A. Linton and R. A. Wilson: The maximal subgroups of the Fischer groups Fi_{24} and Fi'_{24}. Proc. London Math. Soc., **63**, 113–164 (1991)
122. S. A. Linton: The maximal subgroups of the Thompson group. J. London Math. Soc., **39**, 79–88 (1989), and **43**, 253–254 (1991)
123. D. Livingstone: On a permutation representation of the Janko group. J. Algebra, **6**, 43–55 (1967)
124. F. Lübeck: Small degree representations of finite Chevalley groups in defining characteristic. LMS J. Comput. Math., **4**, 135–169 (2001)
125. H. Lüneburg: Die Suzukigruppen und ihre Geometrien. Springer (1965)
126. R. Lyons: Evidence for a new finite simple group. J. Algebra, **20**, 540–569 (1972), and **34**, 188–189 (1975)

127. K. Magaard: The maximal subgroups of the Chevalley groups $F_4(F)$ where F is a finite or algebraically closed field of characteristic $\neq 2,3$. Ph. D. thesis, Calif. Inst. Tech. (1990)

128. S. S. Magliveras: The subgroup structure of the Higman–Sims simple group. Bull. Amer. Math. Soc., **77**, 535–539 (1971)

129. G. Malle: The maximal subgroups of $^2F_4(q^2)$. J. Algebra, **139**, 52–69 (1989)

130. E. Mathieu: Mémoire sur le nombre de valeurs que peut acquérir une fonction quand on y permut ses variables de toutes les manières possibles. J. Math. Pures Appl., **5**, 9–42 (1860)

131. E. Mathieu: Mémoire sur l'étude des fonctions de plusieurs quantités, sur la manière de les former et sur les substitutions qui les laissent invariables. J. Math. Pures Appl., **6**, 241–323 (1861)

132. E. Mathieu: Sur la fonction cinq fois transitive de 24 quantités. J. Math. Pures Appl., **18**, 25–46 (1873)

133. K. McCrimmon: A taste of Jordan algebras. Springer (2004)

134. J. E. McLaughlin: A simple group of order 898,128,000. In: R. Brauer and C.-H. Sah (eds.) The theory of finite groups. Benjamin (1969)

135. K. Meyberg: Eine Theorie der Freudenthalschen Tripelsysteme. I, II. Indag. Math., **30**, 162–174, 175–190 (1968)

136. W. Meyer and W. Neutsch: Über 5-Darstellungen der Lyonsgruppe. Math. Ann., **267**, 519–535 (1984)

137. W. Meyer, W. Neutsch, and R. Parker: The minimal 5-representation of Lyons' sporadic group. Math. Ann., **272**, 29–39 (1985)

138. M. K. Mikdashi: On the covering group of the Mathieu group M_{12}. Ph. D. thesis, Birmingham (1971)

139. G. A. Miller: On the supposed five-fold transitive function of 24 elements and 19!/48 values. Mess. Math., **27**, 187–190 (1898)

140. S. P. Norton: F and other simple groups. Ph. D. thesis, Cambridge (1976)

141. S. P. Norton: The construction of J_4. In: B. Cooperstein and G. Mason (eds.) The Santa Cruz conference on finite groups. Amer. Math. Soc. (1980)

142. S. P. Norton: Anatomy of the Monster. I. In: R. T. Curtis and R. A. Wilson (eds.) The Atlas of finite groups: ten years on. Cambridge Univ. Press (1998)

143. R. A. Parker: The computer calculation of modular characters (the meat-axe). In: M. D. Atkinson (ed.) Computational group theory (Durham, 1982). Academic Press (1984)

144. R. A. Parker and R. A. Wilson: Constructions of Fischer's Baby Monster over fields of characteristic not 2. J. Algebra, **229**, 109–117 (2000)

145. D. Passman: Permutation groups. Benjamin (1968)

146. S. J. F. Rogers: Representations of finite simple groups over fields of characteristic zero. Ph. D. thesis, Birmingham (1997)

147. M. Ronan: Lectures on buildings. Academic Press (1989)

148. A. J. E. Ryba: A new construction of the O'Nan simple group. J. Algebra, **112**, 173–197 (1988)

149. A. J. E. Ryba: Matrix generators for the Held group. In: M. C. Tangora (ed.) Computers in algebra. Dekker (1988)

150. L. J. Rylands and D. E. Taylor: Constructions for octonion and exceptional Jordan algebras. Des. Codes Cryptogr., **21**, 191–203 (2000)

151. B. E. Sagan: The symmetric group. Representations, combinatorial algorithms, and symmetric functions. 2nd ed., Springer (2001)

152. J.-P. Serre: Représentations linéaires des groups finis. Hermann (3rd edition, 1978) (English translation of 2nd edition, Springer, 1977)
153. C. C. Sims: The existence and uniqueness of Lyons' group. In: T. Gagen, M. P. Hale and E. E. Shult (eds.) Finite groups '72. North-Holland (1973)
154. C. C. Sims: How to construct a Baby Monster. In: M. J. Collins (ed.) Finite simple groups II. Academic Press (1980)
155. P. E. Smith: A simple subgroup of M? and $E_8(3)$. Bull. London Math. Soc., **8**, 161–165 (1976)
156. P. E. Smith: On certain finite simple groups. Ph. D. thesis, Cambridge (1975)
157. L. H. Soicher: A new existence and uniqueness proof for the O'Nan group. Bull. London Math. Soc., **22**, 148–152 (1990)
158. T. A. Springer and F. D. Veldkamp: Octonions, Jordan algebras and exceptional groups. Springer (2000)
159. R. Steinberg: Variations on a theme of Chevalley. Pacific J. Math., **9**, 875–891 (1959)
160. M. Suzuki: On a class of doubly transitive groups. Ann. of Math., **79**, 514–589 (1964)
161. M. Suzuki: A simple group of order $448,345,497,600$. In: R. Brauer and C.-H. Sah (eds.) The theory of finite groups. Benjamin (1969)
162. D. E. Taylor: The geometry of the classical groups. Heldermann (1992)
163. J. G. Thompson: A simple subgroup of $E_8(3)$. In: N. Iwahori (ed.) Finite groups. Sapporo and Kyoto, 1974. Japan Soc. for the Promotion of Science (1976)
164. J. Tits: Quaternions over $\mathbf{Q}(\sqrt{5})$, Leech's lattice and the sporadic group of Hall–Janko. J. Algebra, **63**, 56–75 (1980)
165. J. Tits: Les "formes réelles" des groupes de type E_6. Sém. Bourbaki, exp. 162 (1957/8)
166. J. Tits: Moufang octagons and Ree groups of type $^2F_4$. Amer. J. Math., **105**, 539–594 (1983)
167. J. Tits and R. M. Weiss: Moufang polygons. Springer (2002)
168. H. van Maldeghem: Generalized polygons. Birkhäuser (1998)
169. H. Weyl: Classical groups. Princeton Univ. Press (1946)
170. H. Wielandt: Finite permutation groups. Academic Press (1964)
171. R. A. Wilson: On maximal subgroups of the finite classical groups. Preprint, 1982.
172. R. A. Wilson: The complex Leech lattice and maximal subgroups of the Suzuki group. J. Algebra, **84**, 151–188 (1983)
173. R. A. Wilson: The maximal subgroups of Conway's group $\cdot 2$. J. Algebra, **84**, 107–114 (1983)
174. R. A. Wilson: The maximal subgroups of Conway's group Co_1. J. Algebra, **85**, 144–165 (1983)
175. R. A. Wilson: The geometry and maximal subgroups of the simple groups of A. Rudvalis and J. Tits. Proc. London Math. Soc., **48**, 533–563 (1984)
176. R. A. Wilson: The maximal subgroups of the O'Nan group. J. Algebra, **97**, 467–473 (1985)
177. R. A. Wilson: Is J_1 a subgroup of the Monster? Bull. London Math. Soc., **18**, 349–350 (1986)
178. R. A. Wilson: Vector stabilizers and subgroups of Leech lattice groups. J. Algebra, **127**, 387–408 (1989)

179. R. A. Wilson: The geometry of the Hall–Janko group as a quaternionic reflection group. Geom. Dedicata, **20**, 157–173 (1986)
180. R. A. Wilson: The maximal subgroups of the Lyons group. Math. Proc. Cambridge Philos. Soc., **97**, 433–436 (1985)
181. R. A. Wilson: The maximal subgroups of the Baby Monster. I. J. Algebra **211**, 1–14 (1999)
182. R. A. Wilson: An elementary construction of the Ree groups of type $^2G_2$. Proc. Edinburgh Math. Soc., to appear (2009)
183. R. A. Wilson: A simple construction of the Ree groups of type $^2F_4$. J. Algebra, to appear
184. R. A. Wilson: A construction of $F_4(q)$ and $^2F_4(q)$ in characteristic 2 using $SL_3(3)$. Preprint (2009)
185. R. A. Wilson: Octonions and the Leech lattice. J. Algebra, **322**, 2186–2190 (2009)
186. R. A. Wilson: The Conway group and octonions. Preprint (2009)
187. E. Witt: Die 5-fach transitiven Gruppen von Mathieu. Abh. Math. Sem. Hamburg, **12**, 256–264 (1938)
188. S. Yoshiara: The maximal subgroups of the sporadic simple group of O'Nan. J. Fac. Sci. Univ. Tokyo Sect. IA Math., **32**, 105–141 (1985)

Index

abelian group, 6
Adams, J. F., 178
adjoint representation, 132, 242
affine
  general linear group, 22
  group, 21
  subspace, 193
Albert algebra, 148
  integral, 162
alert reader, 44, 64
algebra, 78
  Albert, 148
    integral, 162
  Cayley, 119
  Clifford, 78
  Jordan, 148
  Lie, 174
  non-associative, 119
  octonion, 119
  quaternion, 118
  twisted octonion, 140
almost simple
  group, 22
  type, 85
Alperin, J. L., 105
alternating form, 54
alternating group, 12
alternative law, 120
anabasic transposition, 241
angle, 56
anti-symmetric form, 54
Aschbacher, M., 4, 5, 171, 278
Aschbacher–Dynkin theorem, 85, 87

associate, 247
associative law, 6
Atlas, 278
automorphism, 9
  diagonal, 48, 64, 68
  field, 48, 64, 68
  Frobenius, 43
  graph, 48, 64

Baby Monster, 259
base, 241, 245
base group, 20
basic transposition, 241
Bell, R. B., 105
Benson, C. T., 35
Benson, D. J., 269, 277
bilinear form, 54
binary icosahedral group, 220
binary tetrahedral group, 224
black point, 168
Blackburn, N., 178
block
  of Steiner system, 188
block system, 13
BN-pair, 46, 67, 74, 124
Borel subgroup, 46
  of $E_8(q)$, 176
  of $G_2(q)$, 124
Bray trick, 267
Bray, J. N., 73
brick, 191
Brown, K., 106
building, 46
Burnside, W., 2

R.A. Wilson, *The Finite Simple Groups*,
Graduate Texts in Mathematics 251,
© Springer-Verlag London Limited 2009